AS CHEMISTRY

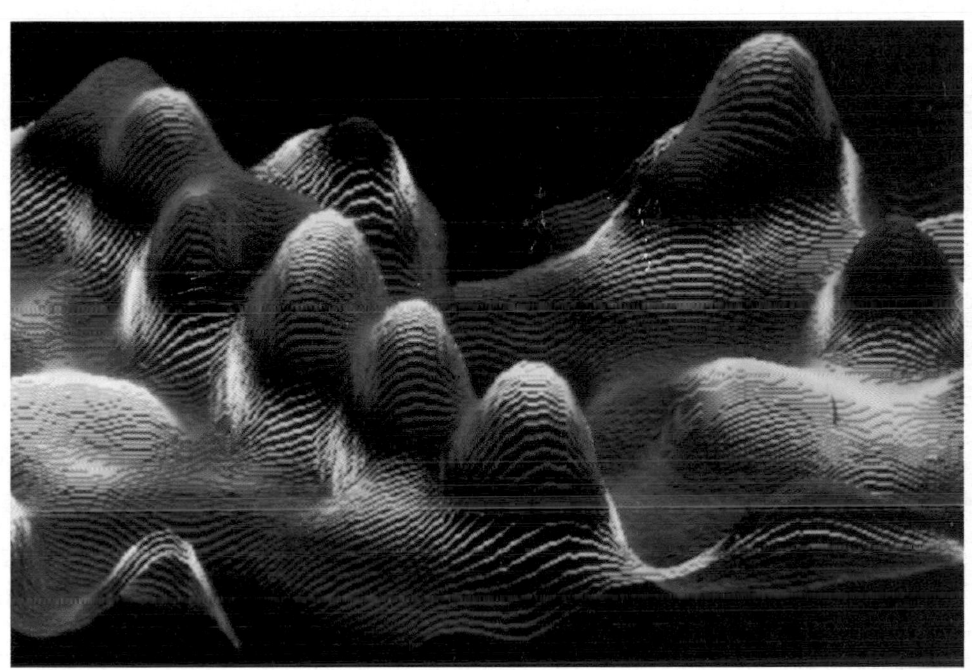

Andrew Hunt

Hodder & Stoughton
A MEMBER OF THE HODDER HEADLINE GROUP

Photo acknowledgements

The publishers would like to thank the following individuals, institutions and companies for permission to reproduce photographs in this book. Every effort has been made to trace ownership of copyright. The publishers would be happy to make arrangements with any copyright holder whom it has not been possible to contact:

Action Plus (165); Andrew Lambert (15 bottom, 16, 17, 19 both, 20, 21, 31, 36, 37 both, 51, 68, 70, 72, 75, 87, 105 both, 137 all, 140, 141, 148, 149, 150, 151, 159, 198, 200 all, 224); Audi (141 left); GSF Picture Library (33 top); Baker Refractories (69 right); Ecoscene/ADR Brown (173)/ Martin Jones (234); Hodder & Stoughton (97, 160 bottom); Holt Studios International (15 top, 163); Institute of Materials/Brian Bansfield at Buehler Krautkramer (71 right); Life File (94, 160 top, 207); Omicron (71 left); Natural History Museum, London (143, 144 bottom); Ruth Nossek (63); Science & Society Picture Library, Science Museum (69 left); Science Photo Library (61, 66 left, 146 bottom)/ Adam Hart-Davis (167)/ Biosym Technologies, Inc. (88)/ Bruce Frisch (171 bottom)/ BSIP Laurent (7)/ Celestial Image Co. (53)/ Damien Lovegrove (174)/ Dept of Physics, Imperial College (57)/ GECO UK (4)/ Geoff Lane, CSIRO (90)/ Geoff Tompkinson (5 bottom, 54, 180)/ James King-Holmes (172)/ JC Revy (126)/ Malcolm Fielding, Johnson Matthey plc (164)/ Martin Bond (162, 233)/ Martin Land (146 top)/ Mehau Kulyk (50)/ NASA, Goddard Space Flight Center (231 bottom)/ Philippe Plailly, Eurelios (66 right)/ Roberto de Gugliemo (144 top)/ Rosenfeld Images Ltd (171 top)/ Scott Camazine (5 top); Sue Cunningham Photographic (231 top)

Orders: please contact Bookpoint Ltd, 130 Milton Park, Abingdon, Oxon OX14 4 SB , UK. Telephone: (44) 01235 827720, Fax: (44) 01235 400454. Lines are open from 9.00–6.00, Monday to Saturday, with a 24 hour message answering service. You can also order through our website www.hodderheadline.co.uk

A catalogue record for this title is available from The British Library

ISBN 0 340 75796 5

First published 2000
Impression number 10 9 8 7 6 5
Year 2006 2005 2004

Cover photo from Science Photo Library.

Illustrated by Jeff Edwards, Hardlines, Peters & Zabransky (UK) Ltd.
Typeset by Wearset, Boldon, Tyne and Wear.
Printed in Italy for Hodder & Stoughton Educational, a division of Hodder Headline, 338 Euston Road, London NW1 3BH.

Contents

Section one	**Studying Chemistry**	1
Section two	**Foundations of Chemistry**	14
Section three	**Physical Chemistry**	49
Section four	**Inorganic Chemistry**	124
Section five	**Organic Chemistry**	177
Section six	**Reference**	238
Answers		247
Index		257

Acknowledgements

I am very grateful for the help and advice of three experienced teachers Del Clark, Noel Dickson and Janet Taylor who helped with the development of this book and made valuable comments and suggestions.

I would also like to acknowledge the suggestions from teachers who commented on the initial plans and draft chapters including Deidre Cawthorne, Nigel Heslop, Lynne Marjoram and Margaret Shears.

I learnt much from my collaboration with Alan Sykes when we wrote the textbook *Chemistry* and related texts. This book draws on ideas from those publications while redeveloping them for the new AS level courses.

I have been thankful for the support of the team at Hodder & Stoughton including Lynda King, Charlotte Litt, Suzanne O'Farrell and Elisabeth Tribe.

I am very pleased with the collaboration with New Media, publisher of the well-known *Chemistry Set* CD-ROM. This has made it possible to produce a CD-ROM and web site to accompany this book. My thanks to Dick Fletcher and David Tymm.

Finally I would like to acknowledge my debt to the many chemistry teachers and chemists I have worked with while contributing to the Nuffield Chemistry, SATIS 16–19 and Salters Chemistry projects. Writing and editing for these projects has clarified my understanding of chemical ideas and helped me to find ways to explain them clearly.

Andrew Hunt

Section one
Studying Chemistry

Contents

1.1 How to use this book

1.2 Why study chemistry?

1.3 Laboratory investigations

1.4 Safety

1.5 Units and measurement

1.6 Key skills

1.1 How to use this book

This is your guide to the first year of an advanced chemistry course. As you start the course you will see new things, hear of new ideas and learn a new language. At first much will seem strange. Like a traveller in a new country you will find it helpful to have a guide book to help you find your way. This book is your guide to the world of advanced chemistry.

The book

We have divided the book into five main sections together with a reference section. You can find your way through the book guided by the coloured strip at the edge of each page. The first page of each section has a table of contents listing the short topics. This is followed by two pages which give you an overview of the main ideas in the section.

There are many links in the text and diagrams to chemical reactions, tests and preparations which you will carry out in a laboratory.

The Test Yourself questions will help you to think about what you are studying. Check your answers at the end of the book to find out how well your understanding of chemistry is developing.

The Review pages at the end of a section will also help you to see what it is that you have to learn and understand.

Icons

There are two icons which appear in this book:

D You will find this icon in the title bar of many of the Test Yourself exercises. It tells you that to answer some of the questions you need to refer to the data tables at the end of the book, or to the data on the CD-ROM or to a book of data.

CD-ROM You will find this icon at the beginning of some sub-sections if the CD-ROM is especially useful for the whole topic. You will also find the icon beside some of the figures which shows that the structures, images or videos on the CD-ROM can bring the diagrams to life. Sometimes you will also find this icon alongside the Test Yourself exercises where the CD-ROM can help you with the learning activities.

The CD-ROM

The CD-ROM which accompanies this text will help you to make your study of chemistry active and rewarding. The main features are:

- a comprehensive database of the properties of all the elements and compounds which feature in AS Chemistry courses together with a data analyser to display the data,
- an interactive Periodic table which helps you to examine periodic patterns and trends,
- three-dimensional models of molecules and crystals which you can rotate and view from different angles,
- short videos of inorganic and organic reactions to remind you of the observations you have made in the laboratory.

The web site

The web site to accompany the book and CD-ROM can be found at *http://www.aschemistry.co.uk*. The site is designed to help you prepare for tests and examinations based on the course specification you are studying. This web site will keep you up-to-date with the requirements for your course and show you how to use this book and the CD-ROM to study and learn each of the modules.

The web site also has links to pages in other web sites with useful information, software and illustrations.

1.2 Why study chemistry?

Here is a summary of the purposes of studying advanced chemistry based on the aims listed in course specifications. How well does this match with what you expect as you start an advanced course in this subject?

Developing knowledge and understanding

Chemistry is one of the sciences which helps us to understand ourselves and all the materials and living things which surround us.

Looking for patterns in chemical behaviour

Part of being a chemist is having a feeling for the ways in which chemicals behave. Chemists get to know chemicals just as people know their friends and family. Some of the patterns are familiar. Copper sulfate, for example, is blue like other copper compounds. The metals sodium and potassium are soft and stored under oil because they react so readily with air and water. Understanding more subtle patterns has made it possible for chemists to identify and make complex molecules such as the green pigment chlorophyll and the poison strychnine.

Working out the composition and structure of materials

New materials only exist because chemists now understand much more about the ways atoms are arranged in crystals and the forces which hold the atoms together. Thanks to this knowledge we can enjoy fibres which breathe but are waterproof, plastic ropes which are 20 times stronger than similar ropes of steel, metal alloys which can remember their shape, and ceramics which are magnetic. Explaining the behaviour of chemicals in terms of structure and bonding is a central theme in modern chemistry.

Controlling chemical change

Four seemingly simple questions lead to the heart of much of the study of chemistry today.

- *How much?* How much of these chemicals do we need to mix to get the product we want and how much of the product will we then get?
- *How fast?* How can we make sure that this reaction goes at the right speed: not too fast and not too slow? What are the factors we can vary to control reaction speeds?
- *How far?* Will these chemicals react completely to make the product we want or will the reaction seem to stop before we get all we want? If it does, what can we do to push the reaction on to get as big a yield as possible?
- *How?* What is going on during this reaction? Which bonds between atoms are breaking and which new ones are forming? If we understand this chemical mechanism, how can we control it to our advantage?

Test yourself

Remind yourself of some patterns in the ways that chemicals behave by recalling, or looking up, what you have learnt already in science.

1. How soluble are nitrates in water?
2. What happens when a more reactive metal (such as zinc) is added to a solution in water of a salt of a less reactive metal (such as aqueous copper sulfate)?
3. What forms at the negative electrode (cathode) during the electrolysis of a solution of a metal salt?
4. What happens on adding an acid (such as hydrochloric acid) to a carbonate (such as calcium carbonate)?
5. What do these compounds look like: sodium chloride, sodium bromide and sodium iodide?

Developing new skills

Chemistry is partly the knowledge and understanding written down in books such as this, but it is also what chemists do. Chemists need thinking skills and practical skills to create new knowledge and to apply existing knowledge to solve practical problems. Today one of the frontiers of the subject is in biochemistry where scientists are learning to understand complex molecules and processes in living things.

Increasingly chemists rely on modern instrumentation to explore structures and chemical change. They also use information technology to store data, search for information and to publish their findings.

Recognising the value of chemistry to society

Synthesis

Chemistry is about making things. Chemists take simple chemicals and join them together to make new substances. This is synthesis.

On a large scale the chemical industry converts raw materials from the earth, sea and air into valuable new products. A well-known example is the Haber process which turns natural gas and air into ammonia – the chemical needed to make fertilisers, dyes and explosives.

On a smaller scale, chemical reactions produce the specialist chemicals we need for perfumes, photography and medicines.

Analysis

A vital task for chemists is analysis to find out what things are made of. Doctors rely on analysis for diagnosing disease. Analysis is essential for checking that our drinking water is pure and our food is safe to eat. People rightly worry about pollution of the natural environment but without chemical analysis we would not know anything about the causes and scale of pollution.

Figure 1.2.1 ► *An analytical chemist using an atomic absorption spectrometer to determine the concentration of a metal in a liquid sample*

Understanding the link between theory and experiment

Scientists use experiments to test their theories. In chemistry this often begins with careful observations of what happens as chemicals react and change. Particularly convincing are the theories which lead to predictions which turn out to be correct when tested by experiment.

Chemists have devised many ingenious techniques for their experiments. In modern chemistry, spectroscopy is especially important. At first spectroscopists just used the light our eyes can see for their experiments but now they have found that they can find out much more by using other kinds of radiation such as ultraviolet and infra-red rays, radiowaves and microwaves.

Learning to enjoy and take an interest in the subject

It was the sight of colourful mineral crystals which attracted William Perkin (1838–1907) to chemistry. Remarkably the laboratory he set up in his home, when an 18 year-old student, led to a discovery which we now celebrate as the start of the modern dyestuffs industry. About a 100 years later it was also crystals which drew Dorothy Hodgkin to the subject. She won a Nobel prize in 1964 for her use of X-rays to tease out the structures of complex molecules such as penicillin and vitamin B12. Harry Kroto shared the Nobel prize for Chemistry in 1996 for his part in the discovery of a new form of carbon called 'buckyballs'. According to Sir Harry, *'science is to do with fun and solving puzzles'*.

Figure 1.2.2 ▼
A computer model of a C_{60} buckyball – 60 carbon atoms making a cage

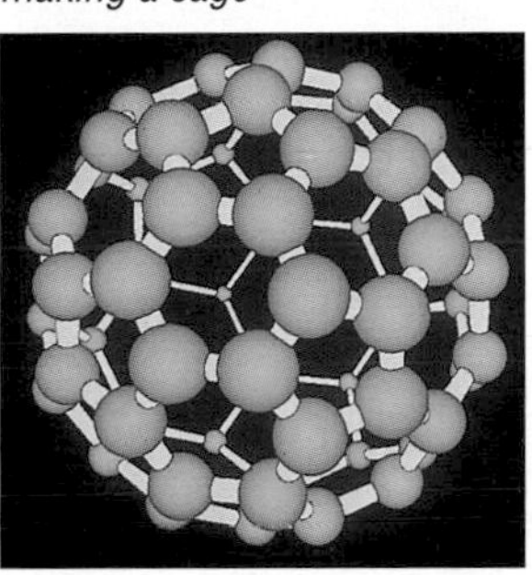

Figure 1.2.3 ◄
Professor Harry Kroto

Some people especially enjoy the practical side of chemistry, getting much pleasure from working with chemicals. They take satisfaction from the good technique which produces high yields of products and accurate results from analysis. Some are fascinated by the theory of chemistry as they discover how models of atoms and molecules can explain the ways in which materials react and change. Yet others are interested in chemistry because of the importance of its applications especially in medicine, pharmacy and dentistry.

1.3 Laboratory investigations

A chemist is familiar with unusual materials which people do not usually encounter in ordinary life. A chemist has to have a feeling for the ways in which these strange substances behave. So chemistry is essentially a practical subject. A laboratory contains the special equipment for handling, analysing and changing chemicals safely.

Studying chemical changes

The theory of chemistry only makes sense when linked to the experimental evidence on which it is based. The theory only comes alive when related to practical investigations.

Chemists' knowledge of reactions is based on many measurements of energy changes, temperature changes and reaction rates. Various forms of spectroscopy are increasingly important tools for detecting alterations to molecules during chemical reactions.

Test yourself

1 Give an example of a reaction which:

a) gives off a gas
b) creates a smell
c) produces a colour change
d) forms a precipitate
e) gets very hot.

2 Why might the results of the following examples of quantitative analysis be important and for whom?

a) the concentration of sugars in urine
b) the concentration of alcohol in blood
c) the percentage by mass of haematite in a rock sample
d) the concentration of nitrogen oxides in the air.

Qualitative analysis

Qualitative analysis answers the question 'What is it?'. In an advanced chemistry course this question can be answered with the help of careful observations while carrying out a series of test tube experiments. Changes include gases bubbling off, smells, new colours appearing, precipitates forming, solids dissolving and temperature changes.

The trick is to know what to look for. Some visible changes are much more significant than others and a good analyst will spot the important changes and know what they mean.

Success also depends on good technique when mixing chemicals, heating mixtures and testing for gases.

Reliable interpretation is based on a thorough knowledge of chemicals and how they behave.

Quantitative analysis

Quantitative analysis involves techniques which answer the question 'How much?'. In advanced courses the most important technique is volumetric analysis based on very accurate ways of measuring the volumes of solutions. Pipettes, burettes and graduated flasks make it possible to measure out volumes of solutions very precisely during a titration. There are correct techniques for using all this glassware which must be mastered for accurate results.

In many commercial laboratories, quantitative analysis is based on techniques such as spectroscopy and chromatography. Even so, the titration remains an important procedure for checking and calibrating these automated methods.

Almost all methods of analysis are based on solutions, so an essential skill is to be able to make up solutions to a required concentration and to dilute them accurately.

Synthesis

Much of the purpose and pleasure of chemistry comes from making new materials such as perfumes, drugs, pigments and dyes. Any synthesis involves mixing chemicals under controlled conditions to allow them to react, followed by steps to separate and purify the required product. Qualitative tests can then check that the required product has formed and quantitative methods can determine the yield and purity of the product.

Chemists have techniques for heating mixtures of chemicals safely so that they do not boil away while reacting. They also have techniques for separation and purification such as distillation and filtration.

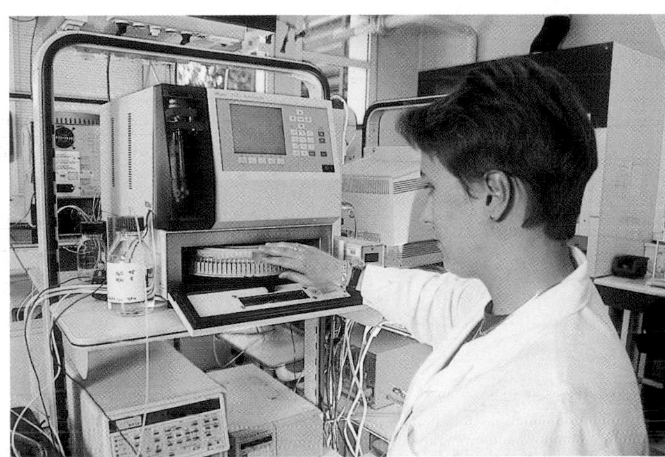

Figure 1.3.1 ◀
This technician is loading an auto-sampler during pharmaceutical drug research

Measuring melting and boiling points helps to identify products and determine how pure they are. Spectroscopy also has an important part to play. The infra-red spectrum of an organic product can be checked against a library of spectra to ensure that the product is as expected.

Planning investigations

Once you have mastered a range of chemical techniques, you will be able to plan your own investigations. This involves using your knowledge of chemistry to identify a question or problem to investigate.

Planning an investigation has to start with gathering information to find out more about the problem from a range of sources including textbooks and computer databases.

Planning involves making a series of decisions: about the reagents, about apparatus and techniques, about the precision of measurements and purity of chemicals, as well as decisions about scale of working, control of variables and safety.

1.4 Safety

Chemistry laboratories are very safe places. This is because knowledge of hazardous chemicals has been carefully documented. In any laboratory there are guidelines for most operations, such as mixing or heating chemicals, the use of fume cupboards, the precautions when using electrical equipment and the disposal of wastes.

The key to health and safety
STOP – THINK – DO

Most accidents arise from human error. People in laboratories should take time to think about what they are doing before they start and pay attention to what they are doing while they work.

Protective equipment such as laboratory coats and safety goggles should be thought of as a last resort or as a back-up to other safety measures.

Hazards and risks

Concentrated sulfuric acid is hazardous because it is corrosive. Anything that can cause harm if things go wrong is a hazard. Figure 1.4.1 shows the symbols used on bottles, jars and packs to identify chemical hazards.

Risk is a measure of the likelihood that someone will actually suffer harm from something hazardous. There is little risk from sulfuric acid when it is properly stored in a locked cupboard. The risk increases when someone carries a bottle of the acid into a laboratory ready for use.

Control of substances hazardous to health

The COSHH regulations (1988) control the use of substances hazardous to health. The regulations define which substances are hazardous including mixtures as well as pure substances. The regulations place the responsibility on employers who have a duty to carry out a risk assessment to prevent or control any exposure of their employees to hazardous substances.

Employees are expected to minimise exposure as far as possible by various means such as:

- substituting a new chemical that is less hazardous
- changing the method of working perhaps by redesigning the apparatus
- using fume cupboards or safety screens to keep people away from the hazardous substances
- providing personal protection such as goggles for eye protection.

Risk assessment

The COSHH regulations lay down procedures for risk assessment which are ultimately the responsibility of the employer. The governors in a school or college are the responsible employers but they delegate the responsibility to teachers. As a student you will be asked to carry out a risk assessment when planning an investigation. Always check your assessment with your teacher before starting any practical work.

1 Start by writing down your plans so that you have listed the chemicals you will use, given the quantities and described the procedures you are going to follow.
2 Use the standard reference sources to identify any hazardous chemicals. Record the hazards and the ways by which you might be exposed to them.
3 Decide what control measure you can take to reduce the risks. You might:
   - decide not to carry out the procedure and use a video or computer simulation instead
   - substitute a less hazardous chemical for one which seems to create too high a risk of harm
   - use a less hazardous form of the chemical such as a more dilute solution
   - change the design of the apparatus or lower the temperature
   - segregate yourself from the hazardous substances using a safety screen or fume cupboard
   - wear personal protection such as gloves.
4 Check that you know how to dispose of any hazardous residues once your practical work is complete.

Test yourself

1 Assess the risk of a practical activity in chemistry. Use these headings:
   - Title of activity
   - Outline of procedure
   - Hazardous substances used or made
   - Quantities used or made
   - Nature of the hazards
   - Sources of information about hazards
   - Control measures and precautions

VERY TOXIC
A substance which if inhaled, ingested or taken in through the skin, may involve extremely serious acute or chronic health risks and even death.

TOXIC
A substance which if inhaled, ingested or taken in through the skin, may involve serious acute or chronic health risk and even death.

NOTE: There are no specific symbols for carcinogen (may cause cancer), mutagen (may cause heritable genetic damage) or teratogen (may cause harm to unborn child). Such substances are labelled 'toxic' or 'very toxic', with a risk phrase to describe the way they may cause harm.

EXTREMELY FLAMMABLE
A liquid having a flash point of less than 0 °C and a boiling point of less than or equal to 35 °C.

HIGHLY FLAMMABLE
A substance which is:
- spontaneously flammable in air
- a solid and may catch fire after brief contact with a flame and keep burning after removal of flame
- gaseous and flammable in air at normal pressure
- liable to emit highly flammable gases when in contact with water or damp air
- a liquid with a flash point below 21°C.

FLAMMABLE
A liquid with a flash point greater than 21°C and less than or equal to 55 °C.

EXPLOSIVE
A substance which may explode under the effect of a flame or heat, or which is more sensitive to shocks or friction than dinitrobenzene.

CORROSIVE
A substance which on contact with living tissues may destroy the tissues.

IRRITANT
A non-corrosive substance which, through immediate, prolonged or repeated contact with the skin or eyes, may cause inflammation or lesions.

DANGEROUS FOR ENVIRONMENT
Materials which may harm the (mainly aquatic) environment.

HARMFUL
A substance which if inhaled, ingested or taken in through the skin, may involve limited health risks.

OXIDISING
A substance which produces a reaction giving off great heat when in contact with other substances, particularly flammable substances.

Figure 1.4.1 ▲
Hazard warning symbols

1.5 Units and measurements

SI units are the internationally agreed units for measurements in science. There are seven base units in the system. All other units are derived from the base units.

Every physical quantity in the system has a symbol. A physical quantity has a value and a unit. In calculations it is good practice to substitute both the value and the unit in formulae as shown in the worked examples in this book. The first six units in this table are the base units used in chemistry. Note that in print the symbols of physical quantities appear in italics but the units do not.

Figure 1.5.1 ▶

Physical quantity	Symbol	Unit	Unit symbol
length	l	metre	m
mass	m	kilogram	kg
time	t	second	s
electric current	I	ampere	A
temperature	T	kelvin	K
amount of substance A	n_A	mole	mol
volume	V	cubic metre	m^3
density	ρ	kilogram per cubic metre	$kg\ m^{-3}$
pressure	p	pascal (newton per square metre)	$N\ m^{-2}$
frequency	n	hertz	Hz

These are the quantities and units to use when substituting in formulae, but note that chemists often prefer other units for mass, volume and pressure.

There are also agreed prefixes for large and small numbers in the SI system.

Figure 1.5.2 ▶

Multiple or submultiple		Prefix	Symbol
1000	10^3	kilo-	k
0.1	10^{-1}	deci-	d
0.01	10^{-2}	centi-	c
0.001	10^{-3}	milli-	m
0.000 001	10^{-6}	micro-	μ
0.000 000 001	10^{-9}	nano-	n

Errors and accuracy

Errors

Random errors cause repeat measurements to vary and to scatter around a mean value. Averaging a number of readings helps to take care of these random errors. Systematic errors affect all measurements in the same way making them all lower or higher than the true value. Systematic errors do not average out. Identifying and eliminating systematic errors is important for increasing the accuracy of data.

Systematic errors can be reduced by using better equipment or improved practical technique.

Accuracy

Precise measurements have a small random error. So data are precise if repeat measurements have values which are close to each other.

Precise measurements may or may not be accurate. As a result of a systematic error, a series of precise measurements may give values which are almost the same but are not the true value.

Accuracy of data is determined by the agreement between a measured quantity and the correct value. In chemical analysis the correct value is often not known and so chemists need to estimate the errors which may have affected their results and set confidence limits on the values they quote.

Confidence limits are the limits around an experimental mean value within which there is a high probability that the true mean lies.

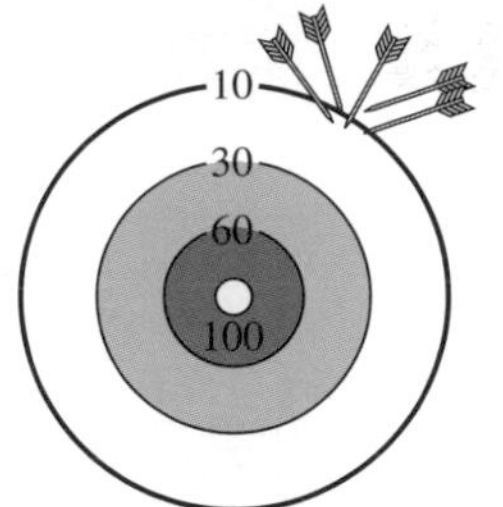

Poor accuracy but good precision

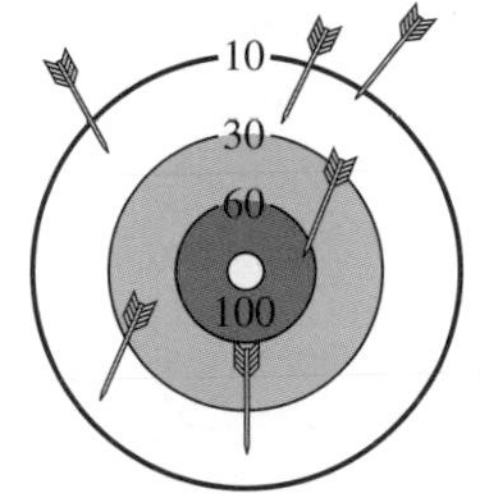

Poor accuracy and poor precision

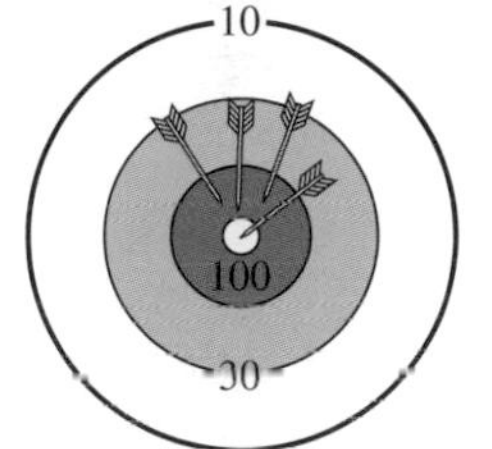

Good accuracy and good precision

***Figure 1.5.3* ▲**
Accuracy and precision – an analogy from archery

Significant figures

Measuring instruments vary in their accuracy and there is uncertainty associated with any measurement. The number of significant figures quoted for a measurement should show the degree of uncertainty.

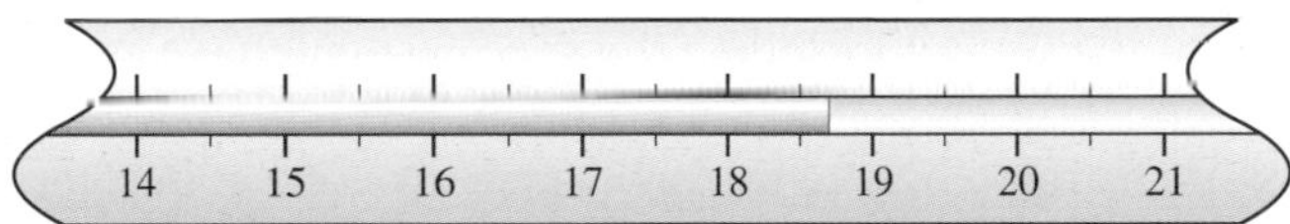

***Figure 1.5.4* ▲**
A thermometer scale. Is the temperature 18.7 °C or is it 18.6 °C or 18.8 °C?

Looking at Figure 1.5.4 there is little doubt that the temperature is between 18.5 °C and 19.0 °C. In this example, three significant figures are justified. Quoting the temperature to three significant figures as 18.7 °C shows that there is uncertainty in the final figure.

Putting values into standard form (see page 13) provides the clearest way of indicating the number of significant figures. This removes any doubt about zeros which sometimes indicate the position of the decimal point, as in 0.0056 g which is a very small mass quoted to two significant figures. This is clearer when the value is in the form 5.6×10^{-3}. In other values the zeros are included to show the number of significant figures. When writing the distance of 3500 m as 3.50×10^3 m it become clear that this is a quantity quoted only to three significant figures.

Test yourself

1 How many significant figures are there in these quantities?

a) 0.005 g
b) 24.0 cm^3
c) 35.5 g mol^{-1}
d) 3000 s

1.6 Key skills

The Chemical Industries Association has described how the workplace is changing. Now and in the future, employees have to be ready to work in teams to run projects, design systems and solve problems. A person may be managing a project on one day but be a subordinate the next. People at work need the key skills of communication, application of number and uses of information technology. Working with others in teams is crucial, so too is the ability to keep learning.

Communication

As you study chemistry you will talk about the subject as well as reading and writing about what you are learning. Talking, listening, reading and writing will all help you to make sense of the ideas.

One way to learn a new topic is to take notes by extracting ideas from a passage of text and representing them in different forms such as tables, charts and concept maps.

As you read, it helps if you can recognise that there are different styles of text. There is text which describes a procedure as a series of steps. There are narrative passages which tell a story, perhaps a story about an episode in the history of chemistry. There are passages which describe the properties and uses of elements and compounds. Other passages outline processes such as the refining of raw materials and manufacture of chemicals. There is text which explains ideas. There are passages which explore issues debating alternative points of view.

As well as reading, you have to be able to write about the subject to show that you have understood chemical concepts. You have to learn the conventions of the subject including the specialist language and the symbols. Chemists often use a combination of text and images to explain their ideas and learning how to do this effectively is an important aspect of becoming familiar with the subject.

Test yourself

1 Look through this book to identify different types of text. Identify and compare the structures of these types of text and learn to use them in your own writing. Notice when chemists use tables, diagrams, graphs and charts to clarify their writing. Find examples of:

- a step-by-step procedure
- a narrative telling a story
- a description of the properties of a chemical
- an account of a process
- an explanation of an idea
- a discussion of an issue.

Application of number

On an advanced course you must be able to work with numbers. The skill of applying numbers means that you know you can use maths to solve a problem. You should be able to answer questions such as 'What data do I need?', 'How do I obtain the data?', 'How do I organise the data and carry out calculations to an appropriate level of accuracy?' and 'How do I present my findings effectively?'. You must also know how to interpret the findings and explain what they mean, taking into account possible sources of error.

It is a big help if you can estimate the results of calculations (without a calculator) as a quick check that your answers are reasonable. When quoting a numerical result always check that the number of significant figures can be justified by referring to the accuracy of the data used in the calculation.

Some chemical ideas are applied most effectively by using a mathematical formula, so it is important to be able to rearrange formulae.

Any physical quantity, such as a mass, volume or temperature, has a numerical value and a unit. Checking that the units in a calculation are consistent is essential for a correct result.

Chemists have to think in three dimensions, appreciating angles and shapes when exploring crystal structures and shapes of molecules. They have devised ways of representing three-dimensional objects in two dimensions.

Standard form

Mathematicians and scientists use standard form to write very large or very small numbers. Standard form is based on powers of 10. So 1200 in standard form becomes 1.2×10^3.

The Faraday constant, for example = 96 480 C mol^{-1}
= $9.648 \times 10\,000$ C mol^{-1}
= 9.648×10^4 C mol^{-1}

which is now in standard form.

Information technology

As well as the general use of information technology for word processing and handling information with spreadsheets, there are other uses of IT which are very helpful in chemistry.

- There are databases which you can use to explore periodic patterns in the properties of elements and compounds.
- Modelling software will allow you to display the shapes and sizes of molecules.
- Sensors with data loggers make it possible for you to collect experimental data automatically and to display them graphically.
- The Internet has a growing number of websites with up-to-date case studies, facts and figures which complement the information in this book.

Skills of learning, problem solving and collaboration

As well as learning chemistry you will also be finding out how to study effectively. At every stage you will do better if you can set targets, plan the next stage of your studies and successfully meet your targets.

Laboratory investigations, research tasks and industrial visits in a chemistry course all provide opportunities to work effectively with others so that you each contribute something important and distinctive to a shared task. These activities will often involve challenges so that you have to analyse problems, agree what it would mean to solve the problems, suggest and assess possible solutions, take forward the preferred solution and keep under review the progress you are making towards a solution.

Test yourself

2 Visit a library or resource centre to find out more about the IT available which can help you study advanced chemistry. Find out how you can:

- plot a graph or chart to show the changes in properties of elements or compounds in the periodic table from a database on a CD-ROM
- display the shape of a molecule and measure bond lengths and angles
- simulate a process or procedure
- find out about the uses of an element, such as chlorine, from a website.

Section two

Foundations of Chemistry

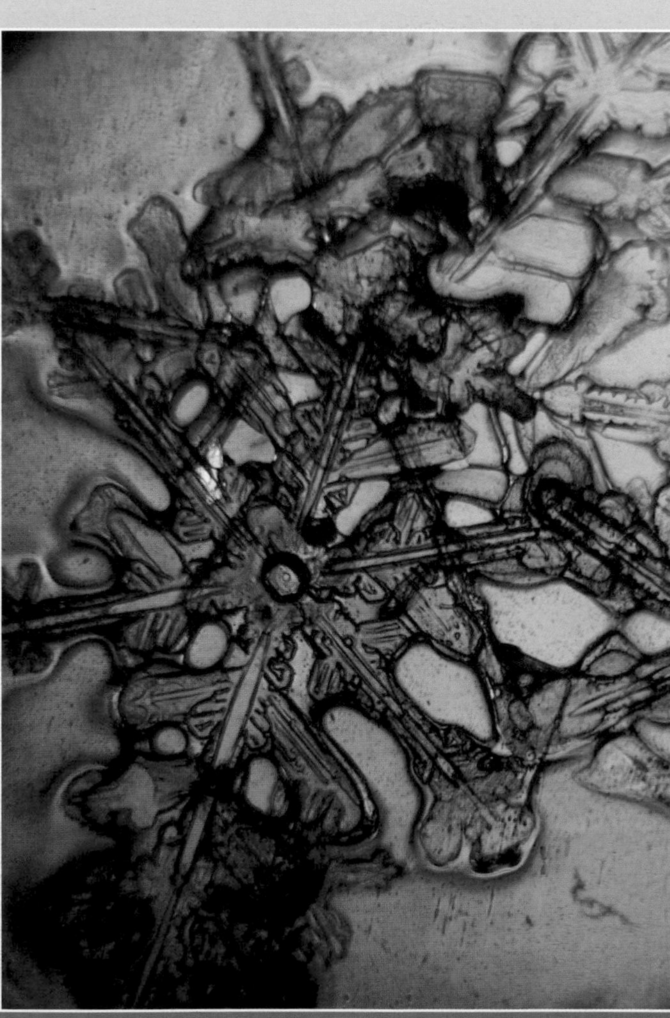

Contents

2.1 Chemicals and where they come from

2.2 States of matter

2.3 Matter and chemical change

2.4 Solutions

2.5 Chemical equations

2.6 Types of chemical changes

2.7 Chemical quantities

2.8 Finding formulae

2.9 Calculations from equations

2.10 Titrations

2.1 Chemicals and where they come from

Chemists are at the forefront of biomedical science, helping us to understand both how our bodies work and how we can use medicines to prevent or treat disease. Chemists help us to understand our environment and have developed the subtle techniques which make it possible to detect hazardous changes in the air we breathe or the water we drink. Chemists too have helped to develop modern materials which have transformed our homes, the way we travel and the games we play.

Atoms

Chemistry is the science of atoms and their magical transformations. Modern chemistry is not so much the science of 100 elements and their compounds, but of the infinite variety of molecules and structures which can be conjured up by building up atoms in different ways.

Chemists today have techniques which allow them to see atoms.

Molecules

Modern methods of analysis allow chemists to see how atoms link together to make molecules, even in complex molecules with many atoms. Today there is great interest in the molecules in plants, especially those plants which have traditionally been used as plant remedies. The periwinkle from Madagascar (*Catharanthus roseus*) is a source of anti-cancer drugs. Doctors use one of them, vinblastine, in the treatment of leukaemias, lung cancer and breast cancer.

Figure 2.1.1 ▲
The Madagascar periwinkle

Raw materials

The major raw materials for the chemical industry are fossil fuels (especially oil and natural gas), metallic ores and minerals, air and water.

The chemical industry in the UK is the country's fifth largest industrial sector. The industry grew on the basis of coal and a range of minerals such as salt and limestone which were available in the UK on a large scale. Today oil and natural gas from the North Sea are important raw materials for the petrochemical industry.

Purity

Many important chemicals come from the ground. Sometimes the chemicals are found pure but usually they are mixtures. Figure 2.1.2 shows a piece of granite consisting of a mixture of crystals. The feldspar, mica and quartz crystals are the minerals that make up the granite rock.

Chemists use the word pure to mean a single substance not mixed with anything else. So water or oxygen gas can be pure. Air cannot be pure in this sense because it is a mixture of nitrogen, oxygen, carbon dioxide and other gases.

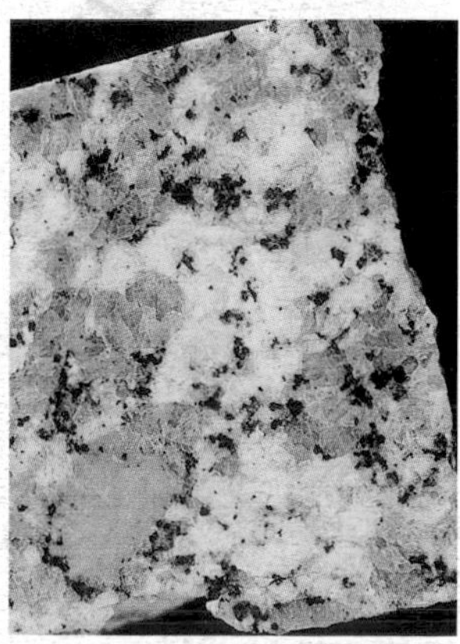

Figure 2.1.2 ▲
The polished surface of granite showing crystals of feldspar and mica

Some rocks are made of only one mineral. An important example is limestone which is quarried on a large scale for use in industry and agriculture. Limestone consists of calcium carbonate which also occurs naturally as chalk and marble.

Crude oil is a very complex mixture of chemicals, mostly hydrocarbons. The chemical industry has to separate, purify and process the chemicals in oil before it can use them as starting points for manufacturing pharmaceuticals, dyes and plastics.

The sea is another complex mixture consisting mainly of water and salts. Extracting pure elements, such as magnesium and bromine, from sea water is another challenge for the chemical industry.

Chemical plants

The genetic modification of crops is opening up new possibilities for chemistry. In time there could be less emphasis on traditional chemical plants with their distillation columns, reactors and furnaces often working at high temperatures and pressures. Genetic manipulation would make it possible to grow the raw materials for chemistry on the farm. Already there is an experimental form of cotton which produces the fibre already coloured with the blue dye for denim clothes.

The biodegradable polymer marketed as Biopol is the first polyester material to be produced commercially from bacteria. The bacterial cells form the polymer as an energy store as they grow for the same reason that animal cells produce fat.

Bioengineers have isolated the genes from the bacterium and are exploring the possibility of introducing them into crops so that the polymer could be harvested from fields instead of being manufactured in giant fermenters.

Meanwhile fields of oil seed rape are the source of biofuels which may increasingly replace oil and gas as part of the strategy for limiting the consequences of air pollution and global warming.

Figure 2.1.3 ►
Oil seed rape – a source of biofuels

2.2 States of matter

Chemists study the world around them and as they do so, they imagine in their minds what is happening to the atoms, molecules and ions as materials change, mix and react. The starting point is to have a clear picture in mind of what happens to the particles in solids, liquids and gases.

Solids, liquids and gases

Solids

Solids are rigid and keep their own shape. Many solids are dense and crystalline suggesting that the atoms, molecules or ions are packed close together in a regular way. The particles in a solid do not move freely but vibrate about fixed positions.

Figure 2.2.1 ▲
Crystals of copper sulphate

Liquids

Liquids flow and take the shape of any container. Some liquids flow easily, others like oil and treacle are thick and sticky – they are viscous liquids. Liquids, like solids, are hard to compress. The atoms or molecules in a liquid are closely packed but free to move around, sliding past each other.

Gases

Gases quickly spread out to fill the space available. They are much less dense than solids or liquids and are easy to compress. In a gas, the atoms or molecules are far apart. The particles move rapidly in a random way, colliding with each other and with the walls of the container.

In a gas the particles are spread out, so the densities of gases are very low compared with solids and liquids. The particles move rapidly in a random manner, colliding with other particles and the walls of the container. Pressure is caused by particles hitting the walls. Light particles move faster than heavier ones.

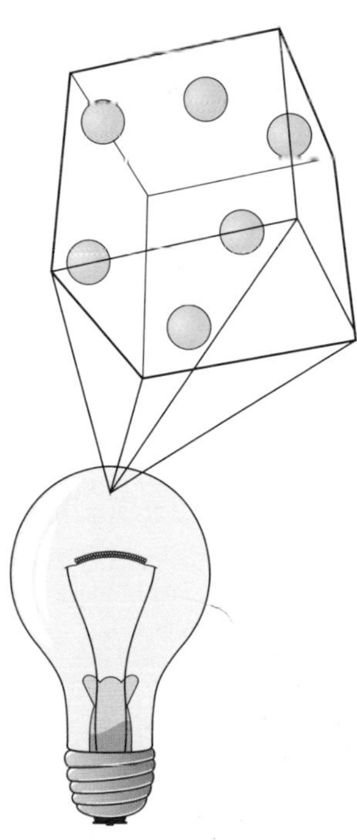

Particles in a solid are packed close together in a regular way. The particles do not move freely, but vibrate about fixed positions

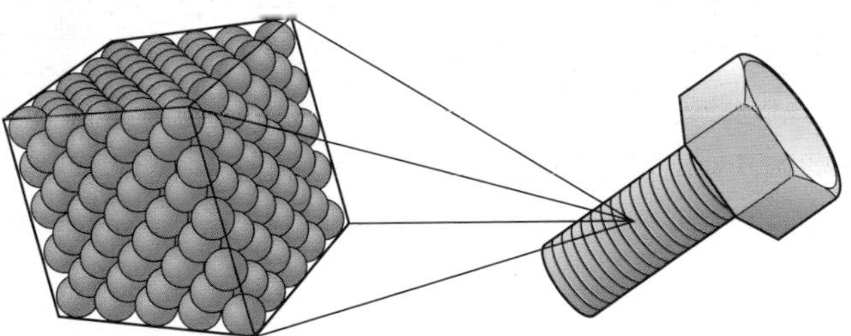

Figure 2.2.2 ▲ *The arrangement of particles in a solid*

The particles in a liquid are closely packed but are free to move around, sliding past each other.

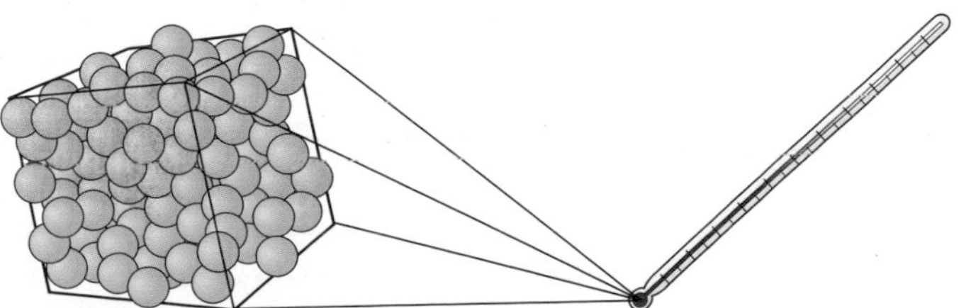

Figure 2.2.3 ▲ *The arrangement of particles in a liquid*

Changes of state

Melting and freezing

Melting is a change of state from a solid to a liquid. Another word for melting is fusion. The melting point for a pure substance is the temperature at which the solid and liquid are in equilibrium. Melting points vary with pressure but only very slightly.

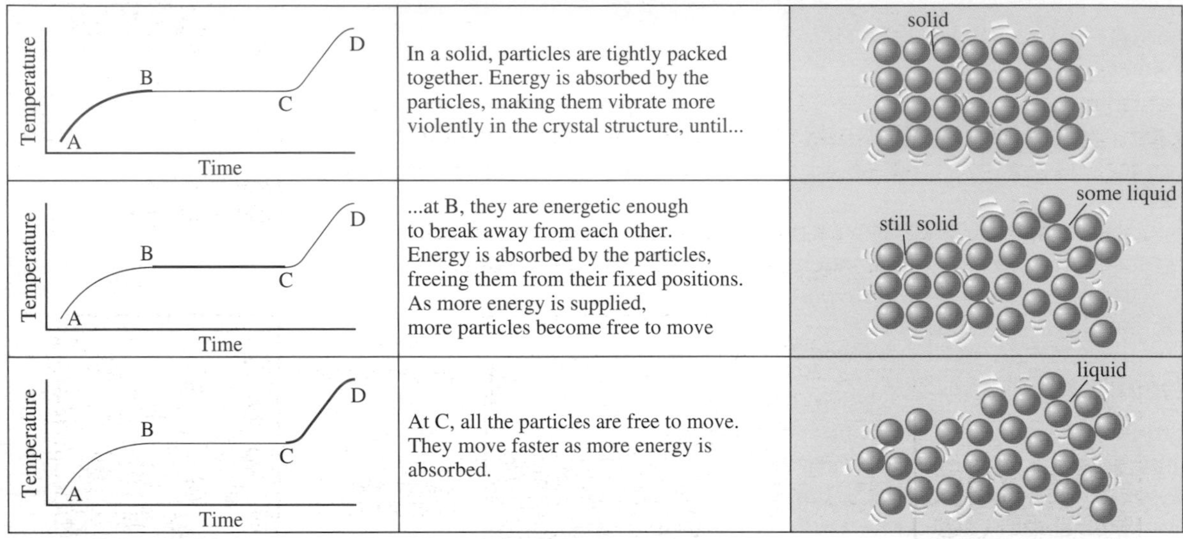

Figure 2.2.5 ▲ *The behaviour of particles in a pure solid as it melts*

Pure compounds have sharp melting points. Measuring melting points is a way of checking the purity and identity of compounds. This technique is especially important in organic chemistry.

Energy is needed to overcome the bonds between particles as a substance melts. The temperature of a pure substance stays constant as it melts. Instead of raising the temperature, the energy goes to overcoming the forces of attraction between the particles.

Pure substances freeze at the same temperature at which they melt. Dissolving a solute in a liquid lowers its freezing point. Adding antifreeze to the water in an engine lowers the freezing point and so prevents the coolant from freezing in winter.

When there is a threat of ice, the highway authorities scatter salt on roads because a mixture of salt and water freezes well below 0 °C.

Evaporation

Evaporation happens at the surface of a liquid as it turns to a gas. Evaporation is an endothermic process (see page 93). Energy must enter from the surroundings to keep a substance at a constant temperature as it evaporates. Liquids therefore feel cold as they evaporate on the skin.

Vapours

Vapours are gases formed by the evaporation of substances which are usually liquids or solids at room temperature. So chemists talk about oxygen gas but water vapour.

Vapours are easily condensed by cooling the substance or increasing the pressure. Physicists sometimes broaden the definition of a vapour to includes gases such as butane, ammonia and carbon dioxide which can be liquefied at room temperature just by increasing the pressure.

Test yourself D

1 Look up the melting and boiling points of these elements and compounds. Are they solid, liquid or gas at room temperature?

decane, eicosane, krypton, gallium, bromobutane, methanal, methanoic acid, silicon tetrachloride, hydrogen fluoride.

Boiling

A liquid boils when it is hot enough for bubbles of vapour to form within the body of the liquid. This happens when the pressure of the vapour escaping from the liquid is equal to the outside pressure.

The boiling point of a liquid varies with pressure. Raising the external pressure raises the boiling point. Boiling points are usually measured at atmospheric pressure. The normal boiling point is the temperature at which the vapour pressure of the liquid equals 1 atmosphere.

Figure 2.2.6 ▲
Iodine subliming

Subliming

Sublimation is the change of a solid directly to a gas on heating. Heating iodine crystals, for example, makes them sublime to a purple vapour which condenses to shiny crystals on a cold surface. This process is used to purify iodine.

Another substance which sublimes is solid carbon dioxide which is called 'dry ice' because it turns to gas at −78 °C without melting.

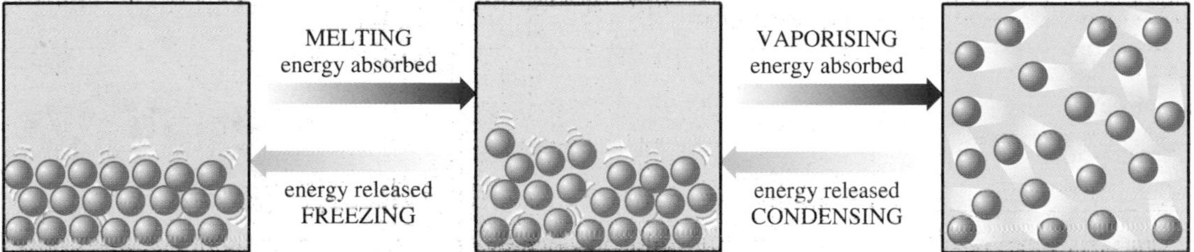

Figure 2.2.7 ▲
The energy changes that accompany changes of state

Liquid crystals

Liquid crystals are a state of matter which is more ordered than a liquid but less ordered than a solid. Although the liquid crystal state was first noticed as long ago as 1888, it is only since the early 1970s that they have been developed for use for digital watches, calculators and portable computer screens.

The molecules of liquid crystals are long and thin. When the solid melts, the liquid retains some but not all of the order of the solid.

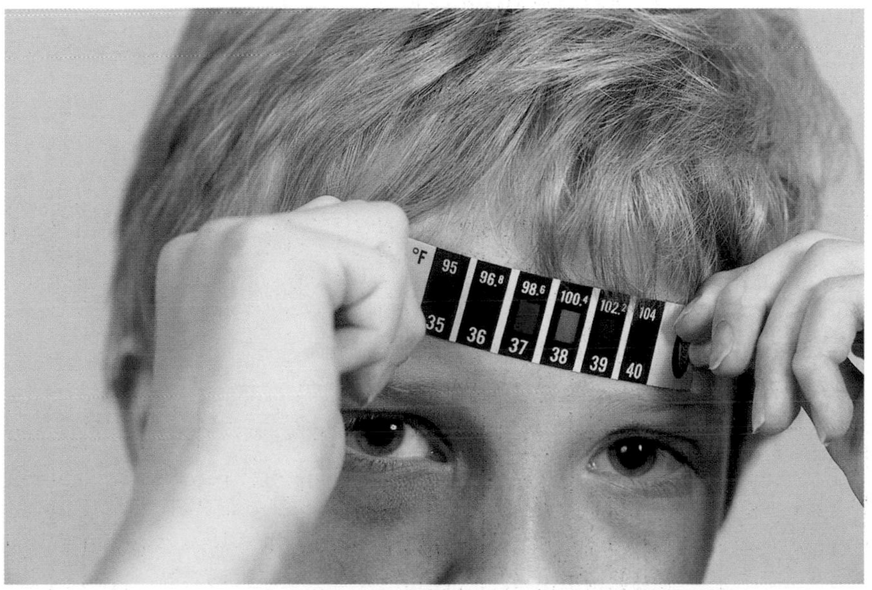

Figure 2.2.8 ◄
A liquid crystal thermometer

2.3 Matter and chemical change

Modern chemistry began once scientists understood the differences between elements and compounds and were able to explain chemical changes in terms of atoms and molecules. This section reminds you of the basic ideas which are fundamental to the study of chemistry.

Elements

Everything is made of elements, the simplest chemicals which cannot be simplified any further by heating or using electricity. There are over 100 elements, but from their studies of the stars, astronomers believe that about 90% of the universe consists of just one element, hydrogen. Another 9% is accounted for by helium, leaving only 1% for all the other elements.

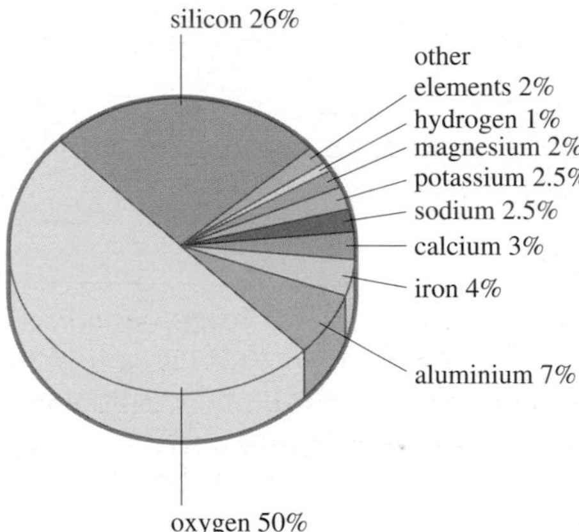

Figure 2.3.1 ►
The proportions of elements in the Earth's crust

Metals and non-metals

Most of the elements, nearly 90 of them, are metals. It is usually easy to recognise a metal by its properties. Metals are shiny, strong, bendable and good conductors of electricity.

Figure 2.3.2 ►
Samples of metals, from left to right, copper, zinc, lead and silver

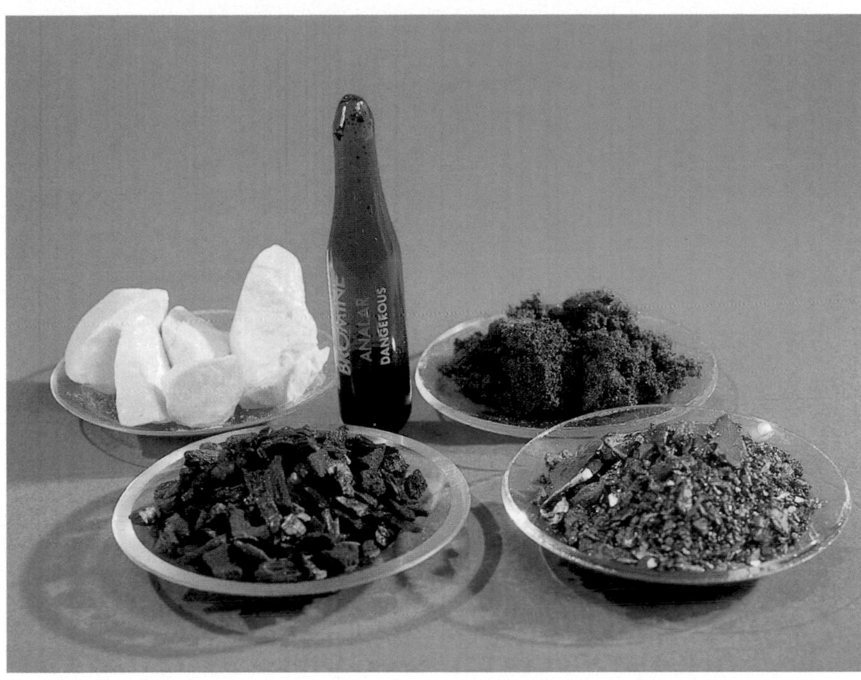

Figure 2.3.3 ◀
Samples of non-metals, sulfur, bromine, phosphorus, carbon and iodine

There are only 22 non-metal elements: this includes a few which are solid at room temperature, such as carbon and sulfur, several gases, such as hydrogen, oxygen, nitrogen and chlorine, and just one liquid, bromine.

Atoms of elements

Each element has its own kind of atom. The atoms consist of protons, neutrons and electrons. The mass of an atom is concentrated in a very small central nucleus consisting of protons and neutrons. The protons are positively charged and the neutrons uncharged.

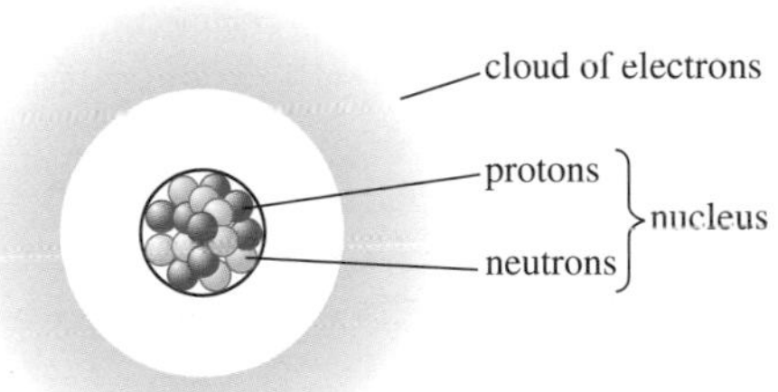

Figure 2.3.4 ◀
Diagram of an atom showing a nucleus surrounded by a cloud of electrons

Note

This book follows the recommendations of the International Union of Pure and Applied Chemistry (IUPAC) when spelling sulfur, sulfate and sulfuric acid instead of sulphur, sulphate and sulphuric acid. IUPAC is the recognised authority for the names of chemicals, for chemical symbols and for the values of chemical data such as relative atomic masses.

Around the nucleus are the electrons. The electrons are negatively charged. The mass of an electron is so small that it can often be ignored. In an atom the number of electrons equals the number of protons in the nucleus. So the total negative charge equals the total positive charge and overall the atom is uncharged.

	Relative mass	Charge
Proton	1	+1
Neutron	1	0
Electron	1/1870 (negligible)	−1

Figure 2.3.5 ◀

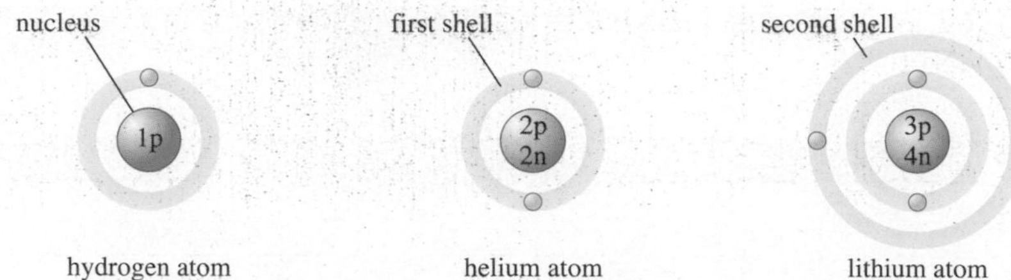

Figure 2.3.6 ▶
Structures of the three simplest atoms showing the electrons in shells

Electrons are arranged in a series of shells around the nucleus. Each shell can only contain a limited number of electrons. The shell nearest the nucleus fills first. When it is full, the electrons go into the next shell, and so on. Figure 2.3.6 shows the structures of the three simplest atoms.

A hydrogen atom has only one proton and no neutrons. It is the simplest and lightest atom, with a relative atomic mass of 1. Helium, with two protons and two neutrons, has a relative atomic mass of 4. The first shell can only hold two electrons, so in a lithium atom the third electron goes into the second shell. The second shell can hold up to eight electrons.

Test yourself

1. Give examples of substances which can be split into elements by heating or by using an electric current (electrolysis).
2. Check that you know the chemical symbols of the elements mentioned in the specification you are studying.
3. Draw up a table to compare metal elements with non-metal elements using the following headings: Property; Metal; Non-metal.
4. Draw diagrams to show the structures of atoms of these elements: beryllium, fluorine, sodium.

Compounds

Compounds form when two or more elements combine. Apart from the atoms of some noble gases (such as helium and neon), all atoms combine with other atoms.

Compounds of non-metals with non-metals

Sugar, water, alcohol, carbon dioxide and all the oily hydrocarbons in petrol are examples of compounds of two or more non-metals.

Most non-metal compounds melt and vaporise easily. They may be gases, liquids or solids at room temperature. They are generally insoluble in water unless they react with water. They do not conduct electricity in solution, even if they do dissolve, unless they react to make ions.

In most such compounds of non-metals with non-metals the atoms combine in small groups to form molecules. Methane in natural gas is an example. Each molecule in methane contains one carbon atom bonded to four hydrogen atoms. The formula of the molecule is CH_4. Figure 2.3.7 shows four ways of representing a methane molecule. Notice that in these diagrams the internal structure of atoms is not shown. The models represent atoms as little spheres.

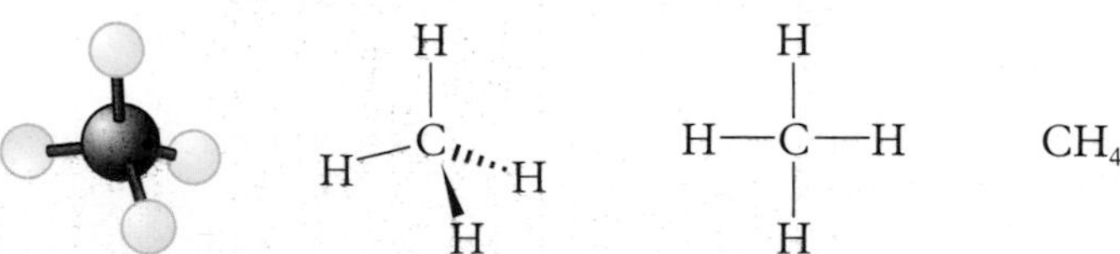

Figure 2.3.7 ▲
Ways of representing a molecule of methane

In many common examples it is possible to work out the likely formula of the molecules knowing how many bonds the atoms normally form (Figure 2.3.10).

Water is a compound of oxygen and hydrogen. Oxygen atoms form two bonds in molecules. Hydrogen atoms form one bond. Thus two hydrogen atoms can bond to one oxygen atom (Figure 2.3.8). The formula of water is H_2O.

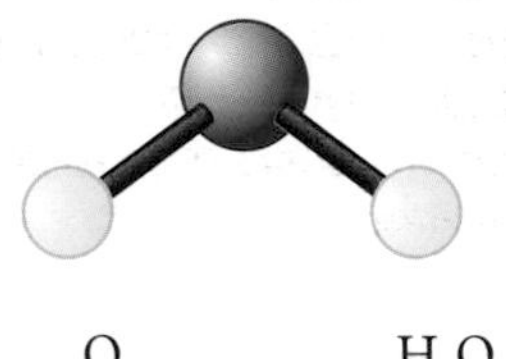

Figure 2.3.8 ▲
Ways of representing a molecule of water

O=C=O

Figure 2.3.9 ▲
Bonding in carbon dioxide showing double bonds between atoms

There are double or even triple bonds between the atoms in some compounds.

Figure 2.3.10 ◀

Element	Symbol	Number of bonds formed
carbon	C	4
nitrogen	N	3
oxygen	O	2
sulfur	S	2
hydrogen	H	1
chlorine	Cl	1

In practice, it is not possible to predict the formulae of all non-metal compounds in this way. The bonding rules in the table cannot account for the formulae of carbon monoxide, CO, or sulfur dioxide, SO_2. These formulae have to be learnt.

Compounds of metals with non-metals

Salt, marble, gypsum and sapphire are all examples of compounds of metals with non-metals.

Compounds of a metal with one or more non-metals conduct electricity when molten or when dissolved in water. They are electrolytes which decompose back into elements when they conduct an electric current. These compounds are electrolytes because they consist of ions.

Sodium chloride (common salt) is an example of an ionic compound. A sodium ion is a sodium atom with a positive charge because it has lost one electron. It has the symbol Na^+. A chloride ion is a chlorine atom with a negative charge because it has gained an extra electron. It has the symbol Cl^-.

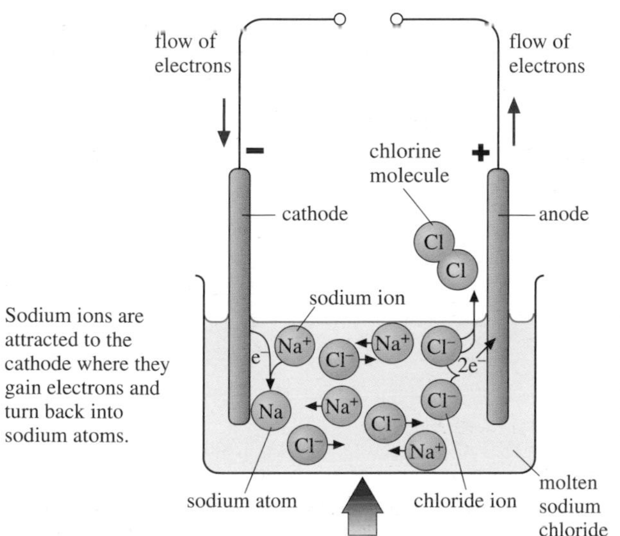

Figure 2.3.11 ◀
Electrolysis of molten sodium chloride splitting the compound into sodium and chlorine

In a crystal of sodium chloride there is an equal number of sodium ions and chloride ions. The formula of sodium chloride is NaCl because for every sodium ion there is one chloride ion.

Test yourself

5 In each of these molecular compounds, name the elements present and work out formula: hydrogen chloride, hydrogen sulfide, carbon disulfide, tetrachloromethane, ammonia.

6 Find examples to illustrate the generalisations that 'Most compounds of non-metals with non-metals are insoluble in water unless they react with water'.

Note

These are the chemical formulae of a few common minerals:

salt, NaCl
marble, $CaCO_3$
gypsum, $CaSO_4$
fluorite, CaF_2

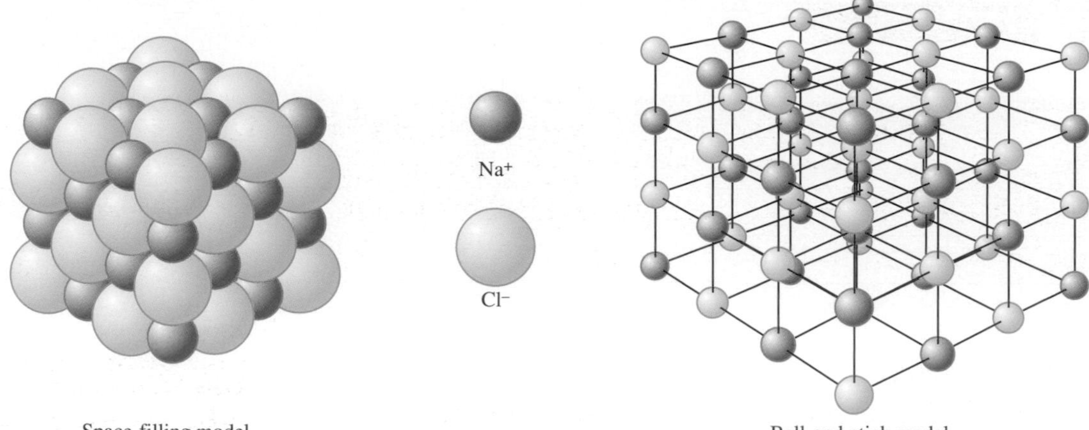

Figure 2.3.12 ▲ *Crystal structure of sodium chloride*

CD-ROM

Not all ions have single positive or negative charges. The formula of magnesium chloride is $MgCl_2$. In this compound, for every magnesium ion there are two chloride ions. All compounds are electrically neutral, so the charge on a magnesium ion must be twice the charge on a chloride ion. The symbol for a magnesium ion is Mg^{2+}.

Look at the table of common ions in Figure 2.3.13.

Figure 2.3.13 ►

Positive ions (cations)			Negative ions (anions)		
Charge	Cation	Symbol	Charge	Anion	Symbol
1+	sodium	Na^+	1−	chloride	Cl^-
	potassium	K^+		bromide	Br^-
	silver	Ag^+		iodide	I^-
	copper(I)	Cu^+		hydroxide	OH^-
	hydrogen	H^+		nitrate	NO_3^-
2+	magnesium	Mg^{2+}	2−	oxide	O^{2-}
	calcium	Ca^{2+}		sulfide	S^{2-}
	zinc	Zn^{2+}		sulfate	SO_4^{2-}
	copper(II)	Cu^{2+}		sulfite	SO_3^{2-}
	iron(II)	Fe^{2+}		carbonate	CO_3^{2-}
3+	aluminium	Al^{3+}	3−	nitride	N^{3-}
	iron(III)	Fe^{3+}		phosphate	PO_4^{3-}

The table shows that:

- metal ions are always positive, while non-metal ions are negative (this is true for all elements except hydrogen)
- some metals can form more than one ion – this is characteristic of transition metals such as copper and iron
- some non-metal ions are compound ions containing more than one type of atom – such as carbonate, sulfate and nitrate ions.

Test yourself

7 What are the formulae of these ionic compounds: potassium iodide, calcium carbonate, sodium sulfate, calcium hydroxide, aluminium chloride?

8 Find examples to illustrate this generalisation 'Compounds of a metal with one or more non-metals conduct electricity when molten or when dissolved in water'.

9 Which of these compounds consist of molecules and which consist of ions: copper oxide, concentrated sulfuric acid, magnesium chloride, lithium fluoride, phosphorus trichloride?

2.4 Solutions

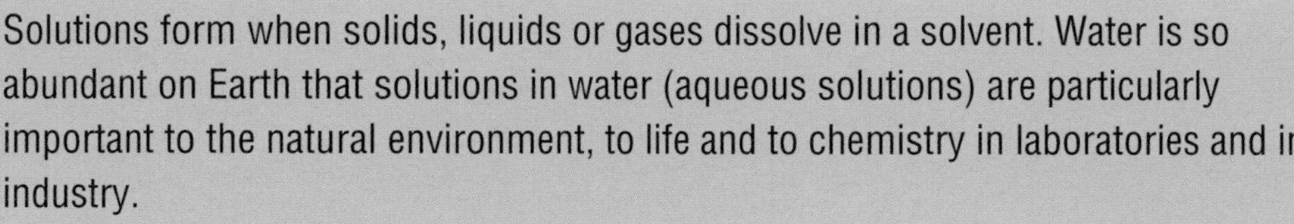

Solutions form when solids, liquids or gases dissolve in a solvent. Water is so abundant on Earth that solutions in water (aqueous solutions) are particularly important to the natural environment, to life and to chemistry in laboratories and in industry.

Patterns of solubility

As a general rule, 'like dissolves like' (see page 89). Water dissolves many ionic compounds and compounds with -OH groups such as alcohols and sugars. Oily solvents, such as paraffin, dissolve other oils and molecular elements or compounds.

Solids generally become more soluble in water as the temperature rises.

Gases become less soluble as the temperature rises. Heating water until it boils, for example, removes dissolved gases from the air. Bubbles of gas appear around the edge of a saucepan before the water boils. The bubbles contain air coming out of solution as the gases get less soluble with the rise in temperature.

Gases get more soluble as the pressure rises. This is how fizzy drinks get their sparkle. Opening a can or bottle of a fizzy drink lowers the pressure and cuts the solubility of the carbon dioxide so that the gas comes bubbling out of solution.

Note

A solute is the substance which dissolves in a solvent to make a solution. In sugar solution the solvent is water and the solute is sucrose. A solution of one or more solutes in water is an aqueous solution (aq).

Saturated solutions

There is a limit to the quantity of substance that will dissolve in water. A solution is saturated when it contains as much of the dissolved substance as possible at a particular temperature.

Soluble or insoluble?

No chemicals are completely soluble and none are completely insoluble. Even so, chemists find it useful to use a rough classification of solubility based on what they see on shaking a little of the solid with water in a test tube:

- very soluble, like potassium nitrate; plenty of the solid quickly dissolves
- soluble, like copper(II) sulfate; crystals visibly dissolve to a significant extent
- sparingly or slightly soluble, like calcium hydroxide; little solid seems to dissolve but the solution becomes quite strongly alkaline
- insoluble, like iron(III) oxide; no sign that any of the material dissolves.

A similar rough classification applies to gases dissolving in water. Ammonia and hydrogen chloride are very soluble. Sulfur dioxide is soluble. Carbon dioxide is slightly soluble. Helium is insoluble.

Test yourself D

1 Why is the solubility of gases such as oxygen and carbon dioxide in water important to living things?

2 With the help of a book of data, classify these solids according to their solubility in water (very soluble, soluble, slightly soluble, insoluble): sodium hydroxide, copper(II) oxide, potassium iodide, sodium chloride, manganese(IV) oxide, zinc sulfate, nickel(II) chloride.

2.5 Chemical equations

Burning, rusting and fermentation are all examples of chemical reactions. Under the right conditions, chemical bonds break and new ones form. This is what happens during a chemical reaction to create new chemicals.

Figure 2.5.1 shows a simple way of demonstrating that when hydrogen burns the product is water. Hydrogen and oxygen (in the air) are both gases at room temperature. When the gases react the changes gives out so much energy that there is a flame. Cooling and condensing the vapour from this flame gives water.

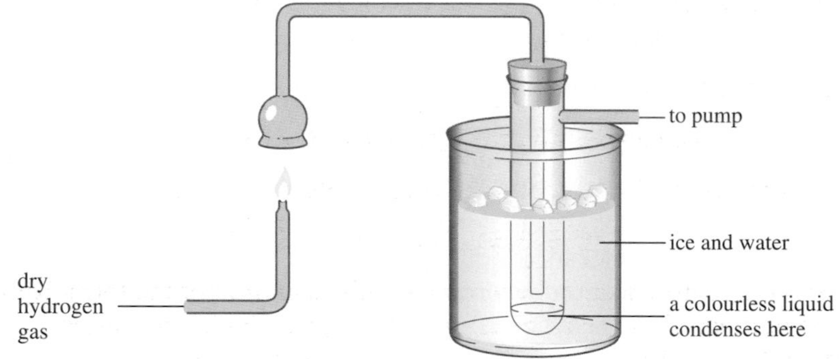

Figure 2.5.1 ► *Demonstration that burning hydrogen produces water. This highly exothermic reaction is used to power rockets for space travel*

Word equations

A word equation describes a chemical change in words. Writing word equations identifies the reactants (on the left) and products (on the right), so it is a useful first step towards a balanced equation with symbols. When hydrogen burns:

$$\underbrace{\text{hydrogen(g)} + \text{oxygen(g)}}_{\text{reactants}} \rightarrow \underbrace{\text{water(l)}}_{\text{product}}$$

Molecular models

In looking at this change chemists imagine what is happening to the molecules. The trick is to interpret the visible changes in terms of theories about atoms and bonding. Models help to make the connection.

We now know that hydrogen molecules and oxygen molecules consist of pairs of atoms. They are diatomic molecules. Figure 2.5.2 shows how molecular models give a picture of the reaction at an atomic level.

Figure 2.5.2 ▼ *Model equation to show hydrogen reacting with oxygen*

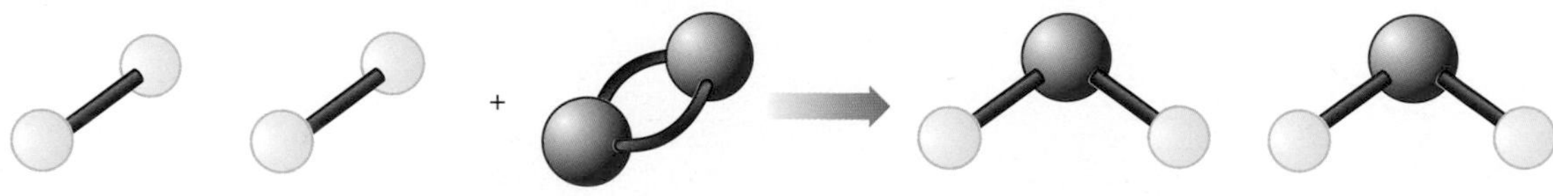

The formula of water is H_2O. Each water molecule contains only one oxygen atom. So one oxygen molecule can give rise to two water molecules provided that there are two hydrogen molecules available to supply all the hydrogen atoms necessary.

There are the same number of atoms on both sides of the equation – the atoms have simply been rearranged.

Chemists normally use symbols rather than models to describe reactions. Symbols are much easier to write or type. State symbols added to a symbol equation show whether the substances are solids, liquids, gases or dissolved in water.

$$2H_2(g) + O_2(g) \rightarrow 2H_2O(l)$$

Balanced symbol equations

Follow a step-by-step procedure to write the equation for a reaction. Before you start you have to know the names and formulae of the reactants and products. As an example consider the burning of methane which reacts with oxygen to form carbon dioxide and water.

Step 1: Write a word equation

methane + oxygen → carbon dioxide + water

Step 2: Write down the symbols for the reactants and products

$$CH_4 + O_2 \rightarrow CO_2 + H_2O$$

Step 3: Balance the equation by writing numbers in front of the formulae so that the number of each type of atom is the same on both sides of the equation. Do not change the formulae to balance the equation.

$$CH_4 + 2O_2 \rightarrow CO_2 + 2H_2O$$

Step 4: Add state symbols

$$CH_4(g) + 2O_2(g) \rightarrow CO_2(g) + 2H_2O(l)$$

Test yourself

1 Use molecular models to show what happens when methane burns in oxygen.

2 Write balanced equations for these changes: magnesium burning in oxygen, sodium reacting with water, calcium hydroxide neutralising hydrochloric acid.

2.6 Types of chemical change

Classifying chemical reactions helps chemists to make sense of all the many changes they study. Learning about chemicals according to their reactions then makes it possible to predict how they behave. Concentrated sulfuric acid, for example, is obviously an acid but it is also an oxidising agent and a dehydrating agent.

CD-ROM

Thermal decomposition

Thermal decomposition is a reaction in which a compound decomposes on heating. An important example for industry and agriculture is the thermal decomposition of calcium carbonate (limestone) to produce calcium oxide:

$$CaCO_3(s) \rightarrow CaO(s) + CO_2(g)$$

Sometimes heating causes decomposition because a compound is stable at room temperature but becomes unstable at a higher temperature. This is the case with the calcium carbonate.

Sometimes heating causes decomposition of a compound which is unstable at room temperature but does not decompose because the rate of reaction is so slow. This is true of the nitrogen oxides. They all tend to decompose into nitrogen and oxygen but only do so on heating.

Oxidation and reduction

Burning is perhaps the commonest example of oxidation. Another example is rusting. At its simplest, oxidation involves adding oxygen to an element or compound.

Reduction is the opposite of oxidation. Metal oxides are reduced during the extraction of metals from their ores. In a blast furnace, for example, carbon monoxide reduces iron oxide to iron.

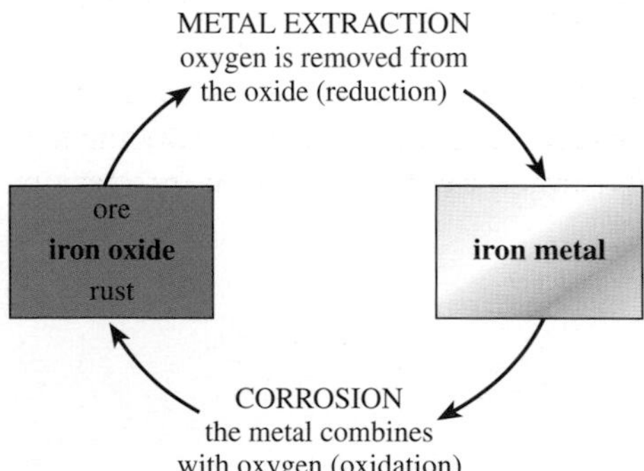

Figure 2.6.1 ►
The cycle of extraction and corrosion for iron

Electron transfer

Magnesium burns brightly in air. The product is a white solid – the ionic compound magnesium oxide, $Mg^{2+}O^{2-}$.

$$2Mg(s) + O_2(g) \rightarrow 2Mg^{2+}O^{2-}(s)$$

During the reaction each magnesium atom gives up two electrons, turning into a magnesium ion:

$$2Mg(s) \rightarrow 2Mg^{2+} + 4e^-$$

Oxygen takes up the electrons from the magnesium producing oxide ions:

$$O_2(g) + 4e^- \rightarrow 2O^{2-}$$

In this way electrons transfer from magnesium atoms to oxygen atoms, turning atoms into ions.

Magnesium atoms also turn into ions when they react with other non-metals such as chlorine, bromine and sulfur (Figure 2.6.2).

$$Mg(s) \rightarrow Mg^{2+} + 2e^-$$

$$Cl_2(s) + 2e^- \rightarrow 2Cl^-$$

Figure 2.6.2 *Electron transfer in the reaction of magnesium with chlorine*

In all its reactions with non-metals, magnesium loses electrons and it atoms turn into positive ions. The non-metals gain electrons and turn into negative ions. These are all examples of redox reactions involving electron transfer. Magnesium is oxidised as it loses electrons. The non metal is reduced as it gains electrons. **Red**uction and **ox**idation always go together, hence the term **redox** reaction.

Note

You may find the mnemonic **oil rig** helpful.

Oxidation	**R**eduction
Is	**I**s
Loss	**G**ain

Test yourself

1 Which element or compound is oxidised and which is reduced in the reaction of:
   a) steam with hot magnesium
   b) copper(II) oxide with hydrogen
   c) aluminium with iron(III) oxide in the thermit process
   d) carbon dioxide with carbon to form carbon monoxide.

2 Write symbol equations to show the transfer of electrons in the reaction of:
   a) sodium with chlorine
   b) zinc with oxygen
   c) calcium with bromine.

Oxidising and reducing agents

An agent is someone or something which gets things done. In spy stories, the main players are secret agents with a mission to make a change. In redox reactions the chemicals with a mission are oxidising and reducing agents.

The term oxidising agent (or oxidant) describes chemical reagents which can oxidise other atoms, molecules or ions by taking electrons away from them. Common oxidising agents are oxygen, chlorine, nitric acid, potassium manganate(VII), potassium dichromate(VI) and hydrogen peroxide.

The term reducing agent (or reductant) describes chemical reagents which can reduce other atoms, molecules or ions by giving them electrons. Common reducing agents are hydrogen, sulfur dioxide and zinc or iron in acid.

It is easy to get into a mental tangle when using these terms. When an oxidising agent reacts it is reduced. When a reducing agent reacts, it is oxidised. This is illustrated by the reaction of magnesium with chlorine (Figure 2.6.3).

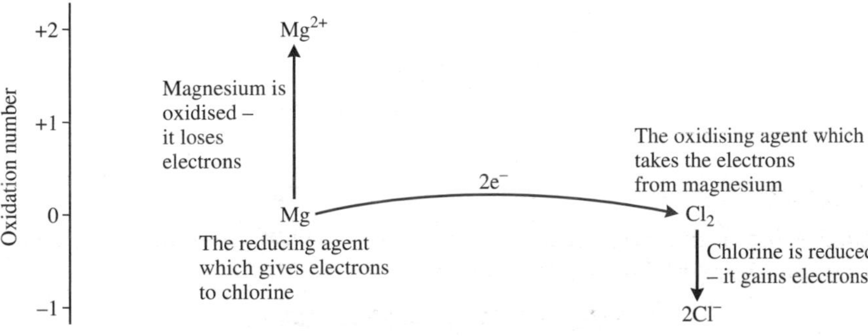

Figure 2.6.3 *Magnesium is oxidised by loss of electrons. It is oxidised by the chlorine, so chlorine is the oxidising agent. At the same time chlorine gains electrons and is reduced by the magnesium. The magnesium is the reducing agent.*

Half-equations

A half equation is an ionic equation used to describe either the gain or the loss of electrons during a redox process. Half equations help to show what is happening during a redox reaction. Two half equations combine to give an overall balanced equation for a redox reaction.

Zinc metal can reduce copper ions to copper. This can be shown as two half equations:

- electron gain (reduction) $Cu^{2+}(aq) + 2e^- \rightarrow Cu(s)$
- electron loss (oxidation) $Zn(s) \rightarrow Zn^{2+}(aq) + 2e^-$

The number of electrons gained must equal the number lost.

$$Cu^{2+}(aq) + 2e^- \rightarrow Cu(s)$$

$$Zn(s) \rightarrow Zn^{2+}(aq) + 2e^-$$

$$Cu^{2+}(aq) + Zn(s) \rightarrow Cu(s) + Zn^{2+}(s)$$

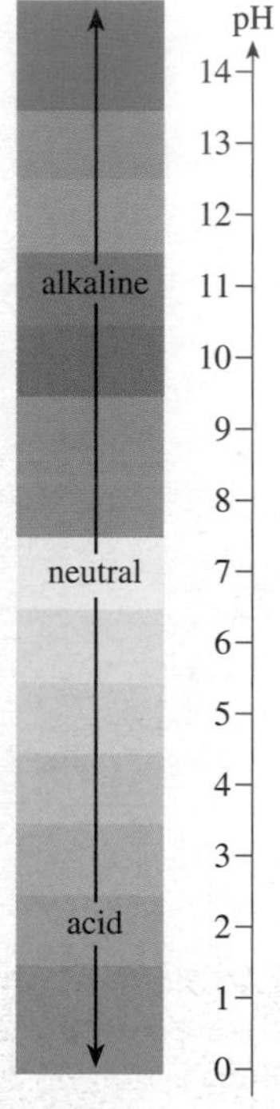

Figure 2.6.4 ▲
The pH scale

Acid–base reactions

Acids and bases neutralise each other to form salts. Sulfuric acid, for example, reacts with sodium hydroxide to make the salt sodium sulfate.

Acids

Acids are compounds with characteristic properties. They:

- form solutions in water with a pH below 7
- change the colours of acid–base indicators
- react with metals such as magnesium to produce hydrogen gas

$$Mg(s) + 2HCl(aq) \rightarrow MgCl_2(aq) + H_2(g)$$

- react with carbonates such as calcium carbonate to form carbon dioxide gas and water

$$CaCO_3(s) + 2HCl(aq) \rightarrow CaCl_2(aq) + CO_2(g) + H_2O(l)$$

- react with basic oxides to form salts and water.

$$CuO(s) + H_2SO_4(aq) \rightarrow CuSO_4(aq) + H_2O(l)$$

Pure acids may be solids (such as citric and tartaric acids), liquids (such as sulfuric, nitric and ethanoic acids) or gases (such as hydrogen chloride which becomes hydrochloric acid when it dissolves in water).

Test yourself

3 Write equations for the reaction of:
   a) calcium oxide with nitric acid
   b) zinc with sulfuric acid
   c) sodium carbonate with hydrochloric acid.

4 Give the names and symbols of the ions formed when these acids dissolve in water:
   a) nitric acid
   b) sulfuric acid.

What makes an acid an acid?

Why do so many different compounds have similar properties? Why do solutions of acids in water behave in the way that they do with indicators, metals, carbonates and bases?

It turns out that acids do not simply mix with water when they dissolve. What happens is that they react to produce aqueous hydrogen ions, $H^+(aq)$. So what acids have in common is that they produce hydrogen ions when they dissolve in water.

$$HCl(g) + water \rightarrow H^+(aq) + Cl^-(aq)$$

The typical reactions of dilute acids in water are the reactions of aqueous hydrogen ions.

With metals: $Mg(s) + 2H^+(aq) \rightarrow Mg^{2+}(aq) + H_2(g)$

With carbonates: $CO_3^{2-}(s) + 2H^+(aq) \rightarrow CO_2(g) + H_2O(l)$

With bases: $O^{2-}(s) + 2H^+(aq) \rightarrow H_2O(l)$

Strong and weak acids

Hydrochloric acid and nitric acid are strong acids. They ionise completely when they dissolve in water.

Ethanoic acid also ionises in water but not so readily. It is a weak acid. In a dilute solution only about one molecule of ethanoic acid in a hundred becomes an ion.

$$CH_3CO_2H(l) + \text{water} \rightleftharpoons CH_3CO_2^-(aq) + H^+(aq)$$

Bases and alkalis

Bases are 'ant-acids' – they are the chemical opposites of acids. Acids give away hydrogen ions; bases take them.

Alkalis are bases which dissolve in water. The common laboratory alkalis are the hydroxides of sodium and potassium, calcium hydroxide (in lime water) and ammonia. Alkalis form solutions with a pH above 7 so they change the colours of acid–base indicators.

In our homes we use alkalis to neutralise acids and to remove grease. Toothpaste is mildly alkaline to neutralise the acids which attack teeth. Milk of magnesia and other ingredients of antacid tablets are designed to neutralise stomach acid.

Manufacturers formulate powerful oven and drain cleaners to remove grease. These cleaners contain sodium or potassium hydroxides. These strong bases are highly 'caustic'. They attack skin. Even dilute solutions of these alkalis can be very hazardous, especially to eyes.

What alkalis have in common is that they dissolve in water to produce hydroxide ions, OH^-. Sodium hydroxide (Na^+OH^-) and potassium hydroxide (K^+OH^-) contain hydroxide ions in the solid as well as in solution. Ammonia produces hydroxide ions by reacting with water, taking a hydrogen ion from each water molecule.

$$NH_3(g) + H_2O(aq) \rightleftharpoons NH_4^+(aq) + OH^-(aq)$$

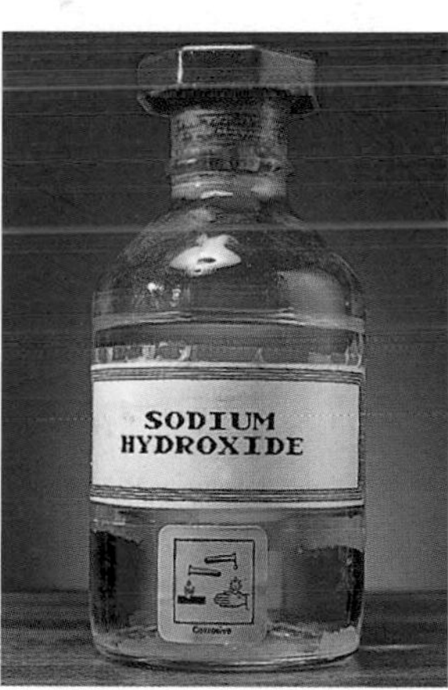

Figure 2.6.5 ▲
The label on a bottle of sodium hydroxide showing the hazard warnings. Sodium hydroxide is highly caustic – hence its older name, caustic soda

Neutralisation reactions

During a neutralisation reaction, an acid reacts with a base to form a salt.

$$HCl(aq) + NaOH(aq) \rightarrow NaCl(aq) + H_2O(l)$$

Mixing the right amounts of hydrochloric acid with sodium hydroxide produces a solution of sodium chloride.

Acids and alkalis neutralise each other because hydrogen ions react with hydroxide ions to form water, which is neutral.

$$H^+(aq) + OH^-(aq) \rightarrow H_2O(l)$$

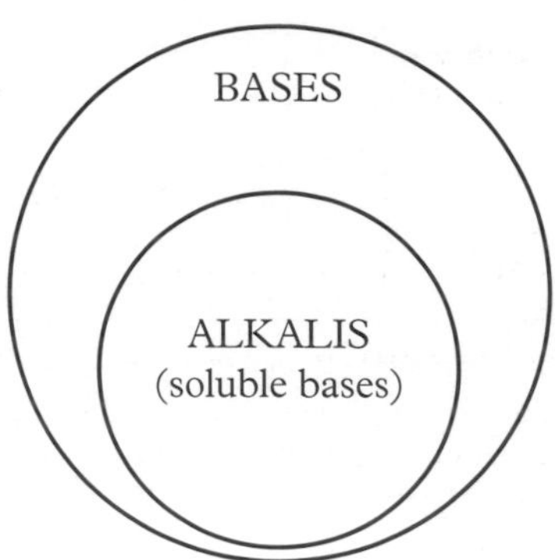

Figure 2.6.6 ▶

5 Copy the Venn diagram in Figure 2.6.6 and complete it by writing the names and formulae of common examples in the correct areas.

Salts

Salts are ionic compounds formed when an acid reacts with a base. In the formula of a salt, the hydrogen of an acid is replaced by metal ions. For example, magnesium sulfate, $MgSO_4$, is a salt of sulfuric acid, H_2SO_4.

Salts therefore have two 'parents'. Salts are related to a parent acid and to a parent base (see Figures 2.6.7 and 2.6.8).

Neutralisation is not the only way to make a salt. Some metal chlorides, for example, are made by heating metals in a stream of chlorine. This is useful for making anhydrous chlorides, such as aluminium chloride (see Figure 2.6.10).

Insoluble salts are conveniently prepared by ionic precipitation.

Parent acid	Salts
hydrochloric acid, HCl	sodium chloride, $NaCl$ calcium chloride, $CaCl_2$ ammonium chloride, NH_4Cl
sulfuric acid, H_2SO_4	sodium sulfate, Na_2SO_4 calcium sulfate, $CaSO_4$ ammonium sulfate, $(NH_4)_2SO_4$
ethanoic acid, CH_3CO_2H	sodium ethanoate, CH_3CO_2Na calcium ethanoate, $(CH_3CO_2)_2Ca$ ammonium ethanoate, $CH_3CO_2NH_4$

Figure 2.6.7 ▶

Parent base	Salts
sodium hydroxide, $NaOH$	sodium chloride, $NaCl$ sodium sulfate, Na_2SO_4 sodium ethanoate, CH_3CO_2Na
calcium hydroxide, $Ca(OH)_2$	calcium chloride, $CaCl_2$ calcium sulfate, $CaSO_4$ calcium ethanoate, $(CH_3CO_2)_2Ca$
ammonia, NH_3	ammonium chloride, NH_4Cl ammonium sulfate, $(NH_4)_2SO_4$ ammonium ethanoate, $CH_3CO_2NH_4$

Figure 2.6.8 ▶

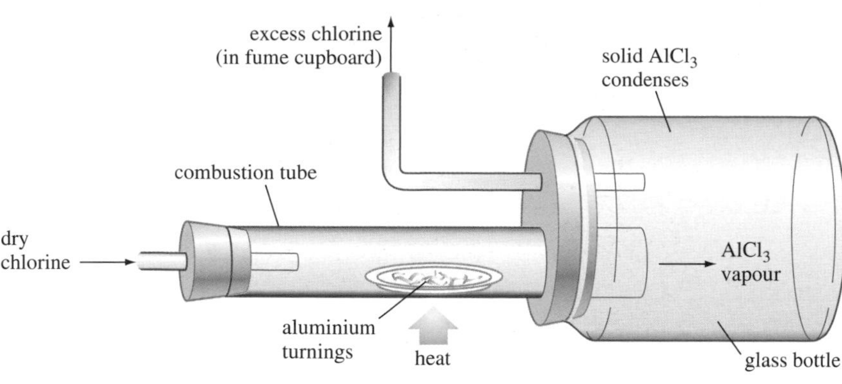

Figure 2.6.10 ▲ CD-ROM
Laboratory apparatus for combining aluminium with chlorine. Note the importance of checking that the chlorine is dry and keeping any moisture from the air out of the apparatus (see page 34). Aluminium chloride sublimes and condenses as a solid in the receiver

Ionic precipitation

A precipitate is an insoluble solid which separates out from a solution during a reaction. A precipitate forms as a greasy scum whenever people wash their hands in hard water. The whitish scum is a precipitate of calcium stearate – a salt of stearic acid. The calcium ions come from the hard water and the stearate ions from the soap.

Tests for anions

Many simple tests depend on the formation of precipitates which can be recognised by their colour. A solution of barium nitrate (or chloride) is the test reagent for sulfate ions (see page 245). Silver nitrate solution is used to test for chlorides, bromides and iodides (see pages 244–245).

Adding silver nitrate to a solution containing chloride ions produces an insoluble white precipitate of silver chloride. On mixing the two soluble salts there are two possible new combinations of ions: silver ions with chloride ions and sodium ions with nitrate ions. Silver chloride is insoluble so it precipitates. Sodium nitrate is soluble so the sodium and nitrate ions stay in solution.

Figure 2.6.9 ▲
Crystals of the mineral fluorite. Fluorite is an insoluble salt consisting of calcium fluoride, CaF_2. The mineral occurs as Blue John in caves in the Peak District of Derbyshire

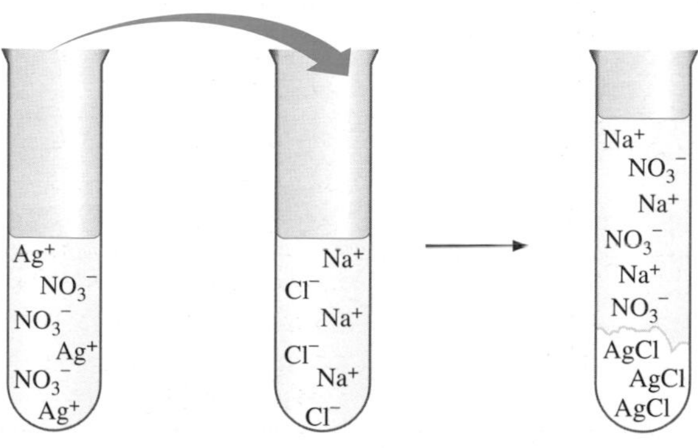

Figure 2.6.11 ◄
The precipitation of silver chloride

Test yourself D

6 Will a precipitate form on mixing these pairs of solutions? If yes, what is the name and formula of the precipitate?

a) zinc sulfate and barium nitrate
b) potassium nitrate and copper(II) sulfate
c) sodium carbonate and calcium chloride
d) lead(II) nitrate and sodium chloride
e) sodium hydroxide and copper(II) sulfate

7 Classify each of the following reactions as: redox, acid–base, precipitation or hydrolysis. Write word and symbol equations for each reaction.

a) zinc burning in chlorine
b) hydrochloric acid reacting with calcium hydroxide
c) hydrogen reacting with copper(II) oxide
d) sulfuric acid reacting with copper(II) oxide
e) barium nitrate solution reacting with sodium chloride solution
f) heating aluminium chloride with water.

Spectator ions

Adding silver nitrate to a solution of potassium chloride produces a precipitate of insoluble silver chloride.

$$\underbrace{Ag^+(aq) + NO_3^-(aq)}_{\text{spectator ions}} + K^+(aq) + Cl^-(aq) \rightarrow AgCl(s) + \underbrace{NO_3^-(aq) + K^+(aq)}_{\text{spectator ions unchanged}}$$

The potassium and nitrate ions remain in solution unchanged. For clarity, chemists omit spectator ions and write an ionic equation.

$$Ag^+(aq) + Cl^-(aq) \rightarrow AgCl(s)$$

Tests for cations

Most metal hydroxides are insoluble in water. The common exceptions are the hydroxides of sodium, potassium and barium. This is the basis of a scheme for identifying common metal ions (see page 244). Adding a solution of sodium hydroxide to a solution of a metal salt normally produces a precipitate of the metal hydroxide. The hydroxides of transition metals, such as copper and iron, can be identified from their colours. The analyst needs more tests to tell the difference between the white hydroxides of metals such as aluminium, calcium and zinc.

Hydrolysis

Hydrolysis reactions use water to break chemical bonds, usually with the help of an acid or alkali to act as a catalyst. Hydro-lysis is based on two Greek words which mean water-splitting.

The reason that moisture must be kept out of the apparatus to make aluminium chloride (see Figure 2.6.10) is that water hydrolyses hot aluminium chloride to aluminium oxide and hydrogen chloride.

Many hydrolysis reactions happen in the gut as the digestive system breaks down food. Hydrolysis splits starch into sugars, proteins into amino acids and fats into fatty acids. There are enzymes to speed up all these reactions.

2.7 Chemical quantities

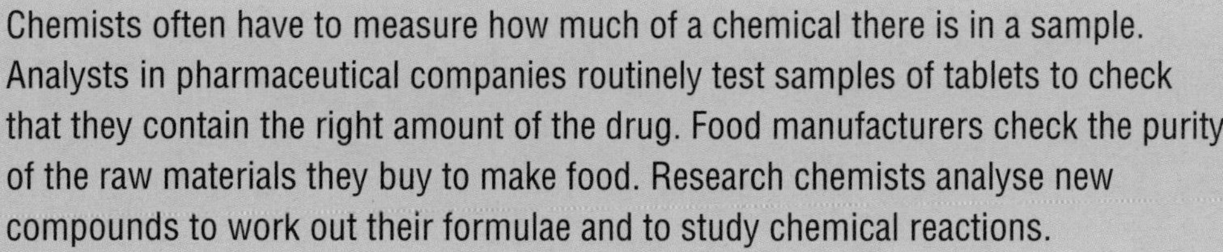
Chemists often have to measure how much of a chemical there is in a sample. Analysts in pharmaceutical companies routinely test samples of tablets to check that they contain the right amount of the drug. Food manufacturers check the purity of the raw materials they buy to make food. Research chemists analyse new compounds to work out their formulae and to study chemical reactions.

Chemical amounts

When finding formulae or working with equations, chemists need to measure amounts containing equal numbers of atoms, molecules or ions. Chemists have balances for determining masses in kilogrammes and graduated glassware for measuring volumes in dm^3 (litres), but there is no measuring instrument for determining chemical amounts directly. Instead chemists first measure masses or volumes and then calculate the chemical amount.

Chemists use the term mole for a 'heap of stuff' containing a standard number of atoms, molecules, ions or any other type of particle. The word 'mole' entered the language of chemistry at the end of the nineteenth century based on the Latin word for a heap or pile.

Relative atomic masses

The key to working with chemical amounts in moles is to know the relative masses of atoms. The accurate technique for determining relative atomic masses is mass spectrometry (see pages 52–54).

Originally chemists measured atomic masses relative to hydrogen, the element with the lightest atoms. In a mass spectrometer it turns out that it is much more accurate to use the most common isotope of carbon as the standard. So now the atomic mass, A_r, of an element is the mean mass of the atoms of the element relative to the mass of atoms of the isotope carbon-12 for which $A_r = 12$ exactly. The values are relative so they do not have units.

Relative molecular mass, M_r

Most non-metals consist of molecules. For example: oxygen, O_2; sulfur, S_8; hydrogen H_2 and phosphorus, P_4.

Most compounds of non-metals with non-metals are also molecular. Examples include water, H_2O, carbon dioxide, CO_2, hydrogen chloride, HCl, and silicon tetrachloride, $SiCl_4$.

$$\text{For ethanol, } M_r(CH_3CH_2OH) = (2 \times \underset{A_r(C)}{12}) + (6 \times \underset{A_r(H)}{1}) + \underset{A_r(O)}{16} = 46$$

Note

Relative atomic mass: the symbol for relative atomic mass is A_r, where the r stands for 'relative'. The values for A_r are average values for the mixture of isotopes found naturally (see pages 54–55). This means that values of relative atomic masses are often not whole numbers. (See page 239 for a table of relative atomic masses.)

Test yourself D

1 How many times heavier are:

a) Mg atoms than C atoms?
b) N atoms than Li atoms?
c) S atoms than He atoms?
d) Fe atoms than N atoms?

Definition

The **relative molecular mass** of the molecules of an element or compound is the sum of the relative atomic masses for the atoms in the molecular formula.

Relative formula mass, M_r

Salts consist of giant structures of ions. To avoid the suggestion that the formulae represent molecules, chemists use the term relative formula mass for ionic compounds and also for other compounds with giant structures such as silicon dioxide, SiO_2.

For anhydrous magnesium nitrate, $M_r[Mg(NO_3)_2]$

$$= \underset{A_r(Mg)}{24} + (2 \times \underset{A_r(N)}{14}) + (6 \times \underset{A_r(O)}{16}) = 148$$

Test yourself D

2 What is the relative molecular mass of:

a) phosphorus, P_4?
b) silicon tetrachloride, $SiCl_4$?
c) sulfuric acid, H_2SO_4?

Molar mass

Amount of substance in chemistry is defined in such a way that the mass of a mole of an element is numerically equal to its relative atomic mass. The relative atomic mass of carbon is 12 so the molar mass of carbon atoms = 12 g mol^{-1}.

The molar mass of hydrogen atoms, $M(H)$ = 1 g mol^{-1}. The molar mass of hydrogen molecules $M(H_2)$ = 2 g mol^{-1}.

Similarly the molar mass of the molecules of an element or compound is numerically equal to its relative molecular mass. The molar mass of ethanol is 46 g mol^{-1}. Likewise the molar mass of an ionic compound is numerically equal to its relative formula mass. The molar mass of magnesium nitrate is 148 g mol^{-1}.

Definition

The **relative formula mass** of an ionic compound is the sum of the relative atomic masses for the atoms in the formula.

Test yourself D

3 What is the relative formula mass of:

a) potassium chloride, KCl?
b) aluminium bromide, $AlBr_3$?
c) iron(III) sulfate, $Fe_2(SO_4)_3$?

Amount in moles

The mole is the SI unit for amount of substance. One mole is the amount of substance that contains as many specified atoms, molecules, ions, electrons and so on as there are atoms in exactly 12 g of the carbon-12 isotope.

Amount of substance is a physical quantity (symbol *n*) which is measured in the unit mole (symbol mol). In chemistry the word 'amount' has this precise meaning. For any pure substance:

$$\text{amount of substance (mol)} = \frac{\text{mass of substance (g)}}{\text{molar mass (g mol}^{-1})}$$

Definition

Molar mass: the mass of one mole of a chemical. The symbol for the molar mass is *M*. The unit is g mol^{-1}. As always with molar amounts, the atom, molecule or ion must be specified.

Figure 2.7.1 ▶ *1 mol amounts of atoms of elements*

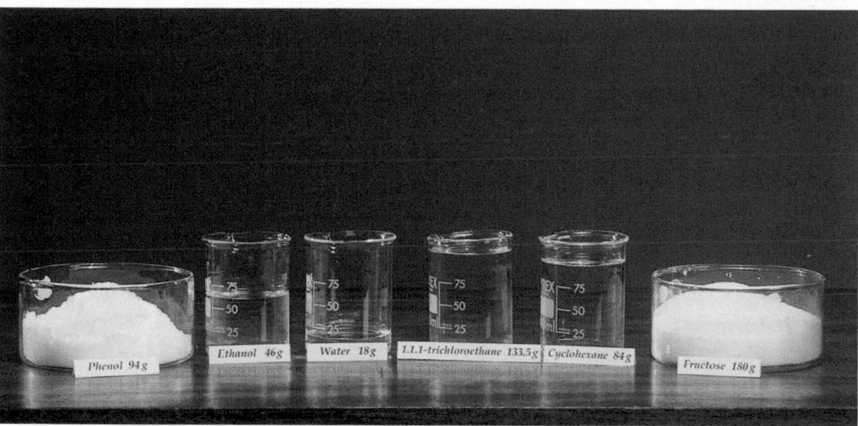

Figure 2.7.2 ▲
1 mol amounts of molecules of compounds

It is important to be precise when measuring amounts in moles. In calcium chloride, $CaCl_2$, for example, there are two chloride ions, Cl^-, combined with each calcium ion., Ca^{2+}. So in one mole of calcium chloride there is one calcium ion mole and two moles of chloride ions.

Avogadro constant

The Avogadro constant is the number of atoms, molecules or ions in one mole of a substance. The Avogadro constant, $L = 6.02 \times 10^{23}\ mol^{-1}$. This vast number is 602 000 000 000 000 000 000 000 particles per mole.

$$\text{amount of substance/mol} \times \text{Avogadro constant/mol}^{-1} = \text{number of specified particles}$$

It is important, as always, to specify the particles. For example 0.25 mol carbon atoms, C, contains 1.50×10^{23} atoms while 0.5 mol oxygen molecules, O_2, contains 3.01×10^{23} molecules.

Figure 2.7.3 ▲
1 mol amounts of some ionic compounds

Test yourself D

4 What is the amount in moles of:

- **a)** 20 g of calcium atoms?
- **b)** 4 g bromine atoms?
- **c)** 160 g of bromine molecules?
- **d)** 6.4 g of sulfur dioxide molecules?
- **e)** 10 g of sodium hydroxide, NaOH?

5 What is the mass of:

- **a)** 0.1 mol iodine atoms?
- **b)** 0.25 mol of chlorine molecules?
- **c)** 2 mol of water molecules?
- **d)** 0.01 mol ammonium chloride, NH_4Cl?
- **e)** 0.125 mol of sulfate ions, SO_4^{2-}?

6 How many moles of:

- **a)** sodium ions are there in 1 mol of sodium carbonate, Na_2CO_3?
- **b)** bromide ions are there in 0.5 mol of barium bromide, $BaBr_2$?
- **c)** nitrogen atoms are there in 2 mol of ammonium nitrate, NH_4NO_3?

7 Use the Avogadro constant to calculate:

- **a)** the number of chloride ions in 0.5 mol of sodium chloride, NaCl
- **b)** the number of oxygen atoms in 2 mol of oxygen molecules, O_2
- **c)** the number of sulfate ions in 3 mol aluminium sulfate, $Al_2(SO_4)_3$.

2.8 Finding formulae

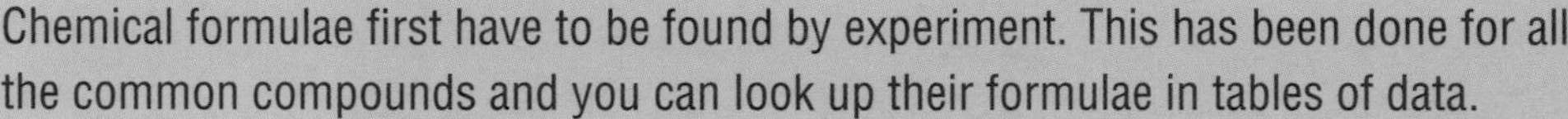
Chemical formulae first have to be found by experiment. This has been done for all the common compounds and you can look up their formulae in tables of data.

Experimental formulae

Empirical evidence is information based on experience or experiment. So chemists use the term empirical for formulae calculated directly from experimental results.

An experiment to find a formula involves measuring the mass of elements which combine in the compound. For carbon compounds chemists use combustion analysis to do this (see page 180).

The empirical formula shows the simplest ratio of the amounts in moles of elements in the compound; it therefore gives the ratios of the numbers of atoms.

Test yourself D

1 What it the formula of a compound in which:

a) 3.6 g magnesium combines with 1.4 g nitrogen?
b) 0.6 g carbon combines with 0.2 g hydrogen?
c) 1.38 g sodium combines with 0.96 g sulfur and 1.92 g oxygen?

Worked example

Analysis of a sample of a salt shows that it consists of 0.378 g iron combined with 1.622 g bromine. What is the formula of the salt?

Notes on the method

The molar masses of the elements come from a table of data.

Recall that:

$$\text{amount of substance/mol} = \frac{\text{mass of substance/g}}{\text{molar mass/g mol}^{-1}}$$

Answer

	iron	bromine
Combining masses	0.378 g	1.622 g
Molar masses of elements	56 g mol^{-1}	80 g mol^{-1}
Amounts combined	$\frac{0.378 \text{ g}}{56 \text{ g mol}^{-1}}$	$\frac{1.622 \text{ g}}{80 \text{ g mol}^{-1}}$
	= 0.00675 mol	= 0.0203 mol
Simplest ratio of amounts	1	3

The formula is $FeBr_3$.

Percentage composition

This is the percentage by mass of each of the elements in a compound. Percentage composition is one way of expressing the results of chemical analysis. The empirical formula of the compound can be calculated from these results.

Worked example

What is the empirical formula of copper pyrites which has the analysis 34.6% copper, 30.5% iron and 34.9% sulfur?

Notes on the method

Follow the procedure in the worked example for finding an empirical formula. The percentages show the combining masses in a 100 g sample.

Answer

	copper	iron	sulfur
Combining masses	34.6 g	30.5 g	34.9 g
Molar masses of elements	64 g mol^{-1}	56 g mol^{-1}	32 g mol^{-1}
Amounts combined	$\frac{34.6\text{ g}}{64\text{ g mol}^{-1}}$	$\frac{30.5\text{ g}}{56\text{ g mol}^{-1}}$	$\frac{34.9\text{ g}}{32\text{ g mol}^{-1}}$
	= 0.54 mol	= 0.54 mol	= 1.09 mol
Simplest ratio of amounts	1	1	2

The formula is $CuFeS_2$.

Test yourself D

2 What is the empirical formula of:

a) an acid with the analysis 2.04% H, 32.46% S and 65.4% O?
b) an alcohol with the analysis 52.2% C, 13.0% H and 34.8% O?

The percentage composition of a compound can be worked out from its chemical formula. This is a guide to people who formulate products such as fertilisers, medicines and cleaning agents.

Worked example

Two common nitrogen fertilisers are urea, $(H_2N)_2CO$, and ammonium nitrate, NH_4NO_3. Compare the percentage of nitrogen in the two compounds.

Notes on the method

Note that when part of a formula is in brackets, the number outside refers to all the atoms in the bracket.

Answer

Relative formula mass of urea = 2 × (2 + 14) + 12 + 16 = 60 of which (2 × 14) = 28 is nitrogen

Percentage of nitrogen in urea $= \frac{28}{60} \times 100\% = 46.7\%$

Relative formula mass of ammonium nitrate = 14 + 4 + 14 + (3 × 16) = 80 of which (2 × 14) = 28 is nitrogen

Percentage of nitrogen in ammonium nitrate $= \frac{28}{80} \times 100\% = 35\%$

Urea contains the higher percentage of nitrogen.

Test yourself D

3 What is the percentage of copper in each of these minerals:

a) cuprite, Cu_2O?
b) malachite, $Cu_2(OH)_2CO_3$?
c) bornite, Cu_5FeS_4

2.9 Calculations from equations

An equation is not just a useful shorthand for describing what happens during a reaction. Chemists also use equations to calculate the correct proportions in which to mix the reactants and the expected yield of product (see page 217).

Calculating masses of reactants and products

Follow these four steps to solve problems based on equations:

1 Write the balanced equation for the reaction.
2 Write the amounts in moles for the reactants and products of interest.
3 Convert amounts in moles to masses.
4 Scale the masses to the quantities required.

Test yourself D

1 What mass of calcium oxide, CaO, forms when 25 g calcium carbonate, $CaCO_3$, decomposes on heating?

2 In a blast furnace, carbon monoxide, CO, reduces iron(III) oxide, Fe_2O_3, to iron. Calculate the mass of carbon monoxide required to reduce 16 tonnes of iron and the mass of iron formed.

3 What mass of sulfur combines with 8.0 g copper to form copper(I) sulfide, Cu_2S?

Worked example

What mass of ethanol forms when 9 g glucose ferments?

Notes on the method

Work out the molar masses, M, in g mol^{-1}.

The carbon dioxide can be ignored here.

Answer

Step 1: $C_6H_{12}O_6(aq) \rightarrow 2C_2H_5OH(aq) + 2CO_2(g)$

Step 2: 1 mol $C_6H_{12}O_6(aq)$ ferments to make 2 mol $C_2H_5OH(aq)$.

Step 3: $M(C_6H_{12}O_6)$

$$= (6 \times 12\text{ g mol}^{-1}) + (12 \times 1\text{ g mol}^{-1}) + (6 \times 16\text{ g mol}^{-1})$$
$$= 180\text{ g mol}^{-1}$$

$$M(C_2H_5OH) = (2 + 12\text{ g mol}^{-1}) + (6 + 1\text{ g mol}^{-1}) + 16\text{ g mol}^{-1}$$
$$= 46\text{ g mol}^{-1}$$

So $(1\text{ mol} \times 180\text{ g mol}^{-1}) = 180$ g glucose ferments to give $(2\text{ mol} \times 46\text{ g mol}^{-1})$ = 92 g ethanol

Step 4: Hence 9 g glucose produces $= \frac{9}{180} \times 92\text{ g} = 4.6$ g ethanol

Gas volumes

The volume of a sample of gas depends on three things:

- the temperature
- the pressure
- the amount of gas in moles.

Avogadro's law

At a fixed pressure and temperature the volume of gas depends only on the number of gas molecules. The type of gas does not matter. This idea was first suggested by the Italian scientist Amadeo Avogadro (1776–1856). He showed that this theory could explain the reacting volumes of gases.

Avogadro's law states that equal volumes of gases under the same conditions of temperature and pressure contain equal numbers of molecules (and so equal amounts in moles).

The molar volume of a gas

What Avogadro's law means is that one mole of any gas occupies the same volume under the same conditions. This is the molar volume of the gas under the given conditions.

The volume occupied by one mole of a gas at 273 K (0 °C) and 1 atmosphere pressure (101.3 kPa) is about 22.4 dm^3 (22 400 cm^3). These are the conditions of standard temperature and pressure (stp).

In a warmish laboratory at 298 K (25 °C) the volume of one mole of gas is about 24 dm^3 (24 000 cm^3) at 1 atmosphere pressure.

$$\text{volume of gas/cm}^3 = \text{amount of gas/mol} \times \text{molar volume/cm}^3\ \text{mol}^{-1}$$

So under laboratory conditions, the volume of gas formed in a reaction can be estimated from this relationship:

$$\text{volume of gas/cm}^3 = \text{amount of gas/mol} \times 24\ 000\ \text{cm}^3\ \text{mol}^{-1}$$

Gay-Lussac's law of combining volumes

The French chemist Joseph Gay-Lussac (1778–1850) showed by experiment that when gases are involved in reactions, the ratios of the volumes of gases are simple whole numbers so long as all measurements are taken at the same temperature and pressure.

dry ammonia
3-way tap
dry hydrogen chloride
100 75 50 25
25 50 75 100
Syringe A
Syringe B

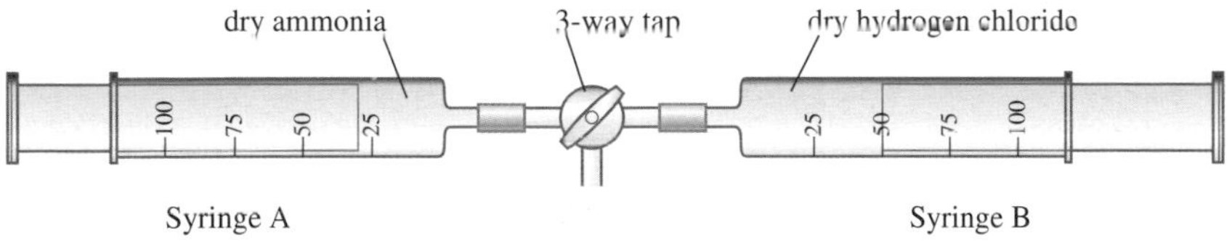

Figure 2.9.1 ◄
Apparatus for measuring the reacting volumes on mixing 30 cm³ dry ammonia with 50 cm³ dry hydrogen chloride. On mixing, a white solid forms. The volume of solid is insignificant. The volume of gas remaining is 20 cm³ which turns out to be excess hydrogen chloride. So 30 cm³ ammonia reacts with 30 cm³ hydrogen chloride. The ratio of the volumes is 1:1

Avogadro's law accounts for Gay-Lussac's observations. If equal volumes of gases contains equal numbers of molecules under the same conditions, it follows that the ratios of the volumes are also the ratios of the number of molecules which are the same as the ratios of the amounts in moles.

$$\begin{array}{ccc} NH_3(g) & + \ HCl(g) & \rightarrow NH_4Cl(s) \\ 1\ \text{mol} & 1\ \text{mol} & \\ 1\ \text{volume} & 1\ \text{volume} & \end{array}$$

Gas volume calculations

Gas volume calculations are straightforward when the reactants and products are all gases. The ratio of the gas volumes in a reaction must be the same as the ratio of the numbers of moles in the equation.

Test yourself

4 What is the amount in moles of gas molecules at room temperature and pressure in:

a) 240 000 cm^3 chlorine?
b) 48 cm^3 hydrogen?
c) 3 dm^3 ammonia?

5 What is the volume of the following amounts of gas at room temperature and pressure?

a) 2 mol nitrogen
b) 0.0002 mol neon
c) 0.125 mol carbon dioxide?

Worked example

What volume of oxygen reacts with 60 cm^3 methane and what volume of carbon dioxide forms if all gas volumes are measured under the same conditions?

Notes on the method

Write the balanced equation.

Note that below 100 °C the water formed condenses to an insignificant volume of liquid.

Answer

The equation for the reaction:

$$CH_4(g) + 2O_2(g) \rightarrow CO_2(g) + 2H_2O(l)$$
$$\text{1 mol} \quad \text{2 mol} \quad \text{1 mol}$$

So 60 cm^3 methane reacts with 120 cm^3 oxygen to form 60 cm^3 carbon dioxide.

Test yourself

6 Assuming that all gas volumes are measured under the same conditions of temperature and pressure, what volume of:

a) nitrogen forms when 2 dm^3 ammonia, NH_3, decomposes completely into its elements?

b) oxygen is needed to react with 50 cm^3 ethane, C_2H_6, when it burns, and what volume of carbon dioxide forms?

The other approach to gas volume calculations is based on the fact that the volume of a gas, under given conditions, depends only on the amount of gas in moles.

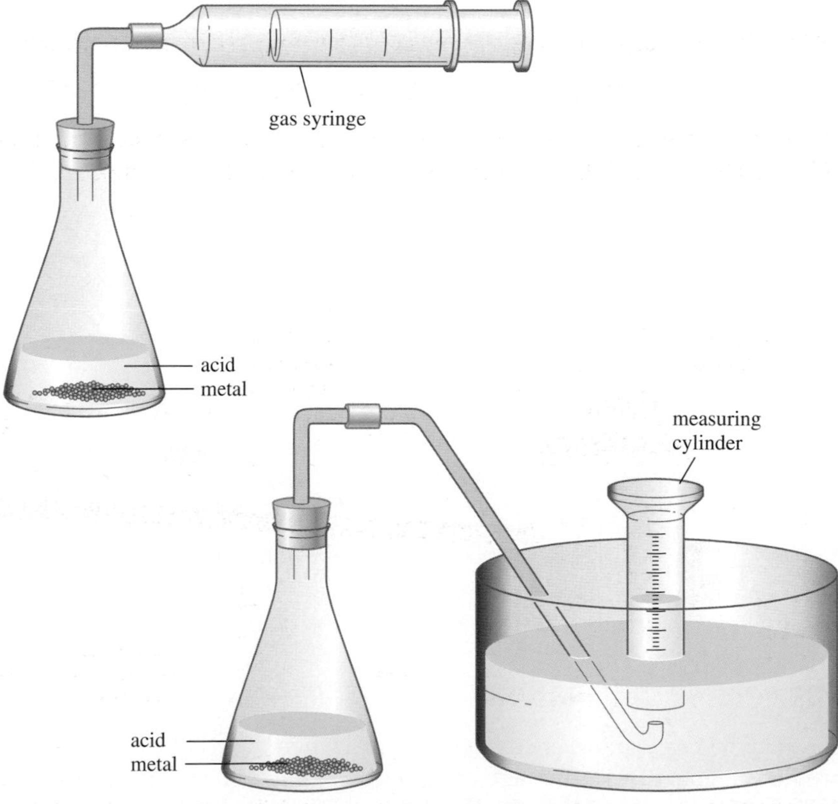

Figure 2.9.2 ► *Apparatus for measuring the volume of gas formed when a metal reacts with an acid*

Worked example

What volume of hydrogen is produced under laboratory conditions when 0.024 g magnesium reacts with excess hydrochloric acid?

Notes on the method

Start by writing the equation for the reaction. Convert the quantity of magnesium to an amount in moles.

Answer

The equation for the reaction is:

$$Mg(s) + 2HCl(aq) \rightarrow MgCl_2(aq) + H_2(g)$$

The amount of zinc = 0.024 g ÷ 24 g mol^{-1} = 0.001 mol

1 mol magnesium produces 1 mol hydrogen.

Volume of hydrogen = 0.001 mol × 24 000 cm^3 mol^{-1} = 24 cm^3

Test yourself D

7 What volume of gas, measured under normal laboratory conditions, forms when:

a) 0.35 g lithium reacts with water?
b) 0.65 g zinc react with excess dilute hydrochloric acid?
c) 0.76 g potassium nitrate, KNO_3, decomposes on heating to potassium nitrite, KNO_2?

Solution calculations

Concentrations of solutions

Concentrations in solution are usually measured in moles per litre of solution (mol dm^{-3}). Chemists talk about the 'molarity' of solutions. A concentration of 1.0 mol dm^{-3} is often written as 1.0 M for short.

$$\text{concentration/mol dm}^{-3} = \frac{\text{amount of solute/mol}}{\text{volume of solution/dm}^3}$$

When ionic crystals dissolve, the ions separate and become independent.

$$CaCl_2(s) + aq \rightarrow Ca^{2+}(aq) + 2Cl^-(aq)$$

So if the concentration of $CaCl_2$ is 0.1 mol dm^{-3}, then the concentration of Ca^{2+} is also 0.1 mol dm^{-3} but the concentration of Cl^- is 0.2 mol dm^{-3}.

Note

There are small volume changes when chemicals dissolve in water so it is important to note that concentrations normally refer to litres of solution not to litres of the solvent.

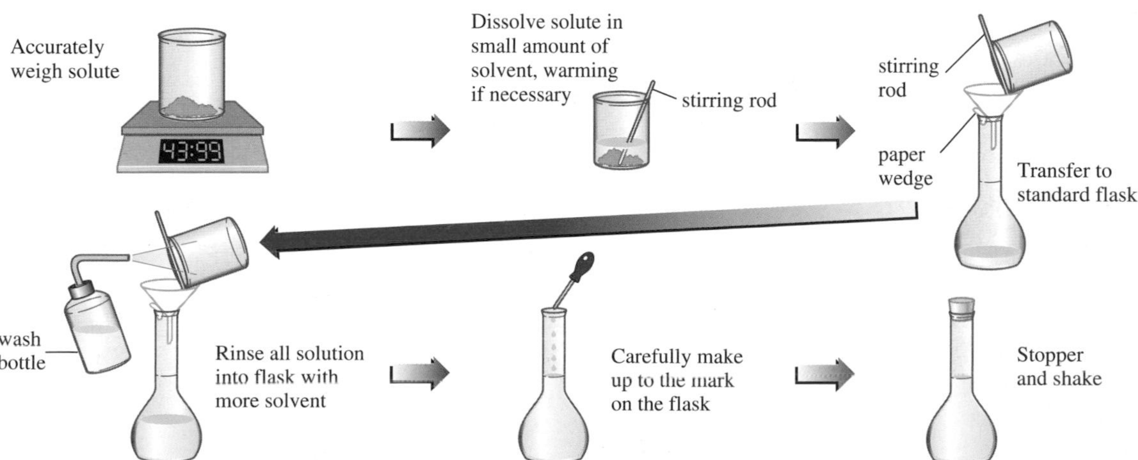
Accurately weigh solute
43:99
Dissolve solute in small amount of solvent, warming if necessary
stirring rod
stirring rod
paper wedge
Transfer to standard flask
wash bottle
Rinse all solution into flask with more solvent
Carefully make up to the mark on the flask
Stopper and shake

Figure 2.9.3 ▲ *Using a graduated flask to prepare a solution with a known concentration*

Test yourself D

8 What is the concentration in mol dm^{-3} of a solution containing:

a) 4.0 g sodium hydroxide, NaOH, in 500 cm^3 solution?
b) 20.75 g potassium iodide in 250 cm^3 solution?

9 What is the mass of solute present in:

a) 50 cm^3 of 2.0 mol dm^{-3} sulfuric acid?
b) 100 cm^3 of 0.01 mol dm^{-3} potassium manganate(VII), $KMnO_4$

10 Calculate the mass of:

a) silver chloride, AgCl, which precipitates on adding excess sodium chloride, NaCl, solution to 20 cm^3 of 0.1 mol dm^{-3} silver nitrate, $AgNO_3$
b) copper(II) oxide, CuO, which dissolves in 25 cm^3 of 1.0 mol dm^{-3} sulfuric acid, H_2SO_4, to make a solution of copper(II) sulfate, $CuSO_4$

11 Calculate the volume at room temperature and pressure of:

a) 1.0 mol dm^{-3} hydrochloric acid, HCl(aq), which a milk of magnesia tablet containing 0.29 g of magnesium oxide can neutralise
b) carbon dioxide given off on adding excess calcium carbonate, $CaCO_3$, to 25 cm^3 of 4.0 mol dm^{-3} hydrochloric acid, HCl(aq).

Worked example

What is the concentration of a solution of silver nitrate made by dissolving 4.25 g of the solid in water and making the solution up to 500 cm^3?

Answer

The molar mass of silver nitrate, $AgNO_3$ = 170 g mol^{-1}

$$\text{Amount of } AgNO_3 \text{ in solution} = \frac{4.25\text{ g}}{170\text{ g mol}^{-1}} = 0.025\text{ mol}$$

$$\text{Volume of the solution} = \frac{500}{1000}\text{ dm}^3 = 0.5\text{ dm}^3$$

$$\text{Concentration of the solution} = \frac{0.025\text{ mol}}{0.5\text{ dm}^3} = 0.05\text{ mol dm}^{-3}$$

Calculations from equations

Calculations from equations that involve solutions are carried out in a similar way to calculations involving masses of solids or the volumes of gases.

Worked example

What is the mass of precipitate on adding excess lead(II) nitrate to 20 cm^3 of 0.5 mol dm^{-3} potassium iodide solution?

Answer

$$Pb(NO_3)_2(aq) + 2KI(aq) \rightarrow PbI_2(s) + 2KNO_3(aq)$$

The lead(II) nitrate is in excess so the amount of precipitate is determined by the amount of potassium iodide.

According to the equation, 2 mol KI produces 1 mol PbI_2(s).

Molar mass of PbI_2(s) = 461 g mol^{-1}

So mass of PbI_2(s) formed from 1 mol KI(aq) = 0.5 × 461 g = 230.5 g mol^{-1}

$$\text{Amount of KI(aq)} = \frac{20}{1000}\text{ dm}^3 \times 0.5\text{ mol dm}^{-3}$$
$$= 0.01\text{ mol}$$

So the mass of precipitate formed = 0.01 mol × 230.5 g mol^{-1}
= 2.3 g

2.10 Titrations

Chemists use titrations to analyse solutions and to investigate reactions. Titrations are common because they are quick, convenient, accurate and easy to automate.

Procedure

A titration only gives accurate results if the reaction is rapid and is exactly described by the chemical equation (meaning that it is stoichiometric).

The analyst adds a measured volume of one solution to a flask with a pipette, then carefully adds the standard solution from a burette. The end point is determined by adding a few drops of an indicator or by using an instrument such as a pH meter.

The calculation

Figure 2.10.1 shows the apparatus for a titration involving a solution A which reacts with solution B. Suppose the equation for the reaction takes the form:

$$n_A A + n_B B \rightarrow \text{products}$$

which means that n_A moles of A reacts with n_B moles of B.

In the laboratory, volumes of solutions are normally measured in cm^3 but they should normally be converted to dm^3 in calculations so that they are consistent with the units used to measure concentrations ($1\ dm^3 = 1000\ cm^3$).

The concentration of B in the flask is c_B mol dm^{-3} and its volume is V_B dm^3.

The concentration of A in the burette is c_A mol dm^{-3}. V_A dm^3 of solution A are added until the indicator shows that the end point has been reached.

The amount of B in the flask at the start $= V_B \times c_B$ mol

The amount of A added from the burette $= V_A \times c_A$ mol

The ratio of these amounts must be the same as the ratio of the amounts shown in the equation.

$$\frac{V_A \times c_A}{V_B \times c_B} = \frac{n_A}{n_B}$$

In any titration all but one of the values in this formula are known. The one unknown is determined from the results.

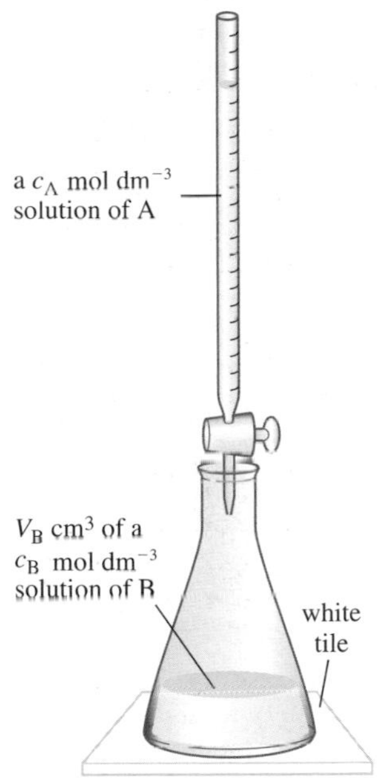

Figure 2.10.1 ▲ *Apparatus for a titration*

Analysing solutions

A titration is often used to measure the concentration of an unknown solution. This is only possible when the equation for the reaction is already known so that the ratio n_A/n_B is known. The analyst also knows the concentration of one of the solutions (the standard solution). The titration gives values for V_A and V_B, so all that remains is to calculate the concentration of the unknown solution.

Definitions

Stoichiometry
The word stoichiometry sounds mysterious but is simply based on Greek words meaning 'element-measure'. Stoichiometry is the basis of quantitative analysis where amounts are measured in moles. A stoichiometric reaction is one which uses up reactants and produces products in amounts exactly as predicted by the balanced equation.

Standard solution
A solution with an accurately known concentration. The direct method for preparing a standard solution is to dissolve a weighed sample of a suitable solid in water and to make the solution up to a definite volume in a graduated flask.

Primary standard
A primary standard is a chemical which can be weighed out accurately to make up a standard solution. A primary standard must be very pure, not gain or lose mass when exposed to the air, be soluble in water and react exactly as described in the chemical equation.

End point
The point during a titration when a colour change shows that enough of the solution in the burette has been added to react exactly with the amount of the chemical in the flask.

Investigating reactions

In titrations to investigate reactions, the problem is to determine the ratio n_A/n_B. The concentrations of both solutions, c_A and c_B, are known. The values of V_A and V_B come from the titration. So the desired ratio can be calculated, indicating what goes on the left-hand side of the equation for the reaction.

Worked example

Lime water is a saturated solution of calcium hydroxide in water. 25.0 cm^3 of 0.04 mol dm^{-3} hydrochloric acid from a burette neutralised 20.0 cm^3 of lime water. What was the concentration of the lime water?

Notes on the method

Start by writing the equation for the reaction.

Since the two volumes appear as a ratio there is no need to convert volumes in cm^3 to volumes in dm^3.

In this case the one unknown is the concentration of the lime water, c_A.

Answer

The equation for the reaction is:

$$Ca(OH)_2(aq) + 2HCl(aq) \rightarrow CaCl_2(aq) + 2H_2O(l)$$

The volume of calcium hydroxide in the flask, $V_A = 20.0\ \text{cm}^3$

Let the concentration of calcium hydroxide be c_A.

The volume of hydrochloric acid added from the burette, $V_B = 25.0\ \text{cm}^3$

The concentration of hydrochloric acid, $c_B = 0.04\ \text{mol dm}^{-3}$

$$\frac{V_A \times c_A}{V_B \times c_B} = \frac{n_A}{n_B}$$

$$\frac{20.0\ \text{cm}^3 \times c_A}{25.0\ \text{cm}^3 \times 0.04\ \text{mol dm}^{-3}} = \frac{1}{2}$$

$$\text{Therefore } c_A = \frac{25.0\ \text{cm}^3 \times 0.04\ \text{mol dm}^{-3}}{2 \times 20.0\ \text{cm}^3} = 0.025\ \text{mol dm}^{-3}$$

The concentration of the lime water was 0.025 mol dm^{-3}.

Test yourself D

1 A 25 cm^3 sample of nitric acid was neutralised by 18.0 cm^3 of 0.15 mol dm^{-3} sodium hydroxide solution. Calculate the concentration of the nitric acid.

2 A 2.65 g sample of anhydrous sodium carbonate was dissolved in water and the solution made up to 250 cm^3. In a titration, 25.0 cm^3 of this solution was added to a flask and the end point was reached after adding 22.5 cm^3 of hydrochloric acid. Calculate the concentration of the hydrochloric acid.

3 A 41 g sample of the acid H_3PO_4 was dissolved in water and the volume of solution was made up to 1 dm^3. 20.0 cm^3 of this solution was required to react with 25.0 cm^3 of a 0.8 mol dm^{-3} solution of sodium hydroxide. What is the equation for the reaction?

Review

This guidance will help you to organise your notes and revision. Check the terms and topics against the specification you are studying. You may find that some topics are not required for your course.

Key terms

Show that you know the meaning of these terms by giving examples. Consider writing the key term on one side of an index card and the meaning of the term with an example on the other side. Then you can easily test yourself when revising. Alternatively use a computer database with fields for the key term, the definition and the example. Test yourself with the help of reports which just show one field at a time.

- Solids
- Liquids
- Gases
- Vapours
- Changes of state
- Melting
- Boiling
- Atom
- Molecule
- Ion
- Acid
- Base
- Alkali
- Salt
- Relative atomic mass
- Relative formula mass
- Relative molecular mass
- Avogadro constants
- Amount of substance
- Mole
- Molar mass
- Molar volume

Symbols and conventions

Make sure that you understand the symbols and conventions which chemists use when writing equations. Illustrate your notes with examples.

- Symbols for atoms, molecules and ions
- Full balanced equations
- Half-equations
- Ionic equations

Patterns and principles

Use tables, charts, concept maps or mind maps to summarise key ideas. Brighten your notes with colour to make them memorable.

- The characteristics of metal elements and non-metal elements
- The characteristics of compounds: non-metals combined with non-metals, metals combined with non-metals
- The charges on metal and non-metal ions
- Types of chemical change: thermal decomposition, oxidation–reduction, acid–base, ionic precipitation, hydrolysis

Laboratory techniques

Draw flow diagrams to show the key steps in these practical procedures.

- Making up a standard solution
- Carrying out acid–base titrations

Calculations

Give your own worked examples, with the help of the test yourself questions, to show that you can carry out calculations to work out the following from given data.

- Empirical formula
- Molecular formula
- Percentage composition
- Masses of reactants and products
- Volumes of gases involved in reactions
- Concentrations of solutions
- Results of acid–base titrations

Key skills

Improving own learning and performance

The start of an AS course is a good time to concentrate on developing your approach to learning. For each topic, look at the specification for your AS course and check how much of this part of the book you need to know and understand for the examinations. Seek advice from your tutor about effective ways of note taking and learning in chemistry. Make a weekly plan for your private study. Use the check lists in the review sections on pages 121–122, 175–176 and 235–237 to keep track of your progress. Reflect on the methods you use to learn the conventions, facts, patterns and principles of chemistry.

Application of number

The various chemical calculations in this part of the course provide good opportunities to practice the separate skills needed for success with application of number.

An investigation which involves a titration is a complex task which needs careful planning. The multi-stage calculations involve amounts and proportion and you may have to rearrange formulae. You also have to interpret the results and relate them to the purpose of the investigation taking into account sources of experimental uncertainty.

Information technology

At the start of your course you can start to learn how to find and select chemical information from a range of sources. Choose an element, for example, and compare what you can find out about the element from a printed book of data, a textbook, a CD-ROM with images, video and data, and a chemical web site.

Also, as suggested in this review section on page 47, you can start to create your own chemical database to provide you with an illustrated glossary of key terms in a form you can use for learning and revision.

Section three

Physical Chemistry

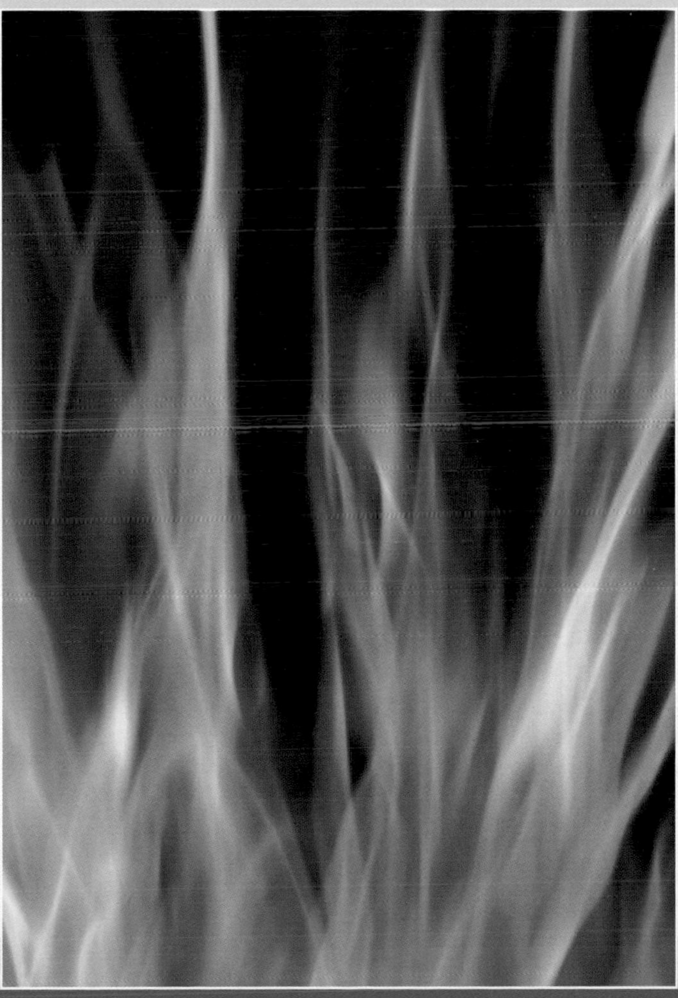

Contents

3.1 What is physical chemistry?

3.2 Atomic structure

3.3 Kinetic theory and gases

3.4 Structure and bonding

3.5 Structure and bonding in metals

3.6 Ionic giant structures

3.7 Covalent molecules and giant structures

3.8 Intermediate types of bonding

3.9 Intermolecular forces

3.10 Structure and bonding in forms of carbon

3.11 Solutions

3.12 Infra-red spectroscopy

3.13 Enthalpy changes

3.14 Reversible reactions

3.15 Chemical equilibrium

3.16 Acid-base equilibria

3.17 Rates of chemical change

3.18 A collision model

3.19 Stability

3.1 What is physical chemistry?

Chemists traditionally divide their subject into three main branches: physical chemistry, inorganic chemistry and organic chemistry. These divisions are still useful but they are perhaps less important than they once were. The theories of physical chemistry help to make sense of all the many changes in inorganic and organic chemistry. Here are some of the main aspects of physical chemistry.

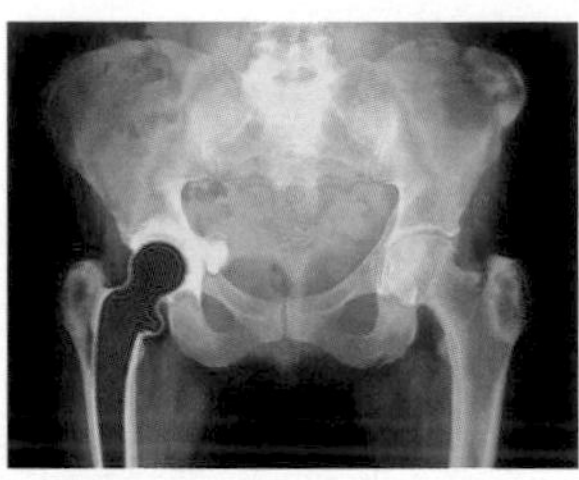

Figure 3.1.1 ▲ Coloured X-ray of a prosthetic hip joint in place

The structure and properties of materials

The starting point for understanding why the materials around us behave as they do is the kinetic theory of matter. This is the theory that solids, liquids and gases consist of small particles in rapid motion. The kinetic theory works particularly well for gases and allows chemists to predict the behaviour of gases in reactions.

Designers have a wide choice of materials thanks to developments in the understanding of structure and bonding in metals, polymers, ceramics and glasses. Scientists now understand the details of how atoms are arranged in many materials (structure) and the forces which hold them together (bonding). This makes it possible to design new materials for special purposes.

Radiation and matter

There is a close connection between light and chemical change. All life, for example, is based on the chemistry of photosynthesis. Sight too depends on chemical changes. This time they take place in the retina when light falls on molecules in the rods and cones.

Over the last hundred years or so, chemists have discovered that they can use many forms of radiation to find out about atoms and the forces between them.

Spectroscopy is the technique which scientists use to analyse radiation. One of the first successes of spectroscopy was the discovery of a new element in the Sun before it had been found on Earth. Thanks to a solar eclipse in 1868, the English astronomer Joseph Lockyer spotted a yellow line in the spectrum of sunlight which could not be matched to any of the spectra of the known elements. The unknown element was called helium – from the Greek word for the Sun, *helios*. Chemists did not discover helium on Earth until 1895.

Spectroscopy is not limited to visible light. Using radiowaves scientists can study the nuclei of atoms, microwaves make it possible to study rotating molecules, with infra-red radiation it is possible to identify particular bonds in molecules, while with the help of UV light spectroscopists can study the electrons in atoms and molecules.

Energy and change

Most useful energy comes from chemical reactions. On a large scale fuels burning in power stations turn water into high pressure steam and drive turbines to generate electricity. On a smaller scale it is the energy available from chemical reactions which gives rise to the 'voltage' of cells and batteries.

Energy is needed to break chemical bonds. Energy is released when new bonds form. So every chemical reaction involves energy changes. Studying these energy changes helps chemists to understand which chemical changes are likely to happen.

Electrochemistry is the study of electricity and chemicals. Electrolysis uses an electric current to split apart chemicals. Several important industrial processes depend on electrolysis, including the manufacture of sodium, magnesium, chlorine and sodium hydroxide.

Chemical cells which produce electricity are in every battery. A huge amount of research and development work is now in progress to develop rechargeable chemical cells. The aim is to design rechargeable cells which are lighter and have a higher storage capacity than the traditional lead–acid cells which power starter motors for car engines.

Figure 3.1.3 ◀
This camera runs on a battery which creates electricity from chemical change

Extent of change

Many reactions are reversible. They go one way or the other depending on the conditions. The study of how far reactions will go and in which direction is an important theme of physical chemistry.

Reversible reactions come to equilibrium. At equilibrium the reaction to make the products is still going on but its effect is cancelled out because the products are reacting to turn back into reactants at the same rate. Despite all this activity at a molecular level, an observer sees nothing. Macroscopically nothing appears to be happening.

Rates of change

Chemical reactions happen at a variety of speeds. Gunpowder explodes very very rapidly, while iron rusts quite slowly.

Raising the temperature speeds up most reactions. Another way to speed up a reaction is to make the chemicals more concentrated in solution or, if they are gases, to compress them by raising the pressure.

One of the frontier areas of modern chemistry is catalysis. Scientists are on the look out for new catalysts to allow faster reactions at lower temperatures. New catalysts can also lead to more efficient processes which make more of the chemicals needed while producing less useless waste.

3.2 Atomic structure

The chemistry of the elements depends on their electrons, especially the electrons in the outer shells. The nuclei of atoms do not change during chemical reactions. Nevertheless chemists are interested in atomic nuclei because they can change too, both during radioactive decay and during the nuclear reactions which occur when atoms are bombarded with high energy particles from a nuclear reactor or particle accelerator to form completely new elements.

Isotopes

Most elements have atoms which are chemically the same but which differ in mass. These are the isotopes (or nuclides) of the element. Hydrogen has three isotopes with relative masses 1, 2 and 3. They all contain the same number of protons and electrons, so they have the same chemical properties. The different masses result from the different numbers of neutrons. Naturally-occurring hydrogen contains 99.9% hydrogen-1. Isotopes are the same on the outside but they have different centres. The electron arrangements are the same so they have the same chemical properties. It is the nuclei on the inside which are different in mass.

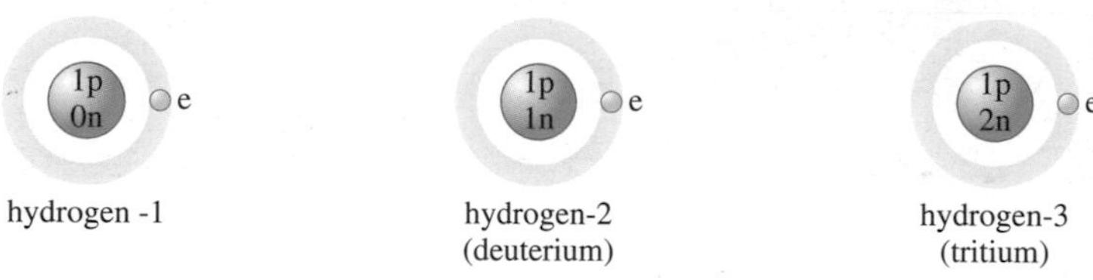

Figure 3.2.1 ▲
The three kinds of hydrogen atoms

Elements from the stars

The elements which make up the Earth and all the living things on the planet formed in a giant star.

At the enormously high temperatures in stars the electrons are stripped away from atoms. The changes in stars which release energy are fusion reactions of the atomic nuclei.

Most stars consist mainly of hydrogen. At 10 million °C or more, the inside of a star is hot enough for the nuclei of hydrogen atoms to join (fuse) to form deuterium and then helium. Next, helium nuclei fuse to form heavier elements, including carbon and oxygen. Fusion reactions of these light elements release huge amounts of energy.

Eventually all the hydrogen in a star is used up. When this happens to small stars (like our Sun) they expand and become red giants, and then gradually fade away as the nuclear fusion reactions stop.

For bigger stars, the story is different. These stars are hot enough to make elements as heavy as iron in their cores. When fusion stops, a dying

Definitions

Mass number
The total number of protons and neutrons (nucleons) in the nucleus of an atom. The alternative term is nucleon number.

Atomic number
The number of protons in the nucleus of an atom. This is equal to the number of electrons in a neutral atom. The alternative term is proton number.

Relative isotopic mass
The mass of the atoms of an isotope relative to the mass of atoms of carbon-12 for which A_r is defined as exactly 12. The values are relative so they do not have units.

Note

There is an agreed shorthand for isotopes based on the element symbol.

$$^{A}_{Z}X$$

mass number — A; atomic number — Z; X — element symbol

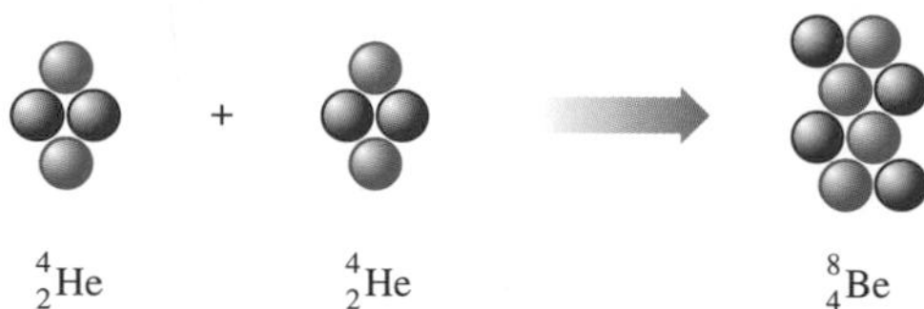

Figure 3.2.2 ◄
Fusion of helium nuclei to form a beryllium nucleus. This diagram shows the protons and neutrons but no electrons

star collapses inward under gravity; it becomes exceedingly hot and the star blows up in a spectacular explosion – a supernova.

The collapsed core of the star becomes a dense mass of neutrons or even a black hole, where gravity is so strong that light cannot escape. Meanwhile the outer layers become so hot that the heaviest elements (beyond iron) can form.

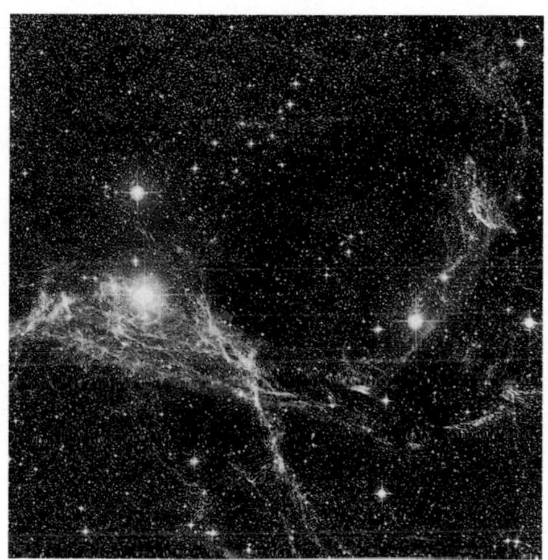

Figure 3.2.3 ◄
The Vela supernova remnant in the Vela constellation

Mass spectrometry

Mass spectrometry is an accurate instrumental technique for determining relative atomic masses and relative molecular masses. Mass spectrometry can also help to measure the abundance of isotopes, to determine molecular structures and to identify unknown compounds. (See page 55 for the use of mass spectrometry to study molecules.)

Inside a mass spectrometer there is a high vacuum so that it is possible to produce and study ionised atoms and molecules including fragments of molecules which do not otherwise exist.

The stages in producing a mass spectrum:

- inject a small sample into the instrument where it vaporises
- bombard the sample with a beam of high energy electrons which turns the atoms or molecules into positive ions by knocking out electrons
- accelerate the positive ions in an electric field
- deflect the moving stream of ions with a magnetic field to focus ions with a particular mass onto the detector
- feed the signal from the detector to a computer which prints out a mass spectrum as the magnetic field steadily changes over a range of values so that it focuses a series of ions with different masses one by one onto the detector.

Test yourself

1 How many protons, neutrons and electrons are there in these atoms or ions?

a) $^{9}_{4}Be$

b) $^{39}_{19}K$

c) $^{235}_{92}U$

d) $^{127}_{53}I^{-}$

e) $^{40}_{20}Ca^{2+}$

2 Write the symbols showing the mass number and atomic number for these atoms or ions:

a) an atom of oxygen with 8 protons, 8 neutrons and 8 electrons

b) an atom of argon with 18 protons, 22 neutrons and 18 electrons

c) an ion of sodium with a 1+ charge and a nucleus of 11 protons and 12 neutrons

d) an ion of sulfur with a 2− charge and a nucleus with 16 protons and 16 neutrons.

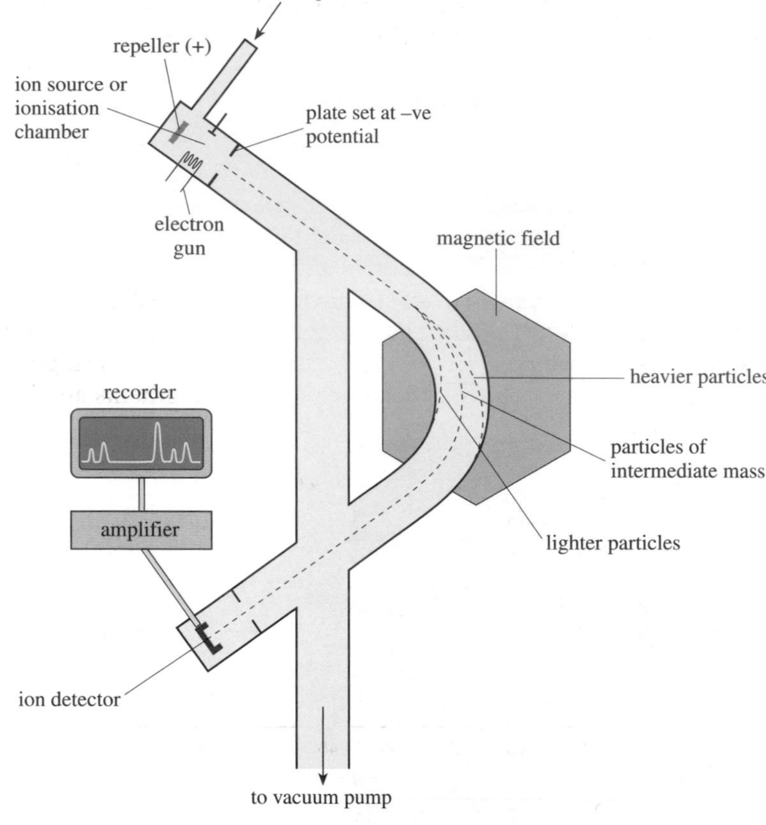

Figure 3.2.4 ► Diagram of a mass spectrometer

Note

Modern mass spectrometers have become very sensitive. It is now possible to detect and identify traces of elements in a specimen down to proportions as low as 10^{-12} g g^{-1}.

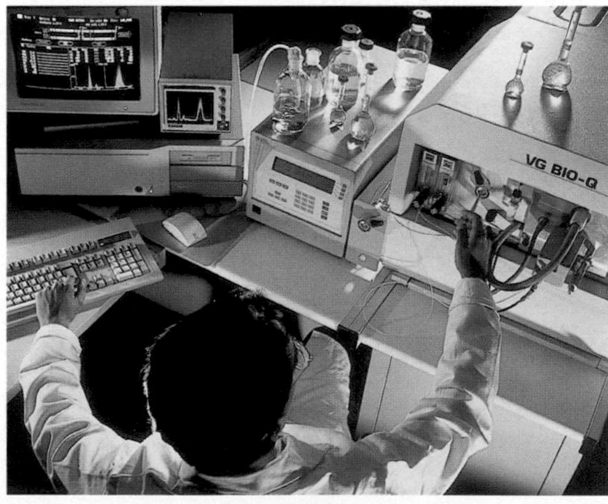

Figure 3.2.5 ► Scientist injecting a sample into a mass spectrometer

Test yourself

3 Use Figure 3.2.6 to estimate the abundances of the two isotopes of magnesium. Calculate the average relative atomic mass of magnesium.

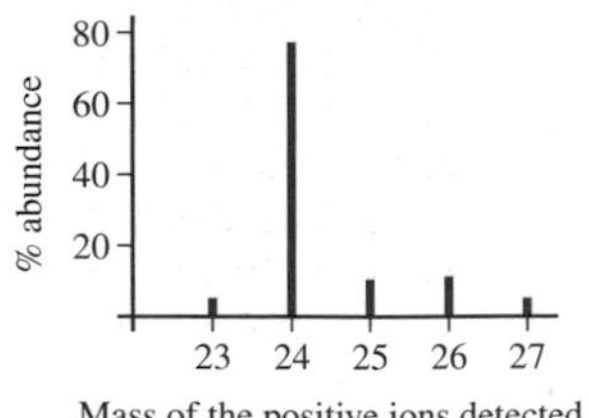

Figure 3.2.6 ▲ Mass spectrum of magnesium

The instrument is calibrated using a reference compound with a known structure and molecular mass so that the computer can print a scale on the mass spectrum.

The mass spectrum for an element shows the relative abundance of the isotopes of the element. This makes it possible to calculate the element's relative atomic mass.

Average relative atomic masses

The reason why the values for relative atomic masses in tables of data are not always whole numbers is that most elements are mixtures of isotopes.

Chlorine, for example, has two isotopes: chlorine-35 and chlorine-37. Naturally-occurring chlorine contains 75% of chlorine-35 and 25% of chlorine-37. So the ratio of chlorine-35 to chlorine-37 is 3:1.

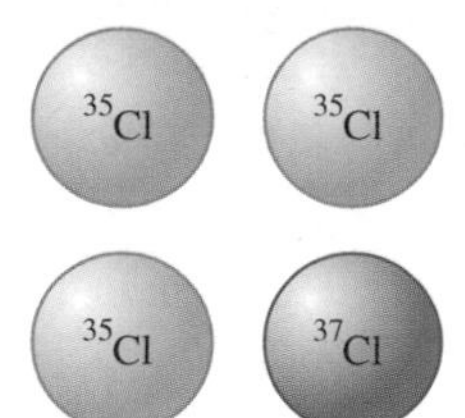

Figure 3.2.7 ▲
On average, for every four chlorine atoms, three are chlorine-35 and one is chlorine-37

The average relative atomic mass of chlorine

$$= \frac{(3 \times 35) + (1 \times 37)}{4} = 35.5$$

Relative molecular masses

A mass spectrometer can also be used to study molecules. After injecting molecules into the instrument, the bombarding electrons not only ionise the molecules but also break them up into fragments. Because of the high vacuum it is possible to produce and study molecular fragments and ions which do not normally exist.

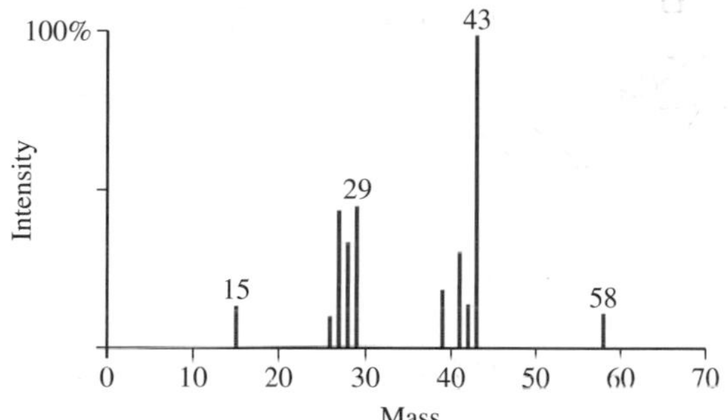

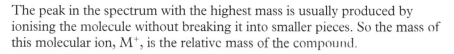

Figure 3.2.8 ◀
The mass spectrum of a hydrocarbon showing the fragmentation pattern

The peak in the spectrum with the highest mass is usually produced by ionising the molecule without breaking it into smaller pieces. So the mass of this molecular ion, M^+, is the relative mass of the compound.

$$\underset{\text{molecule in sample}}{M(g)} + \underset{\text{bombarding electron}}{e^-} \rightarrow \underset{\text{molecular ion}}{M^+(g)} + \underbrace{e^- + e^-}_{\text{electrons}}$$

A chemist who synthesises a new compound can study its fragmentation pattern to identify the fragments from their masses and then piece together likely structures with the help of other methods of analysis such as infra-red spectroscopy (see page 90).

The combination of gas–liquid chromatography (glc) with mass spectrometry is of great importance in modern chemical analysis. First chromatography separates the chemicals in an unknown mixture, such as a sample of polluted water, then mass spectrometry detects and identifies the components.

Ionisation energies

In a mass spectrometer a beam of electrons bombards the sample turning atoms into positive ions. The electron beam has to have enough energy to knock electrons off the atoms of the sample. By varying the intensity of the beam it is possible to estimate the minimum energy needed to remove electrons from the atoms in the sample.

The energy needed to remove an electron from a gaseous atom or ion is its ionisation energy.

The first ionisation energy for an element is the energy needed to

Test yourself

4 Silicon (atomic number 14) has three naturally-occurring isotopes with mass numbers 28, 29 and 30.

a) Write the symbols for the three isotopes of silicon.
b) For each isotope work out the numbers of protons and neutrons in the nucleus.
c) What is the relative atomic mass of silicon which normally consists of 93% silicon-28, 5% silicon-29 and 2% silicon-30?

5 Use Figure 3.2.8 to determine the relative molecular mass of the hydrocarbon.

Test yourself

6 The first five ionisation energies of an element in kJ mol^{-1} are: 738, 1451, 7733, 10 541 and 13 629. How many electrons are there in the outer shell of the atoms of this element? To which group of the periodic table does this element belong?

7 Sketch a graph of log(ionisation energy/kJ mol^{-1}) against number of electrons removed when all the electrons are successively removed from an aluminium atom.

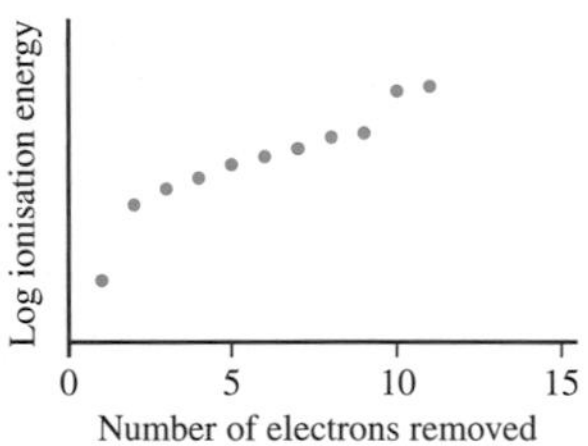

Figure 3.2.9 ▲ *A plot of log(ionisation energy) against number of electrons removed for sodium*

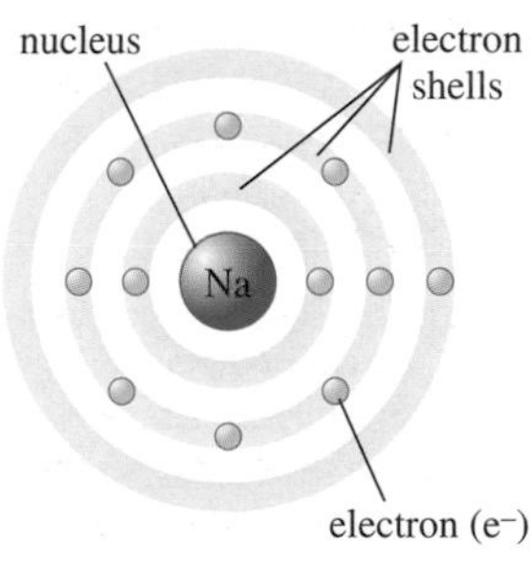

Figure 3.2.10 ▲ *The electrons in shells for a sodium atom*

remove one mole of electrons from one mole of gaseous atoms. Successive ionisation energies for the same element measure the energy per mole needed to remove a second, third, fourth electron and so on.

Scientists measure ionisation energies not only using a mass spectrometer but also by studying the emission spectra of atoms. In these ways it is possible to measure energy changes involving ions with higher charges which do not normally appear in chemical reactions.

$Na(g) \rightarrow Na^+(g) + e^-$	first ionisation energy	$= +\ 496\ kJ\ mol^{-1}$
$Na^+(g) \rightarrow Na^{2+}(g) + e^-$	second ionisation energy	$= +\ 4563\ kJ\ mol^{-1}$
$Na^{2+}(g) \rightarrow Na^{3+}(g) + e^-$	third ionisation energy	$= +\ 6913\ kJ\ mol^{-1}$

There are 11 electrons in a sodium atom so there are 11 successive ionisation energies for this element.

The successive ionisation energies for an element get bigger and bigger. This is not surprising because having removed one electron it is more difficult to remove a second electron from the positive ion formed.

The graph in Figure 3.2.9 is evidence in support of the theory that the electrons in an atom are arranged in a series of energy levels or shells around the nucleus (see page 22). There are big jumps in value each time electrons start to be removed from the next energy level or shell in towards the nucleus.

So for a sodium atom, there is one electron in the outer shell which is furthest from the nucleus. This outer electron is shielded from the full attraction of the positive nucleus by 10 inner electrons. There are eight electrons in the second shell which are closer to the nucleus and only have two inner shielding electrons. The two inner electrons feel the full attraction of the nuclear charge and are closest to the nucleus. These electrons are hardest to remove. (See pages 129–130 for more discussion of shielding and ionisation energies.)

Atomic spectra

Atoms emit radiation when excited by heat or electricity. A spectroscope contains a prism or diffraction grating to split up the radiation according to wavelength revealing a pattern of lines.

Some elements emit radiation in the visible region of the spectrum and these are the elements which give coloured flames in flame tests. Carrying out a flame test and examining the coloured light from the flame with a hand-held spectroscope is a simple demonstration of this type of spectroscopy.

The atomic spectra of elements provide the evidence which chemists need to determine the arrangements of electrons in atoms. Atomic spectra give much more information than ionisation energies.

Applying a high voltage across the electrodes at the ends of a glass tube of hydrogen at low pressure produces a bluish glow. A spectroscope shows that the light from the glow is made up of radiations with particular wavelengths – it is not like a rainbow. A spectroscope produces a line spectrum which can be recorded photographically.

Quantum theory

Quantum theory helped scientists to explain atomic spectra. The basic idea of this theory is that radiation is emitted or absorbed in discrete amounts called 'quanta'. Max Planck (1858–1947), the German physicist, put

Figure 3.2.11 ◄ *The line spectrum for hydrogen in the visible and ultraviolet (UV) regions of the spectrum only*

forward the theory in a paper published in 1900. Albert Einstein (1879–1955) extended the theory to explain what happens when electrons absorb and emit radiation.

The Danish physicist Niels Bohr (1885–1962) took up the quantum theory to explain the lines in hydrogen's atomic spectra. Bohr's theory could account very well for the frequencies of the lines in the spectrum of the hydrogen atom by making these assumptions:

- electrons in a hydrogen atom can only be at certain definite energy levels
- a quantum of light (a photon) is emitted or absorbed when an electron jumps from one energy level to another
- the energy of the light quanta (photons) equals the difference, ΔE, between the two energy levels
- the frequency of the radiation emitted or absorbed is related to the energy of the photon by $\Delta E = h\nu$ where h is Planck's constant and ν is the frequency of the radiation.

Quantum theory explains why atomic emission spectra consist of a series of sharp lines. Each line in an emission spectrum corresponds to an energy jump of a definite size as electrons drop back from a higher energy level to a lower energy level.

The lines of the Balmer series (see Figure 3.2.12) are in the visible part of the spectrum. The bigger the energy jumps, the higher the frequency of electromagnetic radiation emitted.

The energy gaps between energy levels get smaller as they get further from the nucleus. As a result, in each series, the differences between the energy jumps get smaller and smaller until they converge. They converge at the high frequency (big jump) end. The biggest jump in the Lyman series is for an electron dropping back from the very edge of an atom to the lowest energy level (see Figure 3.2.13). The size of this jump is the energy needed to ionise a single electron in a hydrogen atom. So the ionisation energy of hydrogen can be calculated from the frequency of the convergence limit of the Lyman series since (according to quantum theory) $\Delta E = h\nu$.

Electrons in energy levels

There is only one electron in a hydrogen atom, making it the simplest atom with the simplest spectrum. Study of the spectra from other atoms shows that the pattern of energy levels becomes more complex as the number of electrons increases.

The main energy levels or shells divide into sub-levels labelled s, p, d and f. The labels s, p, d and f are left over from early studies of atomic spectra which used the words sharp, principal, diffuse and fundamental to describe different series of lines.

These terms now have no special significance.

Test yourself

8 The energy jump for one electron from the lowest energy level to the outermost energy level in a hydrogen atom is 2.18×10^{-18} J. Calculate:

a) the frequency of the convergence limit for the Lyman series

b) the ionisation energy of hydrogen in kJ mol^{-1}.

Note

Chemists use the capital Greek 'delta', Δ, for a sizeable change or difference. They use the small Greek letter 'delta', δ, for a small quantity or change (see page 81).

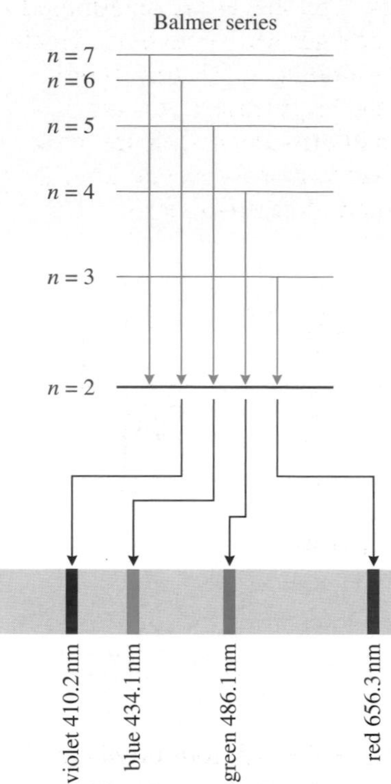

Figure 3.2.12 ▲
An electron jumps between energy levels in the hydrogen atom giving rise to the Balmer series in the emission spectrum. The series of lines are named after the people who first discovered or studied them

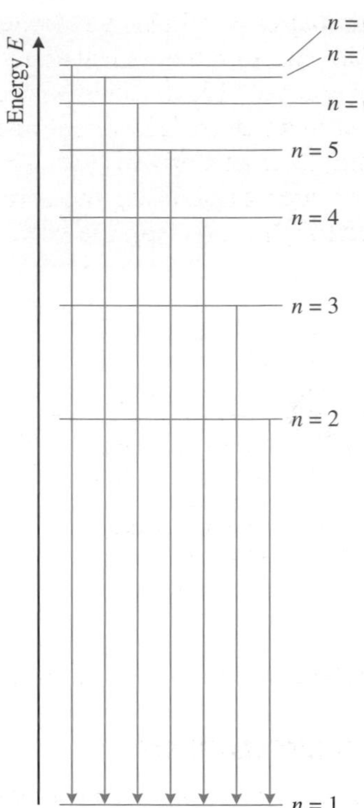

Figure 3.2.13 ▲
The Lyman series for a hydrogen atom. In this series, the transitions of the electron are to n = 1. The wavelengths of the spectral lines are in the ultra-violet region of the spectrum

Figure 3.2.14 ►
The energies of atomic orbitals in atoms. The terms 'energy level' and 'orbital' are often used interchangeably. In a free atom, the orbitals in a sub-shell have the same energy

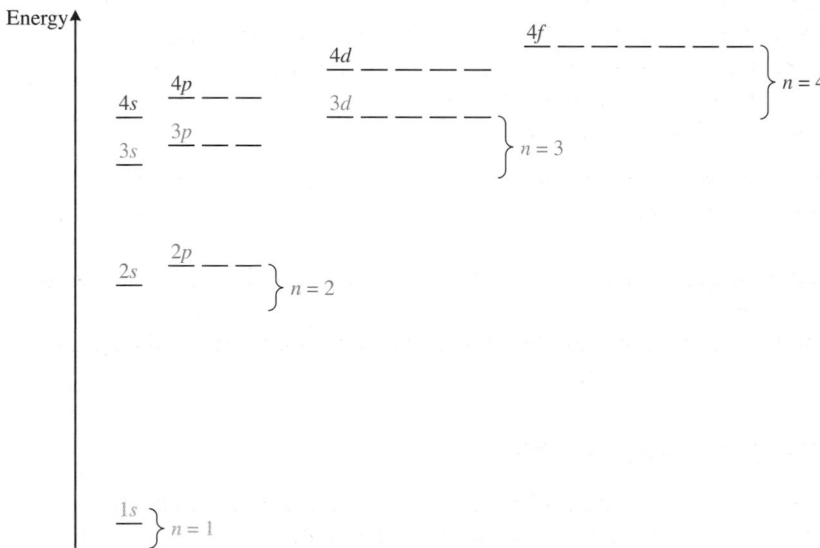

The sub-levels are further divided into atomic orbitals.

Each orbital is defined by its:

- energy
- shape
- direction in space.

The shapes of orbitals are derived from theory. The shapes are determined by solving a mathematical equation (the Schrödinger wave equation) which makes it possible to calculate the probability of finding an electron at any point in an atom. The one orbital in the first shell is spherical. It is an example of an s orbital (1s). The four orbitals in the second shell are made up of one s orbital (2s) and three dumbbell shaped p orbitals. The three p orbitals ($2p_x$, $2p_y$, $2p_z$) are arranged at right angles to each other.

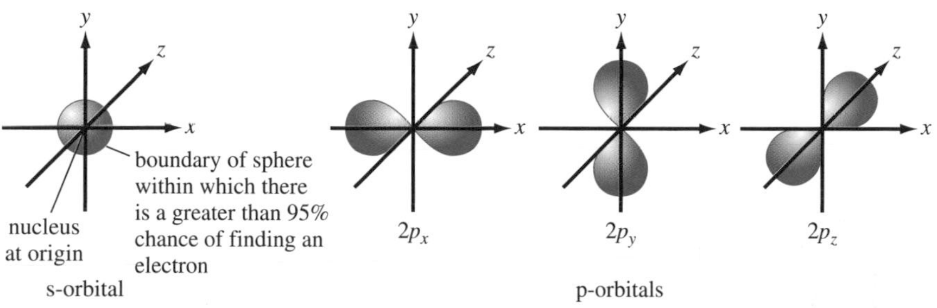

Figure 3.2.15 ▲
The shapes of s and p atomic orbitals

Quantum numbers

Each electron in an atom is unique. Theory shows that just four numbers are all that is needed to identify each electron in an atom. These quantum numbers label the energy level or orbital occupied by an electron in an atom.

The fourth quantum number introduces the idea of 'electron spin'. Spin is the property of electrons which accounts for their behaviour in a magnetic field.

Electrons behave like tiny magnets. In a magnetic field they either line up with the field or against the field and there is a small energy difference between the two alignments.

An atomic orbital can only hold two electrons and they must have opposite spins. Arrows pointing up or down represent electrons with opposite spins in energy level diagrams.

The four quantum numbers are:

- the principal quantum number which indicates the main shell
- the second quantum number which identifies the sub-shell – s, p, d or f
- the third quantum number which shows the orbital within a sub-shell – p_x, p_y or p_z
- the fourth quantum number which states the alignment of the spin – spin up or spin down.

Electron configurations

The electron configuration of an atom describes the number and arrangement of electrons in an atom of an element.
The electrons in an atom fill the energy levels according to a set of rules. The three rules are:

- electrons go into the orbital at the lowest available energy level
- each orbital can only contain at most two electrons (with opposite spins)
- where there are two or more orbitals at the same energy, they fill singly before the electrons pair up.

Note

Newton's equations of motion can be solved to predict the precise orbits of moons and planets. Newton's theories do not work for very small particles such as electrons in atoms. Today theoretical chemists base their explanations on a development of quantum theory called quantum mechanics. The equations of quantum mechanics only give the probabilities of finding particles at various places. Atomic orbitals are the results of quantum mechanical calculations; they show the probabilities of finding electrons in particular regions. So chemists now use the 'fuzzy' term orbital instead of the 'crisp' term orbit.

Note

The first shell ($n = 1$) with one orbital can hold two electrons, the second shell ($n = 2$) with four orbitals holds up to eight electrons and the third shell ($n = 3$) with nine orbitals holds up to 18 electrons. Thus the maximum number of electrons in a shell is given by $2n^2$.

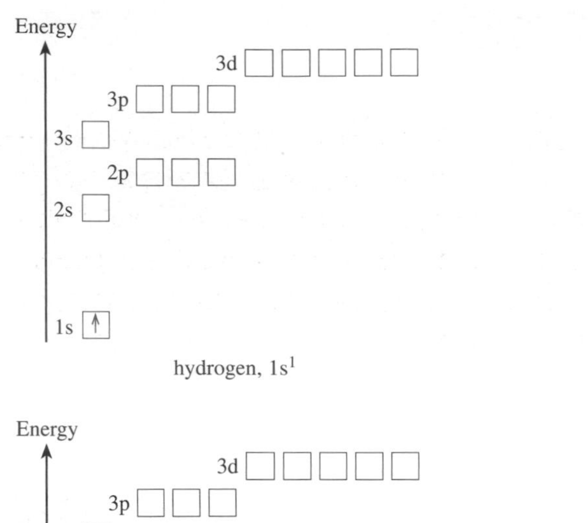

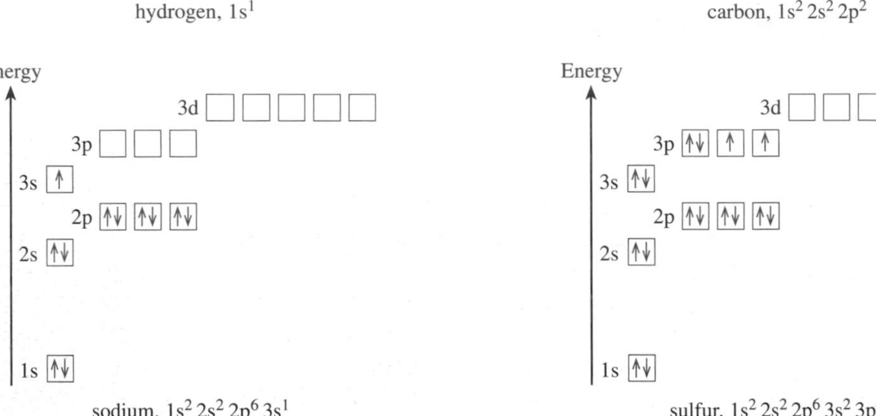

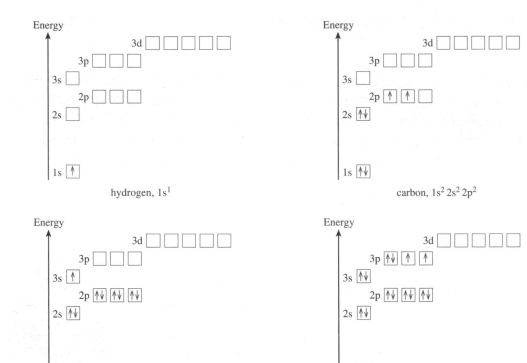

Figure 3.2.16 ▲
Electrons in energy levels for a series of atoms to show the application of the building-up principle

Note

Chemists sometimes use the term 'aufbau principle' for these rules from the German word meaning 'build up'. This is a reminder that electron configurations build up from the bottom, starting with the lowest energy level.

The electron configuration helps to make sense of the chemistry of an element. The electrons in the outer shell largely determine the chemical properties of an element. Elements in the same group of the periodic table have similar properties because they have the same outer electron configuration. There are trends in properties down a group because of the shielding effect of the increasing number of full inner shells.

There are several common conventions for writing electron configurations (see Figure 3.2.17).

A shortened form of electron configuration uses the symbol of the previous noble gas to stand for the full inner shells. According to this convention the electron configuration of sodium is [Ne]$3s^1$.

Test yourself

9 Write out the electron configurations of these atoms:

a) beryllium
b) oxygen
c) silicon
d) phosphorus.

10 Identify the elements with these electron configurations in their outermost shells:

a) $2s^2$
b) $3s^2 3p^1$
c) $3s^2 3p^5$

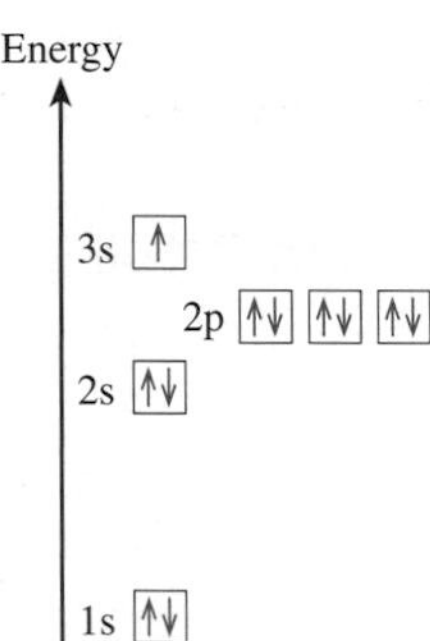

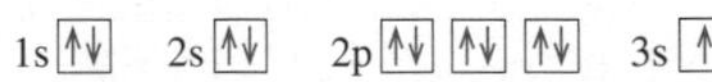

Figure 3.2.17 ►
Representations of the electron configuration of sodium, $1s^2 2s^2 2p^6 3s^1$

3.3 Kinetic theory and gases

Scientists have developed a model to explain the properties of gases which gives a good account of how real gases behave. This model is based on a picture of gases consisting of molecules in rapid random motion. The theory is quantitative and explains not only the behaviour of gases but also the factors which determine the rate and extent of chemical change (kinetics and equilibrium). Real gases do not behave exactly as the model predicts but the deviations give chemists some insight into the weak attractive forces between molecules (intermolecular forces).

Gas volumes

In the late eighteenth century, the hot air balloon flights of the Montgolfier brothers stimulated scientists to study the behaviour of gases. Two of these scientists were French: Joseph Gay-Lussac (1778–1850) and Jacques Charles (1746–1823). They were particularly interested in the variation of the volumes of gases with temperature. Jacques Charles put his theories to the test and in 1783 made the first ascent in a hydrogen balloon.
Over a hundred years earlier, the Irish chemist Robert Boyle (1627–1691) had investigated the effect of pressure on the volume of gases.

We now know that the volume of a sample of gas depends on three things:

- the temperature, T
- the pressure, p
- the amount of gas in moles, n

Figure 3.3.1 ▲
Montgolfier balloon ascent

The effect of temperature on gas volumes

Gases expand as the temperature rises. They expand in a regular way as shown in Figure 3.3.2.

Figure 3.3.2 shows that the volume of a gas is proportional to the temperature on the Kelvin scale.

So the volume, V, of a fixed amount of gas is proportional to the temperature, T, on the Kelvin scale so long as the pressure, p, stays constant.

$$V \propto T \text{ at constant } p$$

This is sometimes called Charles's law.

Note

Scientists measure gas volumes in cm^3, dm^3 and m^3.

$1\ dm^3 = (10\ cm)^3 = 1000\ cm^3 = 1$ litre

$1\ m^3 = (100\ cm)^3 = 10^6\ cm^3$

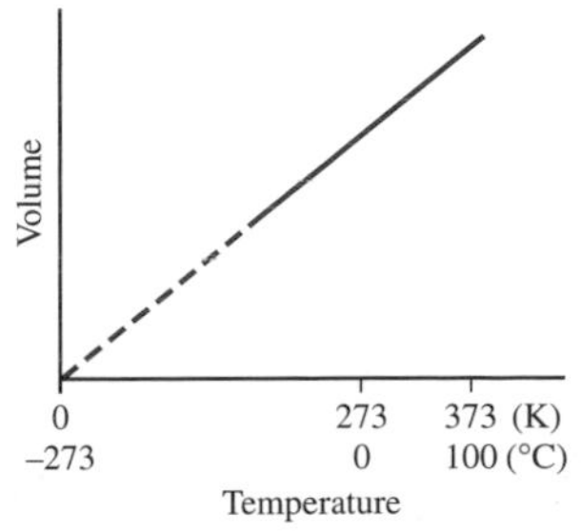

Figure 3.3.2 ◄
A plot of the volume of a sample of gas against temperature at constant pressure is a straight line which on extrapolation cuts the temperature axis at −273.15 °C

Test yourself

1 What are the values of these temperatures on the Kelvin scale?

a) boiling point of nitrogen, −196 °C
b) boiling point of butane, −0.5 °C
c) melting point of sucrose, 186 °C
d) melting point of iron, 1540 °C

Definition

The **kelvin** is the SI unit of temperature on the absolute or Kelvin scale. The symbol is K. Temperatures above absolute zero are measured in kelvin. On this scale water freezes at 273 K and boils at 373 K.

Temperature *differences* measured on either the Celsius or the Kelvin scales are the same and recorded as kelvin (K).

Temperature in Kelvin = temperature in °C + 273

Absolute zero (0 K) is the temperature at which atoms and molecules in crystals are effectively motionless. It is the lowest temperature on the absolute or Kelvin temperature scale.

Many gases studied at around room temperature appear to behave as if they would have zero volume at absolute zero. In practice real gases turn to liquids and solids and occupy a definite volume before absolute zero is reached.

The effect of pressure on gas volumes

As the pressure goes up, the volume of a gas goes down. Robert Boyle discovered that if he kept the temperature constant and worked with a fixed amount of gas, when he doubled the pressure he halved the volume. This is Boyle's law.

$$p \propto 1/V \text{ at constant } T$$
$$\text{or } pV = \text{constant}$$

Note

One variable is proportional to another when the graph is a straight line through the origin.

Definition

Pressure is defined as force per unit area. The SI unit of pressure is the pascal (Pa) which is a pressure of one newton per square metre (1 N m^{-2}). The pascal is a very small unit so pressures are often quoted in kilopascals, kPa.

When studying gases, the standard pressure is atmospheric pressure which is $101.3 \times 10^3 \text{ Nm}^{-2} = 101.3 \text{ kPa}$.

Standard pressure for definitions in thermochemistry is now 1 bar which is $100\,000 \text{ Nm}^{-2} = 100 \text{ kPa}$

The effect of the amount of gas on its volume

If the pressure and temperature are fixed, then the volume of a gas depends on the number of gas molecules present. In other words, the volume of gas depends on the amount of gas in moles. This is Avogadro's law (see page 41).

$$V \propto n \text{ at constant } T \text{ and } p$$

The volume of one mole of a gas (the molar volume) depends on the temperature and pressure. Two sets of conditions are often used to compare one gas with another:

- standard temperature and pressure, stp, (273 K and 1 atmosphere pressure) – under these conditions the volume of one mole of any gas is 22.4 dm^3, and
- room temperature and pressure (298 K and 1 atmosphere pressure) – under these conditions the volume of one mole of any gas is 24 dm^3.

The molar volume is the same for any gas under given conditions of temperature and pressure.

Real and ideal gases

Scientists have in mind an 'ideal gas' which obeys the gas laws perfectly. In practice real gases do not obey the laws under all conditions. Under laboratory conditions, however, there are gases which are close to behaving like an 'ideal gas'. These are the gases which, at room temperature, are well above their boiling points, such as helium, nitrogen, oxygen and hydrogen.

Chemists generally find that the gas laws predict the behaviour of real gases accurately enough to make them a useful practical guide, but it is important to bear in mind that gases such as ammonia, butane, sulfur dioxide and carbon dioxide can show marked deviations from ideal behaviour. These are the gases which boil only a little below room temperature and can be liquefied just by raising the pressure.

Test yourself

2 Show that the ideal gas equation is consistent with:

a) Charles's law when n and P are constant
b) Avogadro's law when P and T are constant.

The ideal gas equation

The behaviour of an ideal gas can be summed up by combining the gas laws into a single equation called the ideal gas equation: $pV = nRT$.

R is the gas constant. The value of R depends on the units used for pressure and volume. If all quantities are in SI units, then $R = 8.314\,\text{J}^{-1}\,\text{mol}^{-1}$.

The ideal gas equation incorporates all the gas laws. For example, with a fixed amount of gas, n is constant and R is a constant. So at constant temperature, T, pV = constant, which is Boyle's law.

Figure 3.3.3 ▲
A can of liquid butane for filling gas lighters. Butane is a gas at room temperature and pressure but can be stored as a liquid under pressure. Butane does not behave as an ideal gas

Measuring molar masses

In the days before mass spectrometry (see pages 53–55) chemists used the ideal gas equation to measure the molar masses of gases and of other substances which evaporate easily.

The method is accurate enough to determine the molecular formula of elements and compounds.

One practical approach is to inject a weighed sample of a liquid into a syringe heated in an oven. Measurements taken include the volume of vapour, the temperature of the vapour and its pressure (the atmospheric pressure). Measurements are converted to SI units and then substituted in the ideal gas equation to find the amount in moles, n.

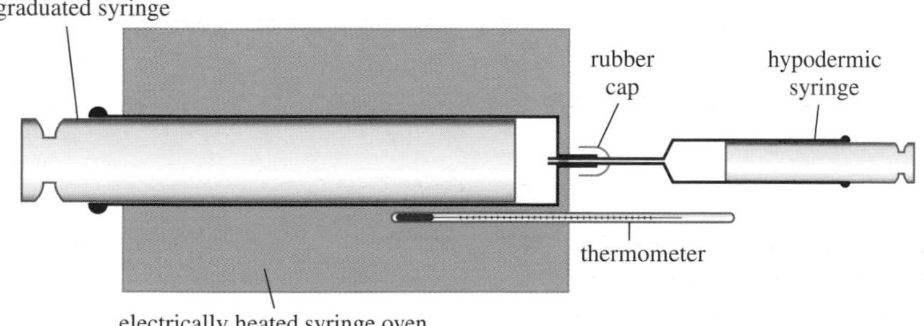

Figure 3.3.4 ◄
Syringe method for determining molar masses of volatile liquids

Worked example

A 0.124 g sample of a liquid with the empirical formula (see pages 38 and 180) C_3H_7 evaporates to give 45 cm^3 vapour at 100 °C and a pressure of 1 atmosphere. What is the molecular formula of the liquid?

Notes on the method

Convert all units to SI units.

Substitute in the equation $pV = nRT$ to find n (the amount in moles).

Answer

Pressure = 101.3×10^3 Nm^{-2}

Volume = 45×10^{-6} m^3

Temperature = 373 K

The gas constant = 8.31 J mol^{-1} K^{-1}

$$n = \frac{pV}{RT} = \frac{101.3 \times 10^3 \text{ Nm}^{-2} \times 45 \times 10^{-6} \text{ m}^3}{8.31 \text{ J mol}^{-1} \text{ K}^{-1} \times 373 \text{ K}} = 1.47 \times 10^{-3} \text{ mol}$$

$$\text{molar mass} = \frac{\text{mass of sample}}{\text{amount in moles}} = \frac{0.124 \text{ g}}{1.47 \times 10^{-3} \text{ mol}} = 84 \text{ g mol}^{-1}$$

A molecular formula is always a simple multiple of the empirical formula (see page 181).

The relative mass of the empirical formula, $M_r(C_3H_7) = (3 \times 12) + (7 \times 1) = 43$.

Even though the vapour of the compound does not behave as an ideal gas, the result is accurate enough to show that the molecular formula is twice the empirical formula. The molecular formula of the compound is C_6H_{14}.

Test yourself

3 A 0.163 g sample of a liquid evaporates to give 65.0 cm^3 of vapour at 101 °C and 10^5 kPa. What is the molar mass of the liquid?

The kinetic theory of gases

The gas laws are based on experimental observations. Scientists rose to the challenge of explaining why gases follow a common set of rules. They began with the picture that a gas consists of molecules in constant random motion colliding with each other. They called their particles-in-motion model the kinetic theory, basing the name on a Greek word for movement.

The scientists made a number of assumptions about the molecules of a gas. Then they were able to derive the ideal gas equation from the model by applying Newton's laws of motion to the collection of particles.

The features of the kinetic theory model are that:

- pressure results from the collisions of the molecules with the walls of the container
- there is no loss of energy when the molecules collide with the walls of a container
- the molecules are so far apart that the volume of the molecules can be neglected in comparison with the total volume of gas
- the molecules do not attract each other

Note

It is important to convert all the quantities to SI units before substituting values in the ideal gas equation. In the equation: p is the pressure in N m^{-2} (pascals), V is the volume in m^3, T is the temperature on the Kelvin scale in K, n the amount in moles and R the gas constant.

- the average kinetic energy of molecules is proportional to temperature on the Kelvin scale.

These assumptions help to explain why real gases approach ideal behaviour at high temperatures and low pressures. At high temperatures the molecules are moving so fast that any small attractive forces between them can be ignored. At low pressures the volumes are so big that the space taken up by the molecules is insignificant.

The theory also helps to explain why real gases deviate from ideal gas behaviour as they get closer and closer to turning into a liquid. As a gas liquefies the molecules get very close together and the volume of the molecules cannot be ignored. Also gases could not liquefy if there were no attractive (intermolecular) forces between the molecules to hold them together.

The Dutch physicist Johannes van der Waals (1837–1923) developed his theory of intermolecular forces (see pages 84–85) by studying the behaviour of real gases and their deviations from the ideal gas equation.

The Maxwell–Boltzmann distribution

The kinetic theory was further developed independently by two physicists: James Maxwell (1831–1879) in Britain and Ludwig Boltzmann (1844–1906) in Austria. In particular they explored the distribution of energies in gases working out how many molecules have a given energy. Figure 3.3.5 shows the distribution of energy for the molecules of a gas under two sets of conditions.

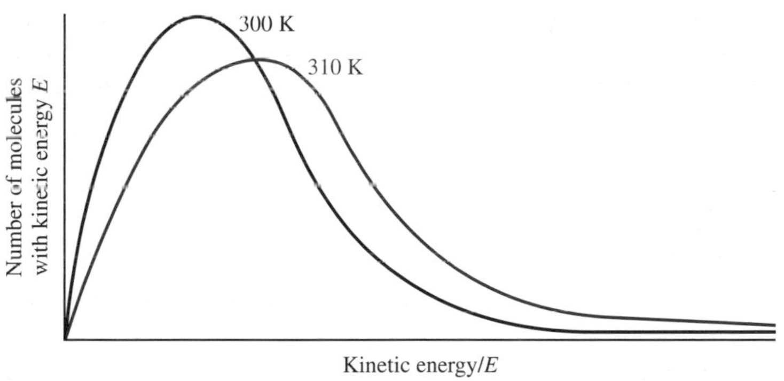

Figure 3.3.5 ▲
The Maxwell–Boltzmann distribution of molecular kinetic energies in a gas at two temperatures

Test yourself

4 Use the kinetic theory model to explain in words (qualitatively) why:

a) the pressure of a fixed amount of gas doubles when the volume is halved at a constant temperature

b) the pressure of a fixed amount of gas at constant volume is proportional to the temperature on the Kelvin scale.

Note

The Boltzmann distribution is important in the collision theory of reaction rates (see pages 117–118) and helps to account for the effects of temperature changes and catalysts on rates of reaction (see pages 118–119).

3.4 Structure and bonding

A central aim of modern chemistry is to explain the properties of elements and compounds in terms of structure and bonding. The structure of a material is a description of the arrangement of the atoms, molecules or ions. Theories of bonding are attempts to account for the forces which hold atoms together. Scientists make extensive use of physics and computer and mathematical models to explain structure and bonding.

Investigating structure and bonding

The word structure has a range of levels of meaning in science. On a grand scale engineers design the structures of buildings and bridges. On the smallest scale scientists explore the inner structure of atoms. In this section the focus is on crystal structure which describes the arrangements of atoms and molecules in crystals.

The regular shapes of crystals suggest an underlying order but until the early part of the twentieth century scientists could only guess at the arrangements of invisible atoms. Sir Lawrence Bragg (1890–1971), working in Cambridge, was the first person to realise that X-rays could be used to investigate crystal structures. His father, Sir William Bragg (1862–1942), helped him by inventing the first instrument to be used to study crystal structures.

X-rays can be used to study structure because their wavelengths are about the same as the distances between the atoms in a crystal. The atoms in the crystal scatter the X-rays producing a diffraction pattern which was originally photographed but can now be recorded electronically. The challenge is to interpret the diffraction pattern and deduce the three-dimensional crystal structure from the pattern of dots.

Many other scientists have followed up the work of the Braggs. One of them, Dorothy Hodgkin (1910–1994) extended the use of X-ray diffraction to unravel the complex structures of biological molecules including

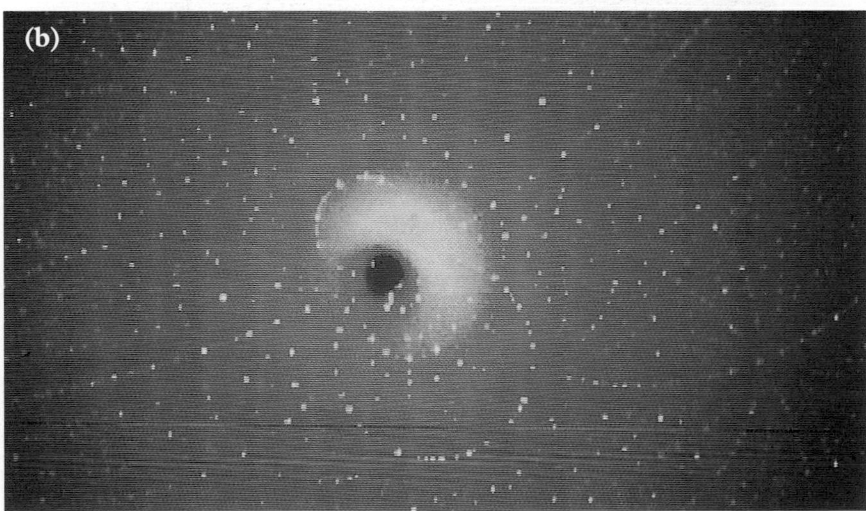

Figure 3.4.1 ▲ ▶
a) Dorothy Hodgkin
b) An X-ray diffraction pattern of lysozyme

penicillin and vitamin B12. Her work on the structure of insulin began in the 1930s but did not reach a conclusion until 1972. Thanks to vast increases in the available computing power, she was able to describe the detailed positions of over 800 atoms in the complex insulin molecule.

A key moment in the advance of molecular biology was the discovery of the structure of DNA in 1954. Francis Crick (born 1916) and James Watson (born 1928) while working in Cambridge used a combination of computing, chemical insight and model building to interpret X-ray diffraction photographs produced by Rosalind Franklin (1920–1958) and Maurice Wilkins (born 1916) in London to elucidate the shape of the double-helical molecules.

Much more recently X-ray diffraction was used to confirm the structure of the new form of carbon, buckminsterfullerene (see page 88).

Definition

Lattice
Chemists use the term lattice to describe an array of atoms or ions in a crystal. So a metal crystal consists of a lattice of metal cations held together by a cloud of mobile electrons.

Two types of structure

Broadly speaking there are two types of structure: giant structures and molecular structures.

Materials with giant structures have crystal structures in which all the atoms or ions are strongly linked by a network of bonds extending throughout the crystal.

Substances with giant structures generally have high melting points and boiling points.

Test yourself D

1 Look up the melting and boiling points of these elements and decide whether they have giant or molecular structures:

beryllium, boron, fluorine, silicon, white phosphorus, sulfur, calcium, cobalt, iodine.

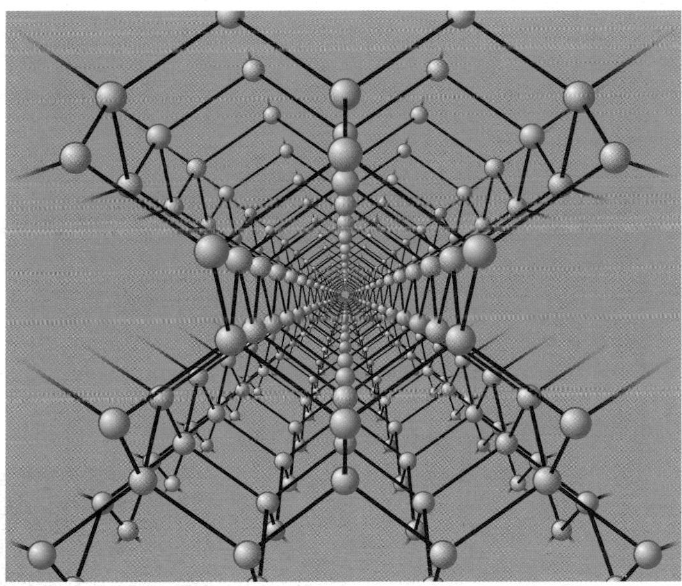

Figure 3.4.2 ◀ A close-up view of the diamond giant structure. Every carbon atom is bonded to four other atoms and the bonding continues throughout the crystal. There are no weak links

Substances with molecular structures consist of small groups of atoms. The bonds linking the atoms in the molecules are relatively strong (intramolecular forces). The forces between molecules (intermolecular forces) are weak. Molecular substances have low melting and boiling points because of the weakness of intermolecular forces.

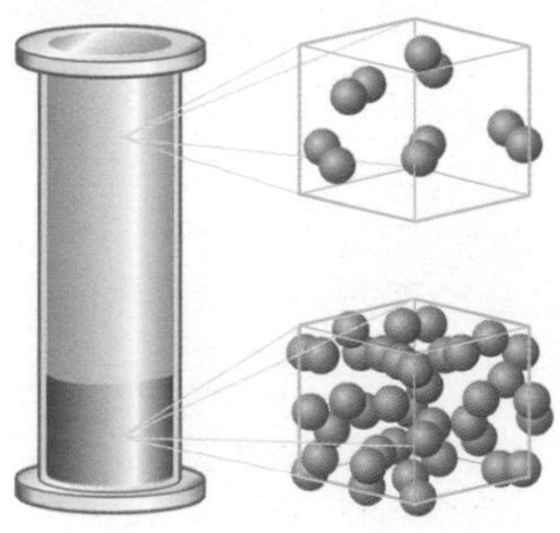

Figure 3.4.3 ▲
Molecules in bromine. Many molecular elements and compounds are liquids or gases at room temperature because little energy is needed to overcome the weak forces between the molecules

Figure 3.4.5 ▲
Metal cables in the electricity grid supported by steel pylons. A reminder that metals are both strong, bendable and able to conduct electricity. Ceramic insulators between the conducting cables and the pylons prevent the electric current leaking away to earth

Three types of strong bonding

The main types of solid materials are metals, polymers, ceramics and glasses. Each of these materials depends for its strength on one of three types of strong bonding.

For each type the strength of the bond depends on electrostatic attractions between positive and negative charges.

Metallic bonding

In a metal crystal each metal atom forms a positive ion by giving up one or two electrons to a 'sea' of negative electrons. Metallic bonding results from the attraction between the positive metal ions and the negative electrons. The electrons are free to drift through the crystal. They do not have fixed positions. In other words they are delocalised.

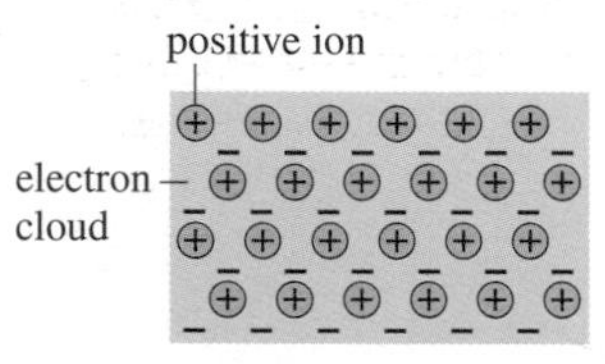

Figure 3.4.4 ▲
Metallic bonding. Metal positive ions held together by a 'sea' of negative electrons

Covalent bonding

A single covalent bond consists of a shared pair of electrons. The bond results from the attraction between the electrons and the positively-charged nuclei of the atoms. The shared electrons are held in place between the two atoms – they are localised. These electrons cannot move and there are no charged particles so covalently bonded materials are normally non-conductors.

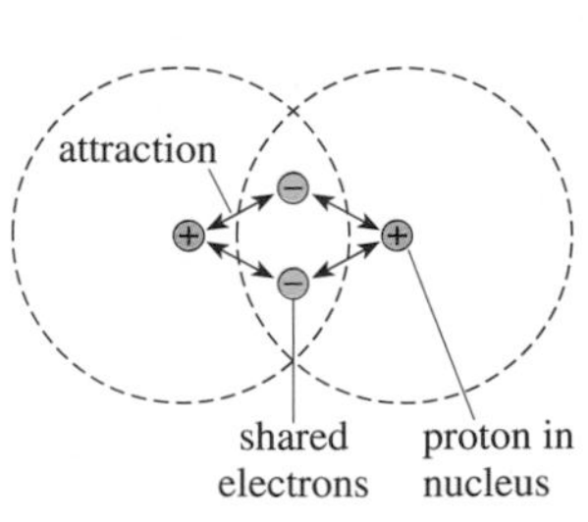

Figure 3.4.7 ▲
A covalent bond between two atoms

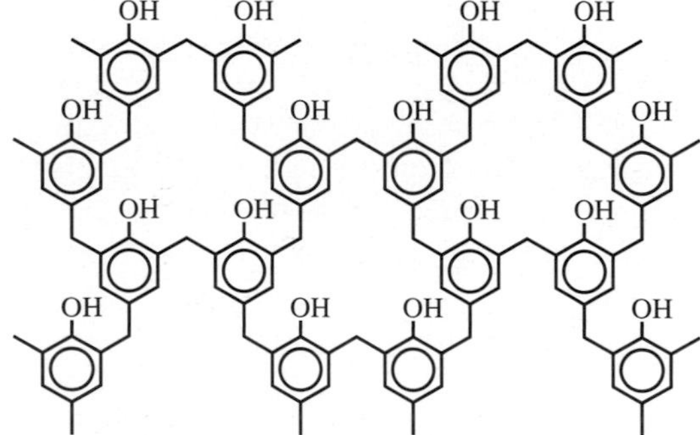

Figure 3.4.6 ▲
A thermosetting polymer in which the long chain molecules are cross-linked so the structure consists of a continuous network of covalent bonds

Ionic bonding

Ionic bonding holds together the ions in the crystals of compounds of metals with non-metals. The bonding is an electrostatic attraction between positive and negative ions.

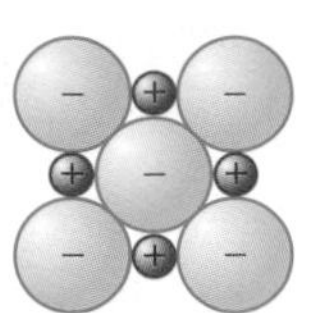

Figure 3.4.8 ▲
Ionic bonding is an electrostatic attraction between positive and negative ions

Figure 3.4.10 ▲
Glass used to make a bridge in the Science Museum. In glass there are two types of bonding: covalent bonding in a giant silicon–oxygen lattice and ionic bonding between charges on the Si-O lattice and metal ions

Figure 3.4.9 ▲
Magnesium oxide bricks in use in a rotary cement kiln. Magnesium oxide is an ionic ceramic with a very high melting point and so it is suitable as a furnace lining. Like most ceramics, these bricks are brittle

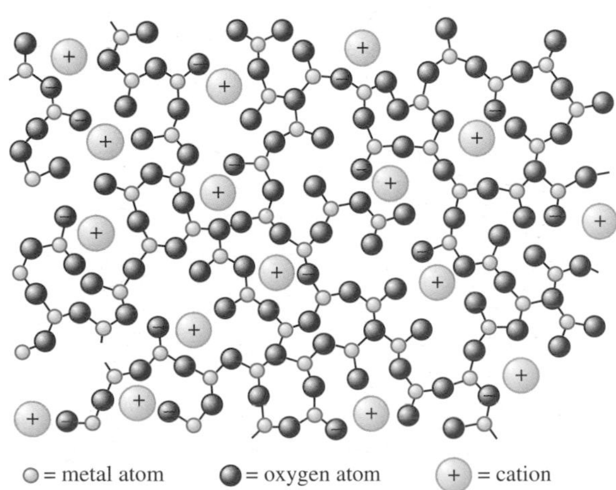

Figure 3.4.11 ▲
Covalent and ionic bonding in glass. Note the irregular arrangement of the atoms and ions. Glass is not crystalline. It is amorphous

Section three Physical Chemistry

3.5 Structure and bonding in metals

Most of the elements are metals – over three-quarters of them. Some metals have given their names to periods of history, for example, the Bronze Age and the Iron Age. People value gold and silver for their rarity and appearance. Advanced technologies depend on the special properties of metals such as platinum, zirconium and titanium. While some metals are useful when pure, many others are better converted to alloys such as bronze, solder and steel. CD-ROM

Metal crystals

Metals consist of a mass of crystal grains – they are polycrystalline. The crystals are hard to see because they are normally very small, usually too small to see without the help of a microscope. There are some exceptions, such as the large zinc crystals visible on the surface of galvanised iron.

Figure 3.5.1 ► *Zinc crystals on the surface of galvanised iron. The atoms in the crystals are arranged in a regular lattice. The crystals grow into each other as the metal solidifies so their shape is not like other crystals*

Often a layer of metal oxide obscures the surface of a metal. Rust on iron and the green coating on weathered copper hide the metals underneath.

Metallurgists study metal crystals by removing any oxide layer, polishing the surface till it shines and then dipping the sample into a solution which etches the surface to show up the boundaries between the crystal grains (see Figure 3.5.2).

Definition

Metallic bonding
Metallic bonding is the strong bonding between atoms in metal crystals (see page 68). Each metal atom in a metal crystal contributes electrons from its outer shell to a 'sea' of delocalised electrons. Only elements with relatively low first ionisation energies form metallic crystals.

Metal structures

X-ray diffraction studies (see page 66) and powerful electron microscopes confirm that metal crystals have regular structures. All metals consist of giant structures of atoms held together by metallic bonding.

Close-packed structures

In many metals the atoms are packed together as closely as possible. In a layer of close-packed spheres each atom has six other spheres touching it.

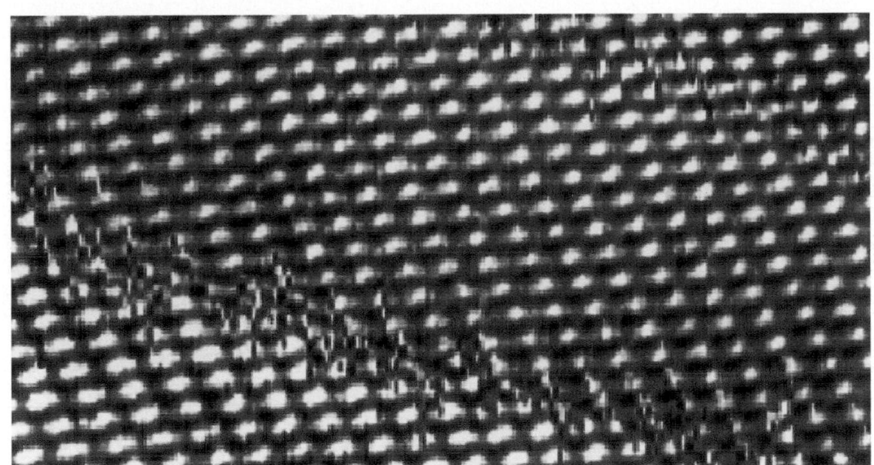

Figure 3.5.3 ▲
Electron microscope photograph showing the regular arrangement of atoms in a gold crystal (magnification 20×10^6)

Figure 3.5.2 ▲
Crystal grains in the structure of copper

In three dimensions, layers of close-packed atoms stack up in two possible ways. In hexagonal close packing the third layer is directly over the first layer (aba). In face-centered cubic close-packing it is the fourth layer which corresponds with the first layer (abca). In the two structures, each atom touches 12 nearest neighbours.

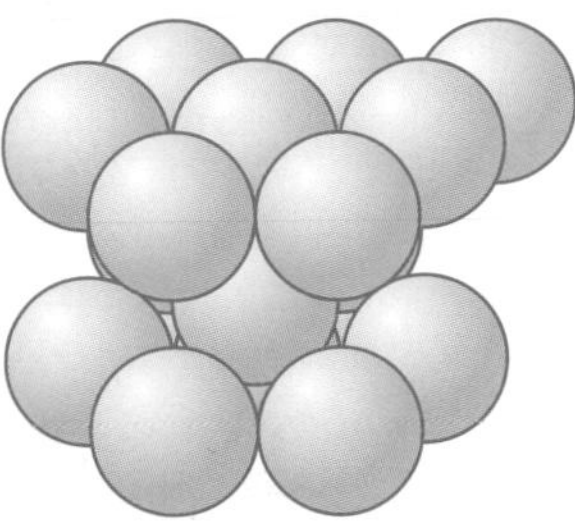

Figure 3.5.4 ▲
Hexagonal close packing of metal atoms. Metals with this structure include magnesium, titanium and zinc

The body-centered cubic structure

Some metals have a structure that is more open than the two close-packed structures. In this structure there is an atom at each corner of a cube surrounding one atom at the centre of the cube.

Metals with the body-centered cubic structure are the group 1 metals lithium, sodium and potassium, also the d-block metals chromium, vanadium and tungsten.

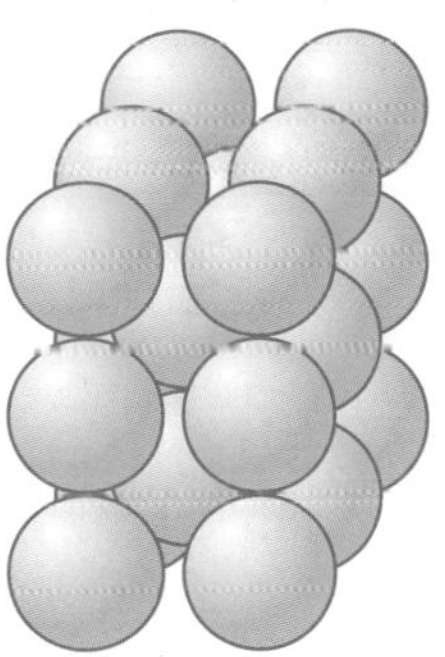

Figure 3.5.5 ▲
Body-centered cubic structure. Each metal atom is surrounded by eight nearest neighbours

Metal properties

Metals can bend without breaking because metallic bonding is not highly directional. Lines or layers of metal atoms can shift their position in a crystal without the bonds breaking. The layers of atoms in a metal can slide over each other so that metals are malleable and ductile.

Metals conduct because the shared bonding electrons can drift through the crystal structure from atom to atom when there is an electric potential difference. They are delocalised electrons.

Test yourself D

1 Find examples of the uses of metals and alloys which illustrate their typical properties. Name the metal or alloy used in each example. Metals:

a) are shiny
b) conduct electricity
c) bend and stretch without breaking
d) have high tensile strength.

2 Examine the patterns of metal properties in the periodic table.

a) Which metals have relatively high densities? Which have relatively low densities?
b) Which metals have relatively high melting and boiling points? Which have relatively low melting and boiling points?

Definition

Delocalised electrons are bonding electrons which are not fixed between two atoms in a bond but are shared between several or many atoms.

3.6 Ionic giant structures

How do chemists explain the extraordinary difference between sodium chloride, which is common salt, and the two elements sodium and chlorine? Sodium is a violently reactive metal while chlorine is a highly toxic gas.

Atoms into ions

Salts such as sodium chloride are compounds of metals with non-metals. Metals, on the left-hand side of the periodic table (see page 239), have just a few electrons in their outer shells. Non-metals, on the right of the table, have almost complete sub-shells. When compounds form between metals and non-metals, the metal atoms lose their outer electrons and become positive ions. The non-metal atoms gain the electrons and become negative ions.

When sodium reacts with chlorine, the sodium atom loses its one outer electron forming an Na^+ ion which now has the same number and arrangement of electrons as the noble gas neon. Each chlorine atom gains one electron forming a Cl^- ion with the same electron configuration as the noble gas argon.

Note

Electron dot-and-cross diagrams provide a balance sheet for keeping track of the electrons when ionic compounds form. They do not describe the mechanisms of the reactions. All the electrons are the same. Chemists use dots and crosses to show which electrons come from the metal atom and which from the non-metal atom.

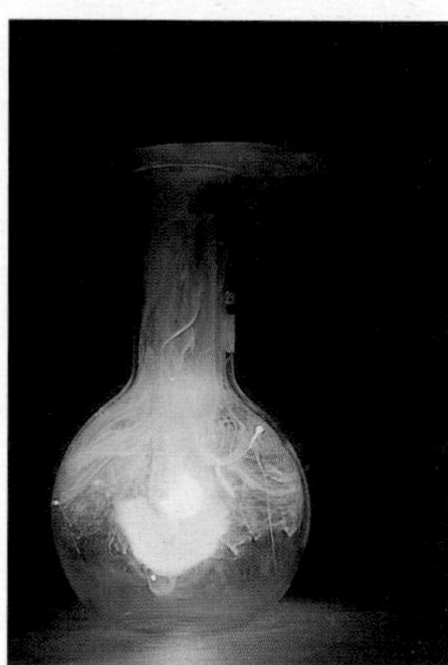

Figure 3.6.1 ▲
Hot sodium reacting with chlorine gas

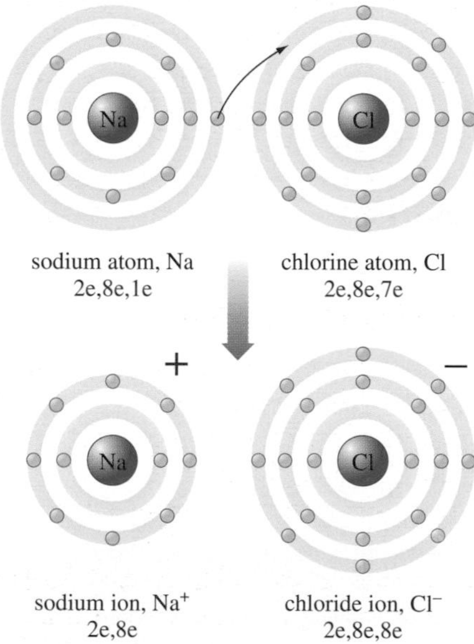

Figure 3.6.2 ▶
Formation of an ion pair of sodium chloride

In many instances when atoms react to form ions they gain or lose electrons in such a way that the ions formed have the same electron configuration as a noble gas. Reactive atoms turn into ions. Sodium chloride is so different from sodium and chlorine because the ions in the salt are chemically entirely distinct from the atoms of sodium and the molecules of chlorine. The ions are much less reactive than the atoms or molecules.

$Na^{\bullet}$	+	$\times\overset{\times\times}{\underset{\times\times}{Cl}}\times$	⟶	Na^{+}	+	$\overset{\times\times}{\underset{\times\times}{\bullet Cl}}\times^{-}$
sodium atom		chlorine atom		sodium ion		chloride ion
(2.8.1)		(2.8.7)		(2.8)		(2.8.8)

Figure 3.6.3 ▲
Dot-and-cross diagram for the formation of sodium and chloride ions showing only the electrons in the outer shells

$Ca^{:}$	+	$\overset{\times\times}{\underset{\times\times}{F}}\times$ $\overset{\times\times}{\underset{\times\times}{F}}\times$	⟶	Ca^{2+}	+	$\overset{\times\times}{\underset{\times\times}{\bullet F}}\times^{-}$ $\overset{\times\times}{\underset{\times\times}{\bullet F}}\times^{-}$
calcium atom		two fluorine atoms		calcium ion		two fluoride ions
(2.8.8.2)		(2.7)		(2.8.8)		(2.8)

Figure 3.6.4 ▲
Dot-and-cross diagram for the formation of calcium and fluoride ions showing only the electrons in the outer shells

Ionic bonding

Ionic bonding is the result of electrostatic forces of attraction between positive metal ions and negative non-metal ions. In a solid salt the ions are built up into a crystal lattice with each positive ion surrounded by negative ions and each negative ion surrounded by positive ions.

The sodium chloride structure

Crystals of sodium chloride are cubes. This reflects the underlying cubic arrangement of the ions (see Figure 2.3.12 on page 24). Each positive ion is surrounded by six chloride ions and each chloride ion is surrounded by six positive ions. This is 6:6 co-ordination.

Many other compounds have the same structure as sodium chloride including: the chlorides, bromides and iodides of Li, Na and K; the oxides and sulfides of Mg, Ca, Sr, Ba; as well as the fluorides, chlorides and bromides of Ag.

The caesium chloride structure

Another salt with a cubic crystal structure is the ionic compound caesium chloride, $Cs^{+}Cl^{-}$. In this structure each positive ion is surrounded by eight nearest neighbours at the corners of a cube and each negative ion is similarly surrounded by eight positive ions.

Other compounds with the caesium chloride structure include CsBr, CsI and NH_4Cl.

In an ionic crystal the ions behave like charged spheres in contact. The structures are only stable if each ion is in contact with its nearest neighbours.

A caesium ion is large enough to have eight chloride ions around it as in the caesium chloride structure. Sodium ions are smaller and only big enough to touch six neighbouring chloride ions as in the sodium chloride structure.

Energy changes

It takes energy to remove electrons from metal ions (see Ionisation Energy on pages 55–56). There are much smaller energy changes (electron

Test yourself D

1 Draw dot-and-cross diagrams for:
  a) lithium fluoride
  b) magnesium chloride
  c) lithium oxide
  d) calcium oxide.

2 With the help of a periodic table, predict the charges on the ions of these elements: caesium, strontium, gallium, selenium and astatine.

3 Why do metals form positive ions while non-metals form negative ions?

4 Work out the numbers of protons, neutrons and electrons in the following pairs of particles. In what ways are the particles in each pair the same and how do they differ?
  a) a sodium ion and a neon atom,
  b) a chloride ion and an argon atom.

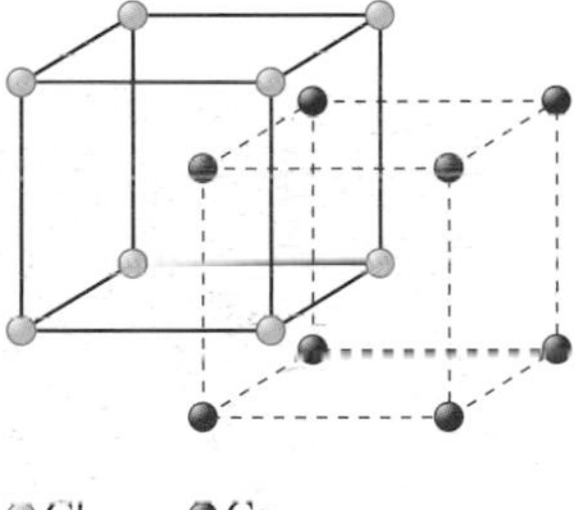

Figure 3.6.5 ▲
Structure of caesium chloride. The structure consists of a simple cubic array of positive ions interpenetrating a cubic array of negative ions. Each caesium ion touches eight chloride ions. Each chloride ion touches eight caesium ions. This is 8:8 co-ordination

Definition

Co-ordination number
The number of nearest neighbours of an atom or ion in a crystal structure.

affinities) on adding electrons to non-metal atoms. So where does the energy come from to turn the atoms into ions? It turns out that ionic crystals are stable because of the large release of energy as the ions, which attract each other, come together to form a crystal lattice.

The strength of the ionic bond, measured in $kJ\ mol^{-1}$, results from the energy given out as millions and millions of positive and negative ions come together to form a crystal lattice. The larger the charges on the ions, the bigger the attractive force between them and the larger the quantity of energy released. The smaller the ions, the closer the charges get to each other, the stronger the force of attraction and the larger the quantity of energy released.

Note

During electrolysis of molten sodium chloride:

- sodium ions are reduced to atoms at the cathode by gaining electrons
- chloride ions are oxidised to chlorine at the anode by losing electrons (see page 23).

Ionic radii

Crystallographers use X-ray diffraction methods to measure the spacing between ions in crystals. They calculate ionic radii from their results. The radius of the positive ion of an element is smaller than its atomic radius because it loses its outer shell when turning into an ion. The radius of the negative ion of an element is larger than its atomic radius as electrons are added to the outer shell.

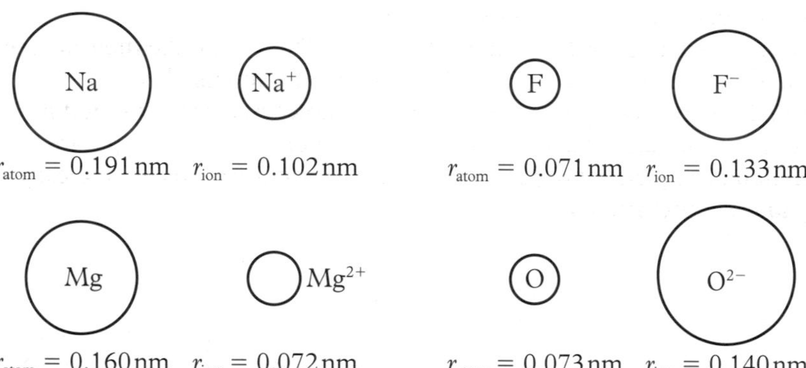

Figure 3.6.6 ▶ *Comparison of the radii of atoms and ions*

Definition

Electron affinity
The energy change per mole on adding electrons to atoms or ions. The gain of the first electron gives out energy but it needs energy to add a second electron to the negative ion.

Properties of ionic compounds

The main properties of compounds with a giant structure are that:

- they are crystalline and hard
- they have high melting and boiling points
- they are often soluble in water (see page 89) but insoluble in non-polar solvents (see page 83)
- they do not conduct electricity when solid but they do conduct when molten or in solution.

Solid salts, such as sodium chloride, are non-conductors. The charged ions cannot move in a solid so there are no moving charges to carry a current.

When sodium chloride melts, the ions are then free to move. Molten sodium chloride is a conductor. Positive sodium ions can move towards the negative electrode (cathode) while negative chloride ions move to the positive electrode (anode). An electric current decomposes an ionic compound such as sodium chloride. This is electrolysis. When the sodium ions reach the electrodes they gain electrons and turn back into sodium atoms. When chloride ions reach the anode they lose electrons turning back into chlorine molecules. So electrolysis reverses the changes which happen when sodium chloride forms from its elements (see page 72).

Test yourself D

5 Look up the melting points of magnesium oxide, calcium oxide, sodium chloride and potassium chloride. Suggest why the melting points of MgO and CaO are much higher than the melting points of NaCl and KCl.

6 Identify the products at the electrode during the electrolysis of:

a) molten potassium bromide
b) molten magnesium chloride.

3.7 Covalent molecules and giant structures

Living things are made up largely of water and a huge variety of compounds of carbon with elements such as hydrogen, oxygen, nitrogen and sulfur. They are all molecular compounds of non-metal elements. These compounds differ greatly from the minerals in the rocks in the crust of the Earth. Most of these minerals consist of giant networks of the two non-metals silicon and oxygen with other elements. So chemists have to understand the structure and bonding of non-metals and their compounds if they are to explain the properties of the organic and inorganic worlds.

Molecular structures

In most non-metal elements the atoms join together in small molecules. Examples of molecular elements are hydrogen, nitrogen, phosphorus, sulfur, chlorine, bromine and iodine.

The covalent bonds holding the atoms together within the molecules are strong so the molecules do not easily break up into atoms. The intermolecular forces (see pages 83–84) are, however, weak so that it is quite easy to separate them. This means that molecular substances are often liquids or gases at room temperature. Molecular solids are typically easy to melt or evaporate.

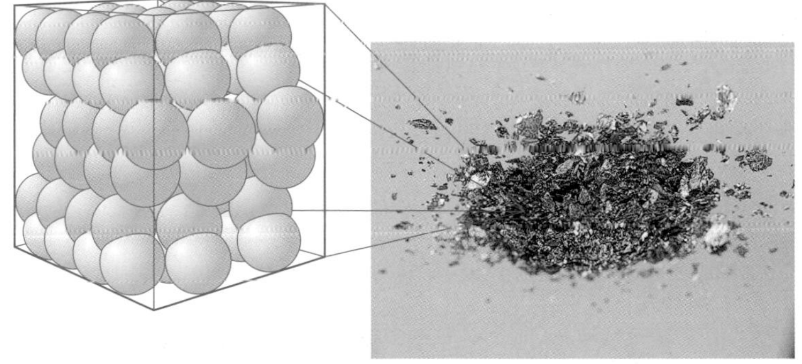

Figure 3.7.1 ◄ *Structure of iodine showing the arrangement of iodine molecules*

Most of the compounds of non-metals with other non-metals are also molecular. This is true of some simple compounds such as water, carbon dioxide, ammonia, methane and hydrogen chloride. It is also true of the many many thousands of carbon compounds which feature in Part 5 (see page 178).

Molecules do not carry electric charge and there are no free electrons in molecular substances; so materials made of molecules cannot conduct electricity. The typical properties of covalent molecular elements or compounds are that:

- they have relatively low melting and boiling points
- the energy needed to separate the molecules is low (see pages 18–19)
- they do not conduct electricity when solid, liquid or gaseous

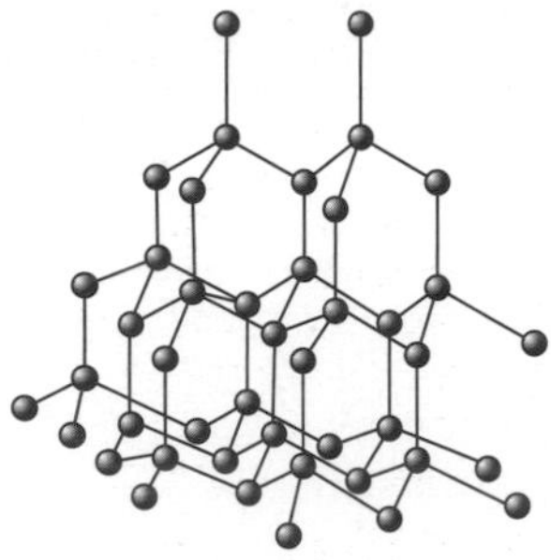

Figure 3.7.2 ▲
Fragment of the diamond giant structure. Note that each carbon atom is linked to four other atoms. This network of bonding extends throughout the giant structure (see Figure 3.4.2 on page 67)

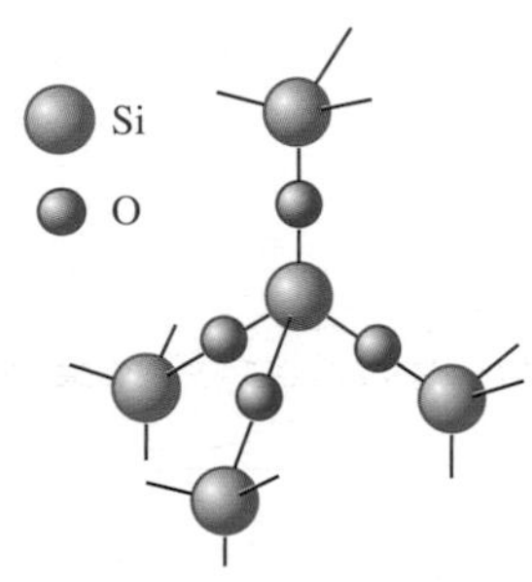

Figure 3.7.3 ▲
A fragment of the giant structure of silicon dioxide in the mineral quartz. Each Si atom is at the centre of a tetrahedron of oxygen atoms. The arrangement of silicon atoms is the same as the arrangement of carbon atoms in diamond but there is an oxygen atom between each silicon atom. Amethyst is crystalline quartz coloured purple due to the presence of iron(III) ions. Sandstone and sand consist mainly of silica

- they are more soluble in non-polar solvents such as hexane than in water, and the solutions do not conduct electricity.

These are not rigid rules. Some covalent compounds are very soluble in water. Sugar and ethanol, for example, dissolve to give solutions which do not conduct electricity. This is because of hydrogen bonding (see pages 85–86).

Hydrogen chloride and sulfur dioxide are also molecular and very soluble. They react with water as they dissolve forming ions (see page 30) and so these solutions do conduct electricity.

Covalent giant structures

A few non-metal elements consist of giant structures of atoms held together by covalent bonding. Examples are carbon and silicon.

The covalent bonds in diamond are strong and point in a definite direction so diamonds are both very hard and have very high melting points.

Diamond does not conduct electricity because the electrons in covalent bonds are fixed (localised) between pairs of atoms.

Some compounds of non-metals with non-metals, such as silicon dioxide and boron nitride, have giant structures with covalent bonding. These compounds are also hard and non-conductors.

Covalent bonding

Covalent bonding holds together the atoms of non-metals in molecules and giant structures.

Covalent bonds form when atoms share electrons. The atoms are held together by the attraction between the positive charges on their nuclei and the negative charge on the shared electrons (see page 68).

The electron configuration of fluorine is $1s^22s^22p^5$. A fluorine atom has seven electrons in the outer shell. When two fluorine atoms combine to form a molecule, they share two electrons. The electron configuration of each atom is then like that of neon, the nearest noble gas.

Chemists draw a line between symbols to represent a covalent bond. So they write a fluorine molecule as F—F. This is the structural formula, showing the atoms and bonding. The molecular formula of fluorine is F_2.

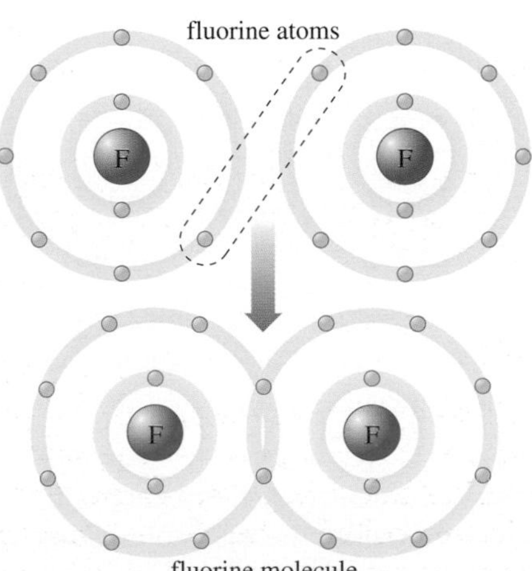

Figure 3.7.4 ►
Covalent bonding in a fluorine molecule

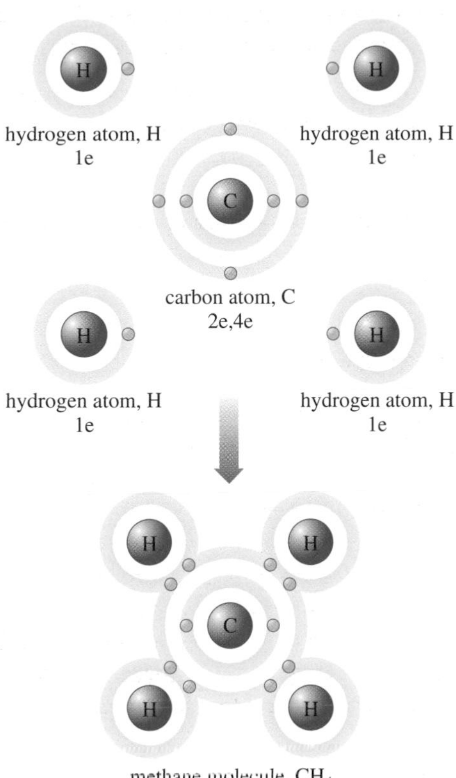

Figure 3.7.5 ◄
Covalent bonding in methane

Test yourself

1 Compare the melting and boiling points of the non-metals boron, fluorine, white phosphorus, germanium and bromine. Which of these elements are molecular and which have giant structures?

2 Compare the formulae and physical properties of carbon dioxide and silicon dioxide. Explain the differences in terms of the structures of the two compounds.

Covalent bonds also link atoms in non-metal compounds. Figure 3.7.5 shows the covalent bonding in methane.

Dot-and-cross diagrams provide a simple way of representing covalent bonding. Only the electrons in outer shells appear in these diagrams.

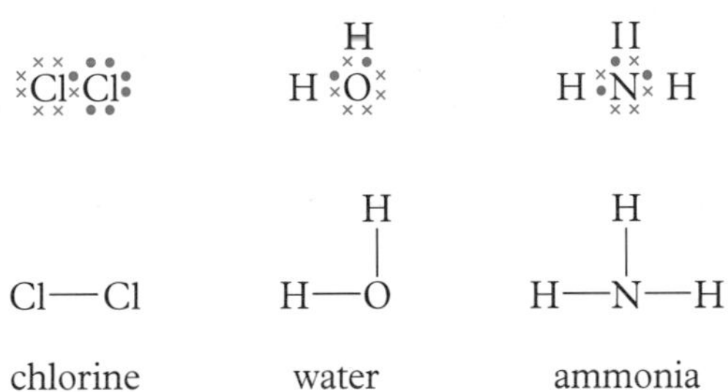

Figure 3.7.6 ◄
Dot-and-cross diagrams to show single covalent bonding in molecules. Also shown is a simpler way of showing the bonding in molecules. A line between two symbols represents a covalent bond

The non-metals common in organic chemistry generally form a fixed number of covalent bonds. This helps to work out the structures of molecules (Figure 3.7.7).

Multiple bonding

One shared pair of electrons makes a single bond. Double bonds and triple bonds are also possible with two or three shared pairs.

There are two covalent bonds between two atoms in oxygen and carbon dioxide. With two electron pairs involved in the bonding, there is a region of high electron density between two atoms joined by a double bond.

Figure 3.7.7 ► The number of covalent bonds formed by atoms

Element	Number of covalent bonds
carbon, C	4
hydrogen, H	1
oxygen, O	2
nitrogen, N	3
halogens, F, Cl, Br, I	1

Figure 3.7.8 ► Examples of molecules with double bonds. Note the number of bonds formed by each atom. How does this compare with the numbers in Figure 3.7.7?

oxygen	carbon dioxide	ethene
O=O	O=C=O	$H_2C=CH_2$

N≡N H—C≡C—H

Figure 3.7.9 ▲ Examples of molecules with triple bonds

Test yourself

3 Draw diagrams showing all the electrons in shells to describe how covalent bonds link the atoms in:

a) hydrogen, H_2
b) hydrogen chloride, HCl
c) ammonia, NH_3

4 Draw dot-and-cross diagrams to show the covalent bonding in:

a) hydrogen bromide, HBr
b) hydrogen sulfide, H_2S
c) ethane, C_2H_6
d) sulfur dioxide, SO_2
e) carbon disulfide, CS_2

5 Identify the atoms with lone pairs in these molecules and state the number of lone pairs:

a) ammonia
b) water
c) hydrogen fluoride.

Lone pairs of electrons

In some molecules there are atoms with pairs of electrons in their outer shells which are not involved in bonding between the atoms in the molecule. Chemists call these 'lone pairs' of electrons.

Lone pairs of electrons:

- affect the shapes of molecules
- form dative covalent bonds
- are important in the chemical reactions of compounds such as water and ammonia.

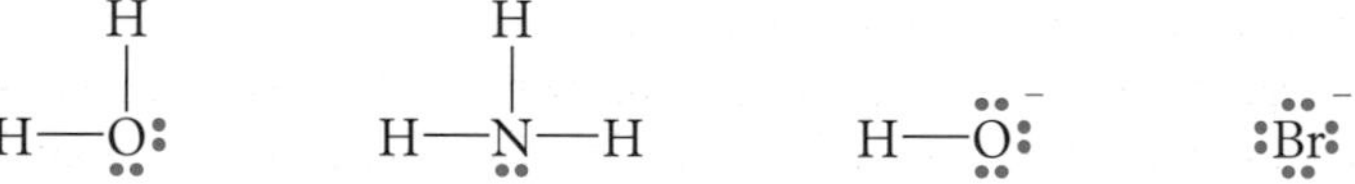

Figure 3.7.10 ▲ Examples of molecules and ions with lone pairs of electrons.

Dative covalent bonds

Two atoms share a pair of electrons in any covalent bond. Usually each atom supplies one electron to make up the pair. Sometimes, however, one atom provides both the electrons. Chemists call this a dative covalent bond because the word 'dative' means 'giving' and one atom gives both the electrons to make the covalent bond. An alternative name is a co-ordinate

$H_3N: + H^+ \longrightarrow NH_4^+$

Figure 3.7.11 ▲ Formation of an ammonium ion

bond Once formed there is no difference between a dative bond and any other covalent bond.

Ammonia forms a dative covalent bond when it reacts with a hydrogen ion to make an ammonium ion.

Dative covalent (co-ordinate) bonding accounts for the structures of carbon monoxide and solid beryllium chloride (see page 144).

> **Test yourself**
>
> **6** Draw dot-and-cross diagrams to show the formation of dative covalent bonds in:
>
> **a)** the H_3O^+ ion
>
> **b)** the BF_4^- ion.

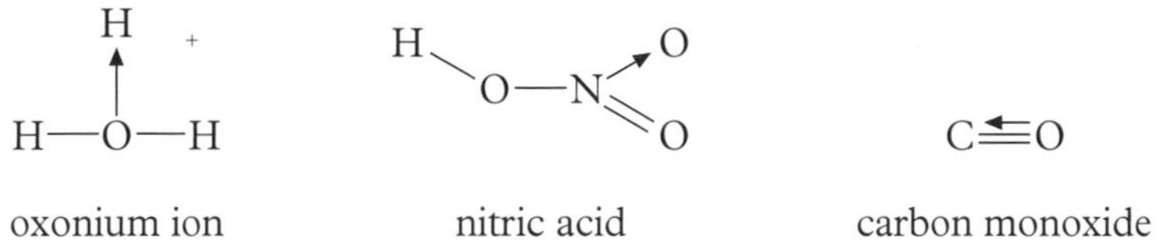

oxonium ion | nitric acid | carbon monoxide

Figure 3.7.12 ◄ *Examples of dative covalent (co-ordinate) bonds*

The octet rule and its limitations

It was the US chemist Gilbert Lewis who first suggested the octet rule in 1916. The rule says that atoms tend to gain, lose or share electrons when they combine with other atoms to acquire a stable octet of electrons. The 'stable octet' is the eight electrons, s^2p^6, corresponding to the outer electron configuration of the nearest noble gas in the periodic table (see Figure 3.7.6, and 3.7.8–10).

There are many exceptions to the octet rule so it is not a safe guide. The rule works pretty well for the elements Li to F in period 2 because there is only a total of four s and p orbitals in the second shell, each of which can hold two electrons. The octet rule also works for ionic compounds of group 1 and 2 metals with the halogens, oxygen and sulfur.

There are some exceptions even in period 2. An example is the molecule BCl_3 when the chloride is a gas. There are only six electrons round the central boron atom (see Figure 3.7.13).

More exceptions arise in period 3 and beyond because d orbitals in the third shell can become involved in bonding. See Figures 3.7.14 and 3.7.16 for molecules with more than eight electrons in the outer shell of the central atom. Chemists sometimes talk about 'expanding the octet' when describing an atom with a share in more than eight bonding electrons.

Shapes of molecules

X-ray diffraction and other instrumental methods make it possible to measure bond angles accurately. The results show that covalent bonds have a definite direction and length.

A theory based on repulsion between electron pairs makes it possible to predict the shapes of molecules and bond angles with surprising accuracy.

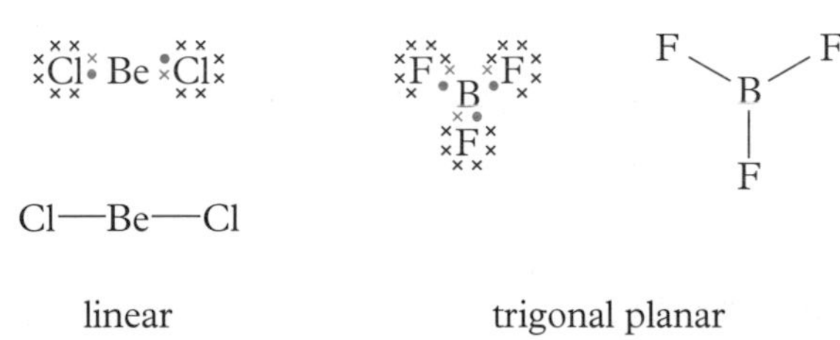

Figure 3.7.13 ◄ *Shapes of molecules with two and three electron pairs around the central atom*

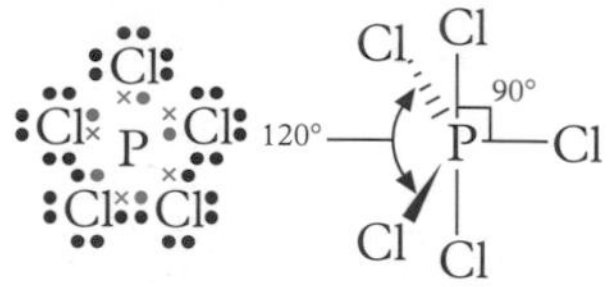

Figure 3.7.14 ▲
A molecule of phosphorus pentachloride vapour. A molecule with five electron pairs around the central atom

Chemists find that they can predict the shapes of molecules and ions by examining the number of bonding and non-bonding lone pairs of electrons in the outer shell of the central atom. The expected shape for a molecule is the one which minimises the repulsion between electron pairs by keeping them as far apart as possible in three dimensions.

Non-bonding lone pairs are held closer to the central atom. The result is that the order of repulsion between electron pairs is:

lone pair–lone pair > lone pair–bond pair > bond pair–bond pair

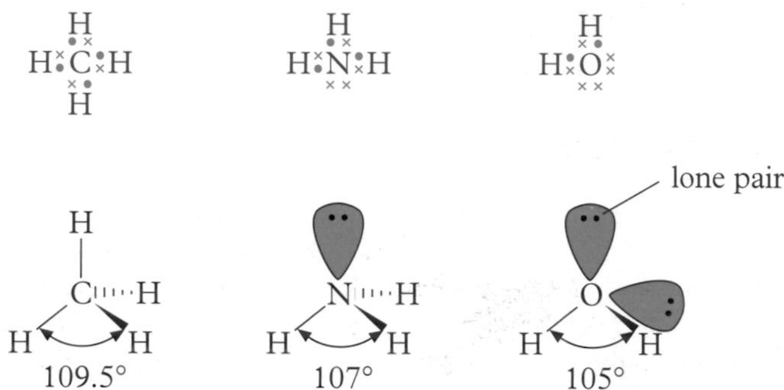

Figure 3.7.15 ►
Bond angles in molecules with four electron pairs around the central atom. The greater repulsion between lone pairs, and between lone pairs and bonding pairs, means that the angles between the covalent bonds decrease as the number of lone pairs increases

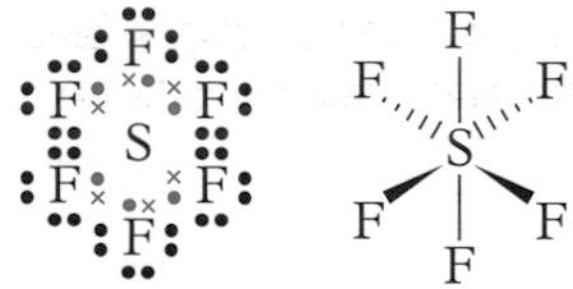

Figure 3.7.16 ▲
An octahedral molecule. An octahedron consists of two pyramids with square bases joined

A molecule with five electron pairs around the central atom takes the shape of two tetrahedrons joined. This is a trigonal bipyramid.

A molecule with six electrons pairs around the central atom is octahedral. An octahedron has eight faces but six corners.

A double bond counts as one region of electrons when it comes to predicting molecular shapes. The same is true for a triple bond.

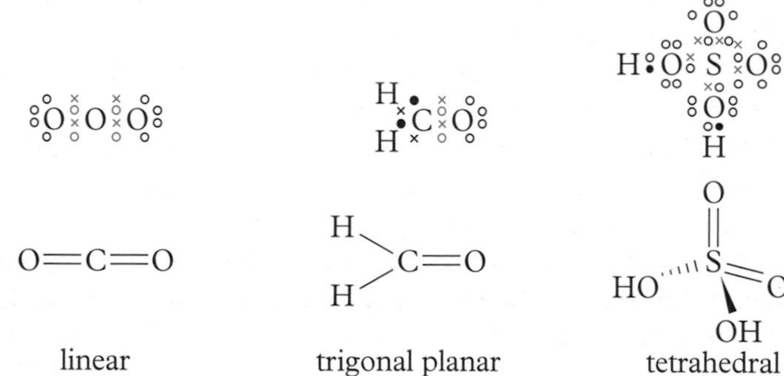

Figure 3.7.17 ▲
Shapes of molecules with multiple bonds

Test yourself

7 Draw dot-and-cross diagrams for these molecules or ions and then predict their shapes:

a) BCl_3 and PH_3
b) CO_2 and SO_2
c) NH_4^+ and NH_2^-
d) SO_4^{2-} and CO_3^{2-}

3.8 Intermediate types of bonding

All the evidence suggests that the bonding in caesium fluoride, Cs^+F^- is ionic. This is a compound between a highly reactive metal and the most reactive non-metal. The bonding in a molecule such as chlorine, where both atoms are the same, is purely covalent. In most compounds, however, the bonding is neither purely ionic nor purely covalent. So how do chemists decide on the type of bonding to expect in a compound?

Polar covalent bonds

The electron pair in a covalent bond is not shared equally if the two atoms joined by the bond are different. The nucleus of one atom attracts the electrons more strongly than the nucleus of the other. This means that one end of the bond has a slight excess of negative charge (δ−). The other end of the bond has a slight deficit of electrons so that the charge cloud of electrons does not cancel the positive charge on the nucleus (δ+).

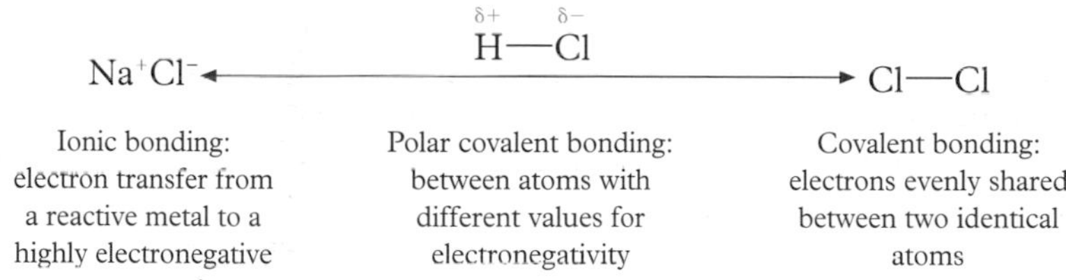

δ+ δ−
H—Cl
Na⁺Cl⁻
Cl—Cl
Ionic bonding: electron transfer from a reactive metal to a highly electronegative non-metal
Polar covalent bonding: between atoms with different values for electronegativity
Covalent bonding: electrons evenly shared between two identical atoms

Figure 3.8.1 ◀
A spectrum from purely ionic to purely covalent bonding

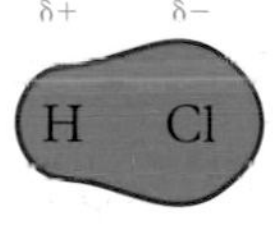

δ+ δ−
H Cl

Figure 3.8.2 ▲
A polar covalent bond in hydrogen chloride

Chemists use electronegativity values as a guide to the extent to which the bonds between them will be polar. The stronger the pulling power of an atom, the higher its electronegativity.

There are two quantitative scales of electronegativity; one devised by Linus Pauling (1901–1994) and the other by Robert Mulliken (1896–1986). However the term electronegativity is generally used to compare one element with another qualitatively so it is enough to know the trends in values across and down the periodic table.

The highly electronegative elements, such as fluorine and oxygen, are at the top right of the periodic table. The least electronegative elements, such as caesium, are at the bottom left.

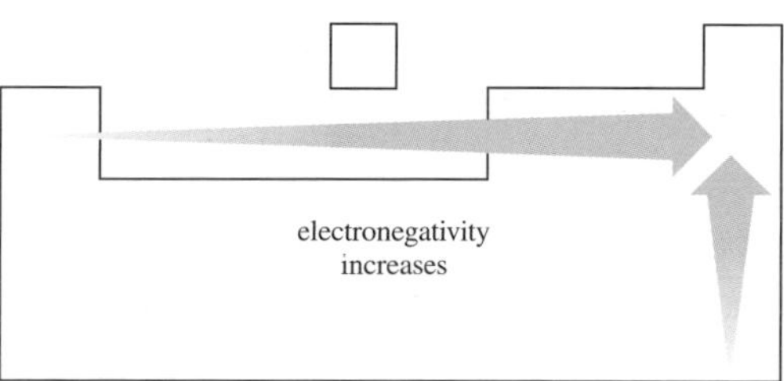

electronegativity increases

Figure 3.8.3 ▲
Trends in electronegativity for s and p block elements. Electronegativity is a measure of the power of an atom in a molecule to pull electrons to itself from other atoms

Note

The symbol δ is the Greek letter 'delta'. Chemists use this symbol for a small quantity or change. They use the symbols δ+ and δ− for the small charges at the ends of a polar bond. They use the capital Greek 'delta', Δ, for larger changes or differences (see page 57).

H
| δ+   δ−
H—C—Br
|
H

H δ+ δ−
  C=O
H

δ+   2δ−
H—C
   |
   H
   δ+

Figure 3.8.4 ▲
Examples of polar bonds between atoms with different electronegativities

The bigger the difference in the electronegativity of the elements forming a bond, the more polar the bond. Oxygen is more electronegative than hydrogen so an O—H bond is polar with a slight negative charge on the oxygen atom and a slight positive charge on the hydrogen atom.

The bonding in a compound becomes essentially ionic if the difference in electronegativity is large enough for the more electronegative element to completely remove electrons from the other element. This happens in compounds such as sodium chloride, magnesium oxide or calcium fluoride.

Polarisation of ions

At the other end of the spectrum of bonding (see Figure 3.8.1), chemists start by picturing purely ionic bonding between two atoms and then consider the extent to which the positive metal ion will distort the neighbouring negative ions giving rise to some degree of electron sharing (that is a degree of covalent bonding).

They find that ionic bonding is favoured if:

- the charges on the ions are small (1+ or 2+, 1− or 2−)
- the radius of the positive ions is large and the radius of the negative ions is small.

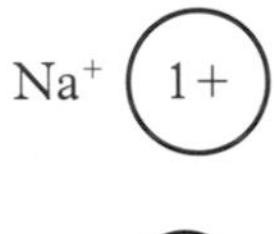

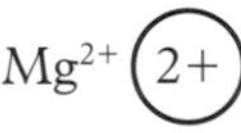

Figure 3.8.6 ▲
Ionic radii for metal ions in the third period

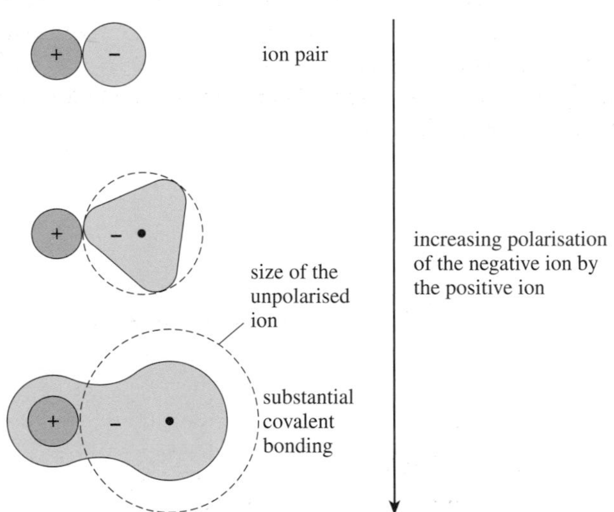

Figure 3.8.5 ▶
Ionic bonding. Ionic bonding with increasing degrees of electron sharing because the positive ion has distorted the neighbouring negative ion. Dotted circles show the unpolarised ions

A spherical anion is non-polar. When a positive ion pulls the electron cloud towards itself it turns the negative ion into a little dipole (with a positive pole and a negative pole). So chemists say that the positive ion polarises the negative ion. The smaller a positive ion and the larger its charge, the greater the extent to which it tends to polarise a negative ion. So polarising power increases along the series: Na^+, Mg^{2+}, Al^{3+}, Si^{4+} as the ions get smaller and more highly charged. Sodium chloride is an ionic, crystalline solid. The bonding in anhydrous aluminium chloride is largely covalent. Silicon chloride is a covalent, molecular liquid. The Si^{4+} ion never exists as a simple ion because its polarising power is so great.

The larger the negative ion and the larger its charge, the more polarisable it becomes. So iodide ions are more polarisable than fluoride ions. Fluorine, which forms the small, single-charge fluoride ion, forms more ionic compounds than any other non-metal. Sulfide ions, S^{2-}, are more polarisable than chloride ions, Cl^-.

Test yourself

1. Use Figure 3.8.3 to predict the polarity of the bonds in these molecules: H_2S, NO, CCl_4, ICl.
2. Put these sets of compounds in order of the character of the bonding, with the most ionic on the left and the most covalent on the right:
   a) NaCl, NaI, NaF, NaBr
   b) Al_2O_3, Na_2O, MgO, SiO_2
   c) LiI, NaI, KI, CsI

3.9 Intermolecular forces

Intermolecular forces are weak attractive forces between molecules. Without intermolecular forces there could be no molecular liquids or solids; real gases would behave more like ideal gases (see page 63); and useful products such as polythene bags and cling-film could not exist.

Weak intermolecular forces arise from electrostatic attractions between dipoles. The attraction can be between:

- molecules with permanent dipoles
- instantaneous dipoles created fleetingly in non-polar atoms or molecules.

Polar molecules

Polar molecules are little electrical dipoles (they have a positive pole and a negative pole). Dipoles tend to line up in an electric field. The bigger the dipole, the bigger the twisting effect (dipole moment). By making measurements with a polar substance between two electrodes it is possible to calculate dipole moments. The units are Debye units named after the physical chemist Peter Debye (1884–1966).

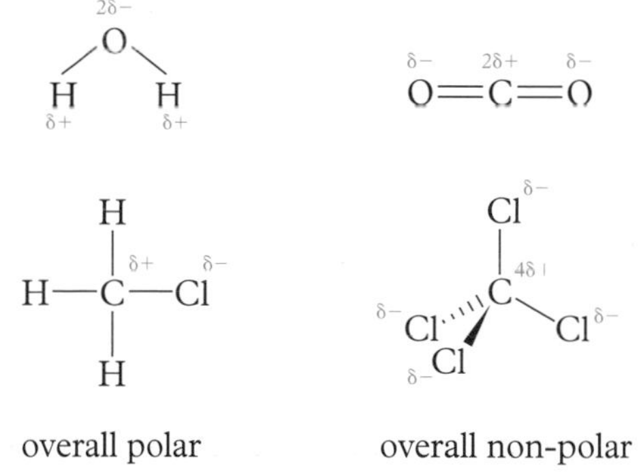

Figure 3.9.1 ◀
Molecules with polar bonds. Note that in the examples on the left the net effect of all the bonds is a polar molecule. In the examples on the right the overall effect is a non polar molecule

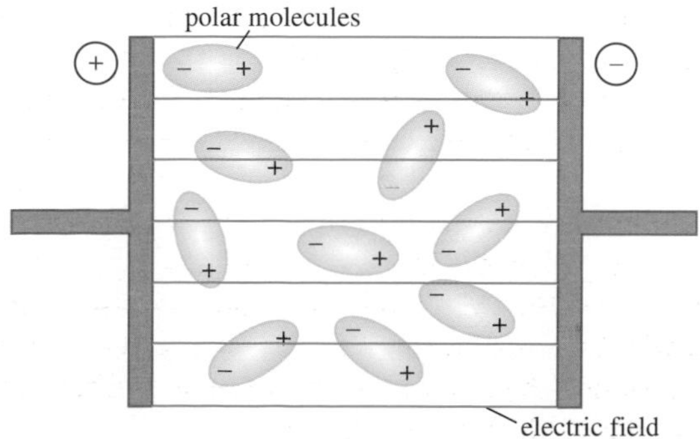

Figure 3.9.2 ◀
Polar molecules in an electric field. The electrostatic forces tend to line up the molecules with the field. The order is disrupted by random movements due to the kinetic energy of the molecules

Figure 3.9.3 ▼

Molecule	Dipole moment (in debye units)
HCl	1.08
H_2O	1.94
CH_3Cl	1.86
$CHCl_3$	1.02
CCl_4	0
CO_2	0

Dipole–dipole interactions

Molecules with permanent dipoles attract each other. The positive end of one molecule tends to attract the negative end of another.

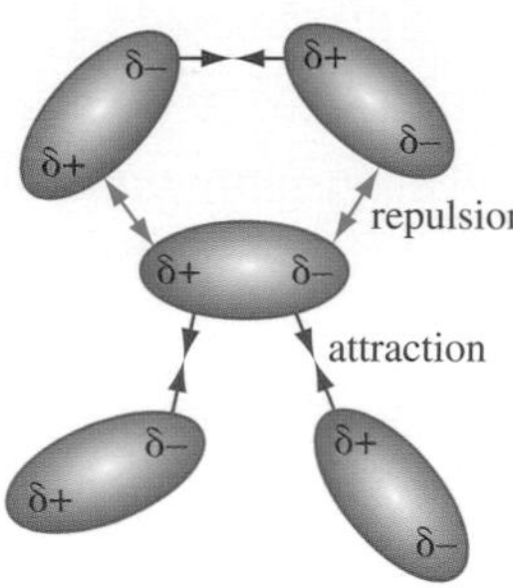
δ−
δ+
δ+
δ−
repulsion
δ+
δ−
attraction
δ−
δ+
δ+
δ−

Figure 3.9.4 ▲
Attractions between molecules with permanent dipoles

Attractions between temporary dipoles

Van der Waals studied gases and showed that the existence of intermolecular forces is one of the reasons why real gases deviate from ideal gas behaviour (see pages 63 and 65).

The problem is to account for the weak attractions between non-polar molecules which are not electrically charged, such as the molecules of iodine, hydrocarbons and the atoms of noble gases. The physicist who developed the theory to explain these forces was Fritz London (1900–1954), so they are sometimes called London forces.

It turns out that when non-polar atoms or molecules meet, there are fleeting repulsions and attractions between the nuclei of the atoms and the surrounding clouds of electrons. Temporary displacements of the electrons lead to temporary dipoles. These instantaneous dipoles can induce dipoles in neighbouring molecules.

The weakest intermolecular forces are the attractions between these instantaneous and induced dipoles which give rise to the tendency for the molecules to cohere. Forces of this kind are roughly a hundred times weaker than covalent bonds.

The greater the number of electrons, the greater the polarisability of the molecule and the greater the possibility for temporary, induced dipoles. This explains why the boiling points rise down group 7 (the halogens) and

Note

Chemists disagree about the definition of the term 'van der Waals forces'. Some use the term for all intermolecular forces, others exclude hydrogen bonding, while another group only uses the term for the weakest attraction between temporary and induced dipoles.

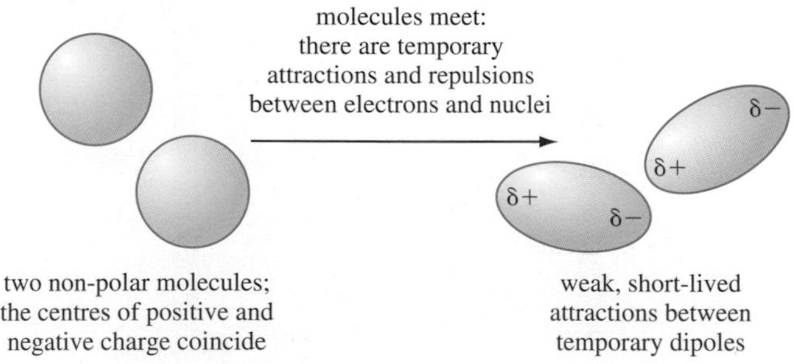
molecules meet:
there are temporary
attractions and repulsions
between electrons and nuclei
δ−
δ+
δ+
δ−
two non-polar molecules;
the centres of positive and
negative charge coincide
weak, short-lived
attractions between
temporary dipoles

Figure 3.9.5 ►
The origins of temporary induced dipoles

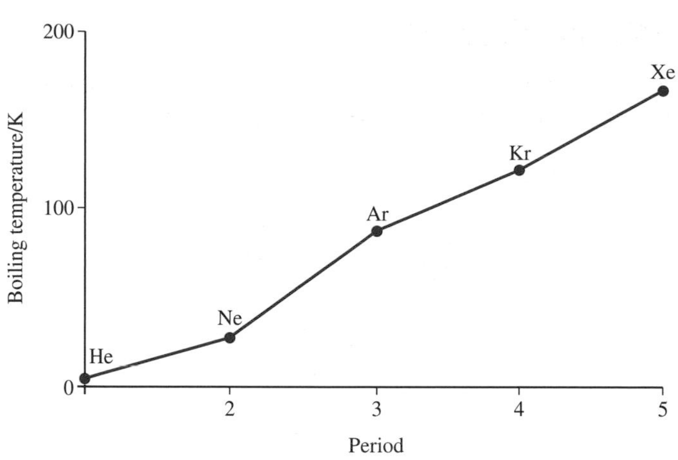

Figure 3.9.6 ◄
The trend in the boiling points of the noble gases with increasing atomic number

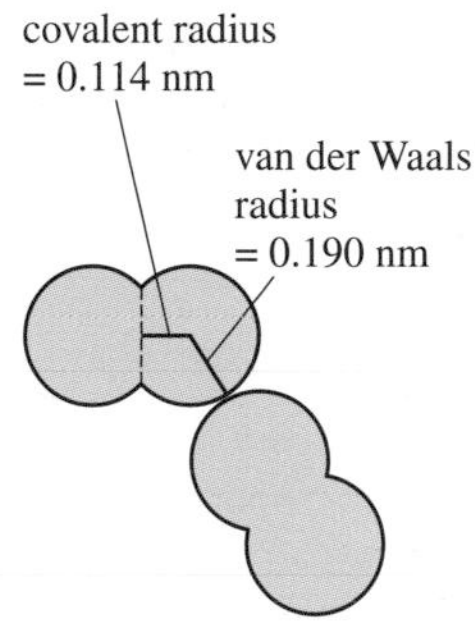

Figure 3.9.7 ▲
Covalent and van der Waals radii for bromine

group 8 (the noble gases). For the same reason, the boiling points in alkanes increase with the increasing number of carbon atoms in these hydrocarbon molecules (see page 197).

Van der Waals radius

The van der Waals radius of an atom is the effective radius of an atom when held in contact with another atom by intermolecular forces.

Atoms do not have a definite size. The apparent size of an atom depends on the way it is bonded to a neighbouring atom, so the ionic radius and the covalent radius of an atom are not the same. Generally the stronger the bonding, the smaller the effective radius. Intermolecular forces are much weaker than covalent bonds, so van der Waals radii are relatively large.

Van der Waals radii determine the effective size of a molecule as it bumps into other molecules in a liquid or gas, and when it packs together with other molecules in a solid.

Hydrogen bonding

Hydrogen bonding is a type of attraction between molecules which is much stronger than other types of intermolecular force, but is still at least 10 times weaker than covalent bonding.

Hydrogen bonding affects molecules in which hydrogen is covalently bonded to one of the three highly electronegative elements fluorine, oxygen and nitrogen.

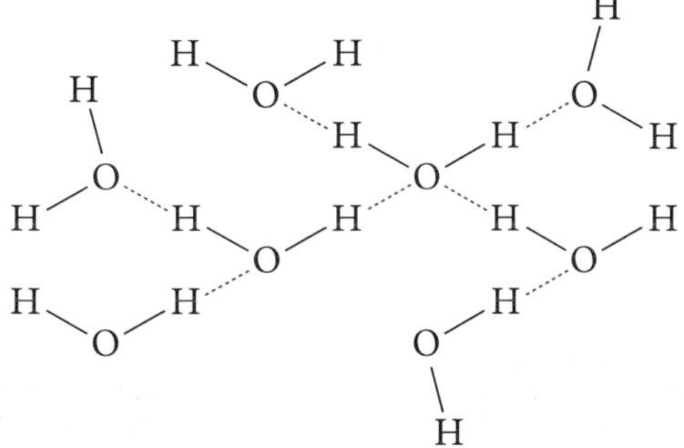

Figure 3.9.8 ◄
Hydrogen bonding in water

Test yourself

1 Considering the shape of these molecules and the polarity of the bonds, divide them into two groups – polar and non-polar: HBr, $CHCl_3$, CCl_4, CO_2, SO_2, C_2H_6.

2 Account for the difference in boiling point between butane, $CH_3CH_2CH_2CH_3$, which boils at −0.5 °C and propanone, CH_3COCH_3 which boils at 56 °C.

Physical Chemistry
Section three

F H H F H H F H F F

hydrogen bond

covalent bond

Figure 3.9.9 ► *Hydrogen bonding in hydrogen fluoride*

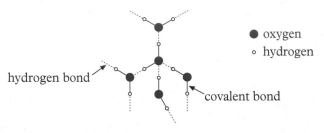

In a hydrogen bond, the hydrogen atom lies between two highly electronegative atoms. It is hydrogen bonded to one of them and covalently bonded to the other. The covalent bond is highly polar. The small hydrogen atom (δ+), which has no inner shells of electrons to shield its nucleus, can get close to the other electronegative atom (δ−) to which it is strongly attracted.

● oxygen
∘ hydrogen

hydrogen bond

covalent bond

Figure 3.9.10 ► *Molecules in ice held together by hydrogen bonding*

Test yourself

3 Which types of intermolecular forces hold together the molecules in: hydrogen bromide, HBr; ethane, CH_3CH_3; methanol, CH_3OH

4 Draw diagrams to show hydrogen bonding between water molecules and:

a) ammonia molecules in a solution of ammonia, NH_3

b) ethanol molecules in a solution of ethanol, CH_3CH_2OH

5 Explain the trend in boiling points in the series: HCl, HBr, HI. Why is the boiling point of hydrogen fluoride, HF, out of line in the series (Figure 3.9.11)?

The three atoms associated with a hydrogen bond are always in a straight line. Hydrogen bonding accounts for:

- the relatively high boiling points of ammonia, water and hydrogen fluoride which are out of line for the trends in the properties of the other hydrides in groups 5, 6 and 7
- the open structure of ice
- the pairing up of bases in a DNA double helix.

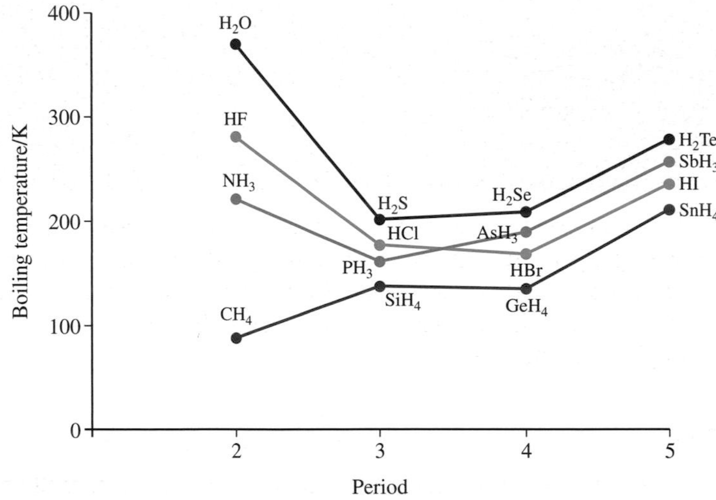

Figure 3.9.11 ► *Boiling points for the hydrides of the elements in groups 4, 5, 6 and 7*

3.10 Structure and bonding in forms of carbon

Carbon can exist in three crystalline forms which nicely illustrate the connection between the properties of materials and their underlying structure and bonding.

Diamond

People have long valued diamonds for their brilliance as gemstones. They have more mundane uses too, mostly as abrasives for cutting and grinding hard materials.

Thanks to its rigid giant structure (see pages 67 and 76), diamond is one of the hardest substances known.

Definition

Allotropes are different forms of the same element in the same physical state. Carbon has allotropes consisting of different giant structures of atoms: diamond and graphite. Carbon also exists in molecular form as the fullerenes.

Figure 3.10.1 ◄
Diamonds in a diamond tipped hole borer

Diamond conducts thermal energy well – five times better than copper. This important property means that diamond-tipped cutting tools do not overheat. The rigidity of the strong and stiff covalent bonds in diamond means that as the atoms get hotter and move faster, the vibrations move rapidly through the giant structure.

Graphite

Graphite is used to make crucibles for casting metals because it melts at the extraordinarily high temperature of 3550 °C. For the same reason, graphite bricks line the walls of industrial furnaces.

The high melting point suggests that graphite has a giant structure. This is confirmed by X-ray diffraction studies which show that the atoms are held together in extended sheets of atoms. The atoms within the layers make up a continuous network of hexagons. Each carbon atom uses just three of its outer electrons to form three normal covalent bonds with other carbon atoms.

Many modern composite materials incorporate graphite fibres for their great tensile strength.

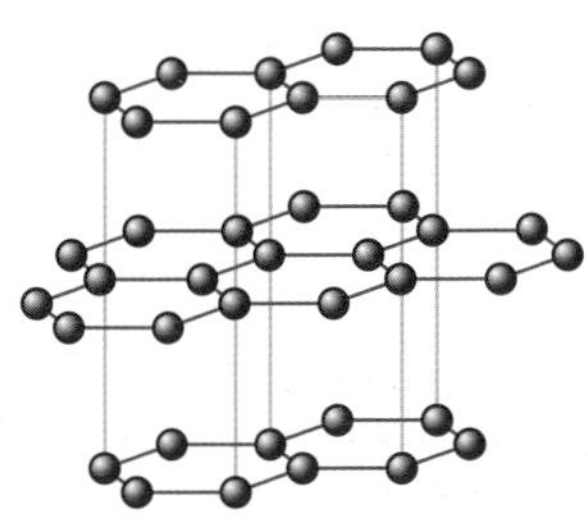

Figure 3.10.2 ▲
The giant structure of graphite. The layers are huge extended sheets of atoms piled on top of each other. The bonding within the layers is strong but the bonding between the layers is relatively weak

Definition

Composites are materials which combine two or more materials to create a new material. A composite combines the desirable properties of its constituents and compensates for their disadvantages.

The combinations are designed to give new materials with better properties especially high strength and stiffness per unit weight.

Parts of aircraft and some sports gear are made from graphite fibres in an epoxy plastic matrix.

Unusually for a non-metal, graphite conducts electricity which accounts for its use in making electrodes. Graphite conducts because each carbon atom in a layer of the structure contributes one electron to a cloud of electrons delocalised over the layer.

Unlike diamond, graphite is soft and feels greasy. One of the important uses of graphite is as a lubricant. Graphite has these properties because there are no strong covalent bonds linking the layers of atoms together. Weak van der Waals forces hold the layers together so they can easily slide over each other.

Fullerenes

Chemists believed that there were just two crystalline forms of carbon until the mid-1980s. Then Harry Kroto and his group at the University of Sussex discovered fullerenes – a group of molecular forms of carbon.

At a molecular level, the fullerenes mimic the geodesic dome invented by the American engineer Robert Buckminster Fuller. When Harry Kroto first identified the new form carbon as C_{60} he and the others in the team wanted to know how the atoms were organised. Several people guessed that the atoms might form a closed cage or tube. The first idea was that the atoms made hexagonal patterns as in graphite, but 60 carbon atoms in a pattern of hexagons makes a flat sheet. The scientists made many models and it was early one morning that one of them hit on the solution by fitting together 12 pentagons and 20 hexagons into a 'football' with 60 corners – one corner for each of the 60 carbon atoms. They named the new molecule buckminsterfullerene – the first of the family of fullerenes to be discovered.

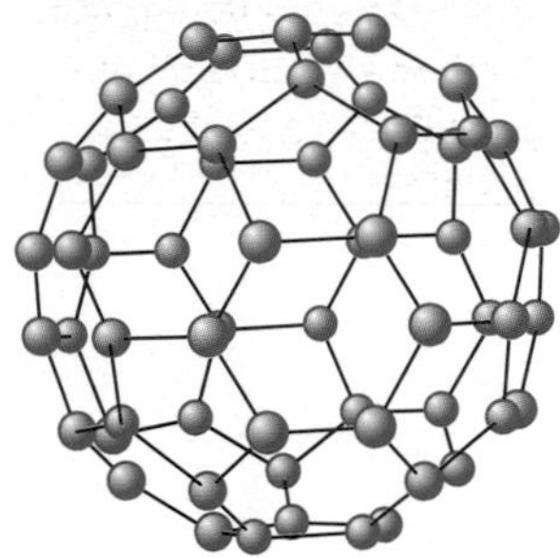

Figure 3.10.3 ▲
Structure of C_{60}. Other fullerenes have the formulae: C_{28}, C_{32}, C_{50}, and C_{70}. Fullerenes exists as tubes as well as spheres

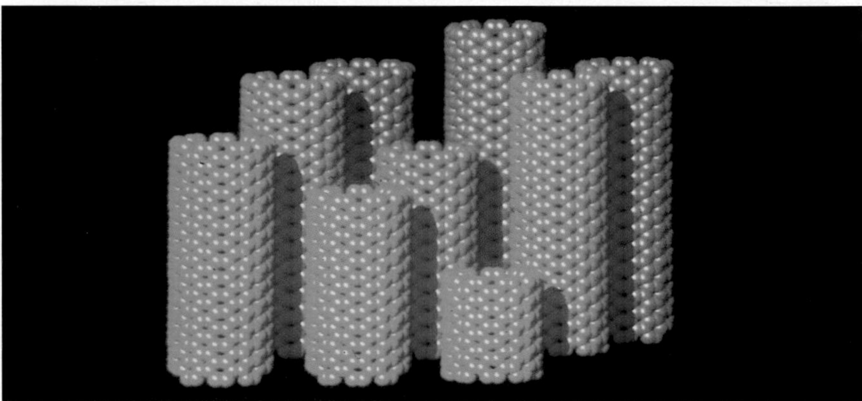

Figure 3.10.4 ▲
The structure of bucky tubes

Test yourself

1 Choose and name an element or compound to match each of the following descriptions. For example, pick one property of the element or compound which can be explained in terms of structure and bonding.

- **a)** a crystalline giant structure of positively- and negatively-charged ions
- **b)** non-polar molecules which are widely spaced and fast moving
- **c)** a layer lattice with delocalised electrons
- **d)** small molecules held together in a crystal structure by hydrogen bonding
- **e)** a sea of electrons around a crystalline giant structure of positive ions
- **f)** polar molecules which are close together but free to move about.

3.11 Solutions

Living on a watery planet and with bodies which consist largely of water, it is no surprise to find that many chemicals dissolve and react in aqueous solution. Water, however, is not the only important solvent. The petrochemical industry supplies many solvents which are non-polar and do not mix with water.

Patterns of solubility

As a general rule 'like dissolves like'. This means, for example, that:

- non-polar liquids mix freely with other non-polar liquids (for example, hydrocarbons mix freely as they do in petrol)
- polar liquids mix with polar liquids (for example ethanol mixes with water)
- polar and non-polar liquids do not mix (oil floats on water).

Similarly highly polar solids such as ionic salts dissolve in water which is polar but not in liquid hydrocarbons which are non-polar.

Non-polar solids such as wax will not dissolve in water but will dissolve in hydrocarbons.

Solutions of ionic salts in water

It is hard to see why the charged ions in a crystal of sodium chloride separate and go into solution in water with only a small energy change. Where does the energy come from to overcome the strong electrostatic attraction between the ions?

The explanation is that the ions are strongly hydrated by the polar water molecules. The water molecules cluster round the ions and bind to them. The energy released as the water molecules bind to the ions is enough to compensate for the energy needed to overcome the ionic bonding between the ions (see pages 69 and 73–74).

Definitions

Miscible liquids are liquids which mix with each other. Water and alcohol are miscible. Oil and water are immiscible.

Hydration takes places when water molecules bond to ions or add to molecules. Water molecules are polar and so they are attracted to both positive ions and negative ions.

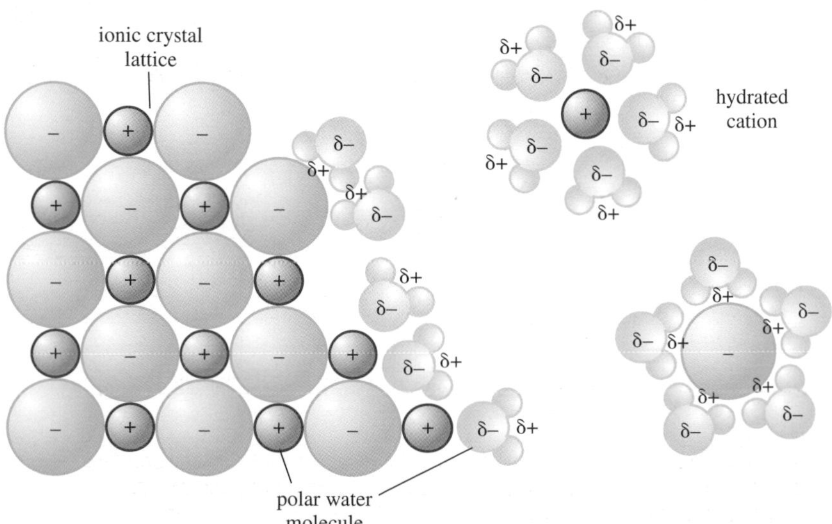

Figure 3.11.1 ◀
Sodium and chloride ions leaving a crystal lattice and becoming hydrated as they dissolve in water. Here the bond between the ions and the polar water molecules is electrostatic attraction

3.12 Infra-red spectroscopy

Infra-red radiation (IR) from a glowing fire makes us feel warm. This is because the frequencies of IR correspond to the natural frequencies of vibrating atoms. Our skin warms up as the molecules absorb IR and vibrate faster.

Note

The units along the bottom of an IR spectrum are wavenumbers (in cm^{-1}). IR spectra typically range from 400 cm^{-1} to 45 000 cm^{-1}. Wavenumbers are easier to work with than wavelengths in the IR region. The wavenumber is the number of wavelengths that will fit into one centimetre.

The bonds which absorb IR strongly are polar covalent bonds such as O—H, C—O and C=O. Bonds like this vibrate in characteristic ways and absorb at specific wavelengths. This allows chemists to look at the IR spectrum of a compound and identify particular groups of atoms.

In a molecule the movements of one bond affect the vibrations of others close by. Even a simple molecule can vibrate in many ways. The aldehyde butanal, C_4H_7CHO, has over 30 ways of vibrating, and the spectrum is still complex even though not all vibrations absorb radiation.

Analysts now have access to databases with IR spectra stored like fingerprints for a large number of pure compounds. Chemists can identify specimens by matching the absorption spectrum of an unknown with one of the known spectra in a database. Matching the IR spectrum of a product with that of a known pure sample can be used to check that the product is pure and free from traces of solvent or by-products.

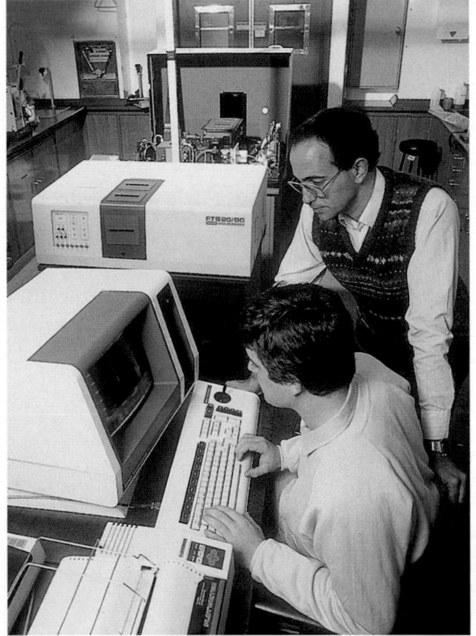

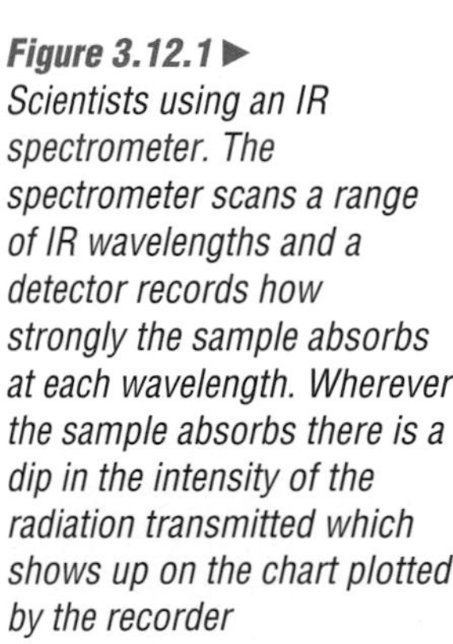

Figure 3.12.1 ▶ *Scientists using an IR spectrometer. The spectrometer scans a range of IR wavelengths and a detector records how strongly the sample absorbs at each wavelength. Wherever the sample absorbs there is a dip in the intensity of the radiation transmitted which shows up on the chart plotted by the recorder*

Exothermic and endothermic changes

Exothermic changes give out energy to their surroundings. Freezing a condensing are exothermic changes of state. Burning is an exothermi chemical reaction. So is respiration in which glucose is oxidised to pr the energy for living things to grow and move.

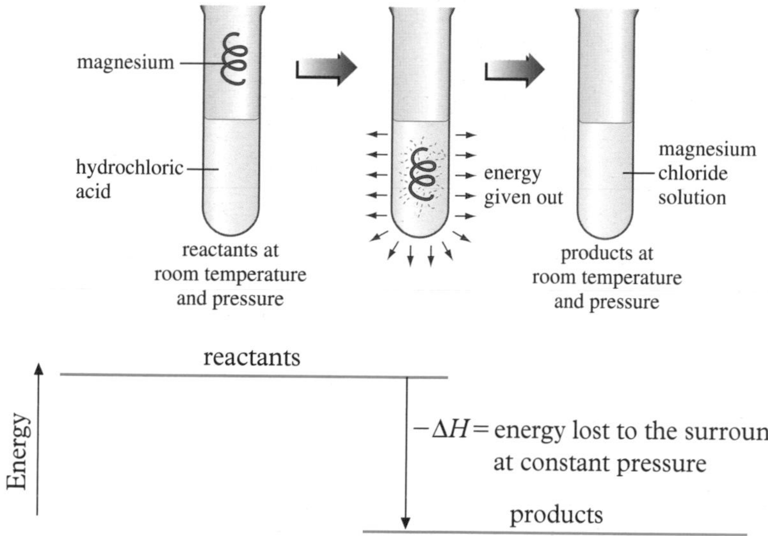

An energy level diagram shows that after an exothermic reaction, the system has less energy than it did at the start. So for an exothermic re the enthalpy change, ΔH, is negative.

Endothermic changes take in energy from their surroundings. Mel and evaporation are endothermic changes of state. Photosynthesis is endothermic: plants take in energy from the Sun to convert carbon di and water to glucose. This is overall the reverse of the chemical chang occurs during respiration.

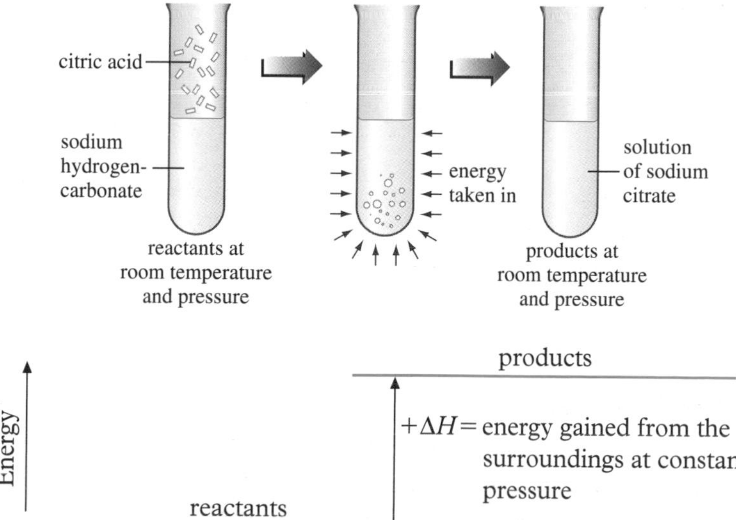

An energy level diagram shows that after an endothermic reaction, the system has more energy than it did at the start. So for an endothermic reaction the enthalpy change, ΔH, is positive.

ethanol

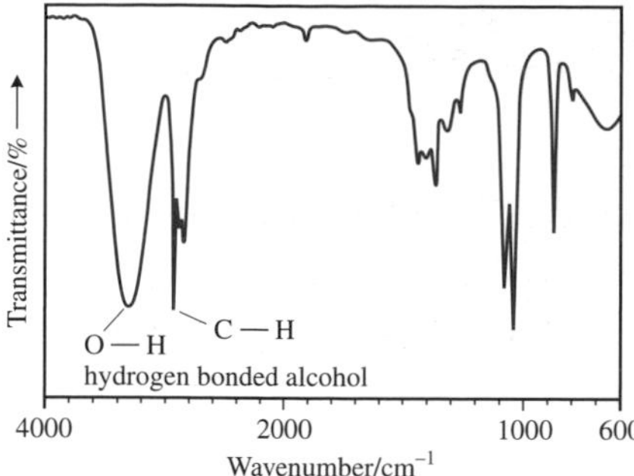

ethanal

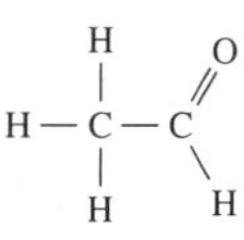

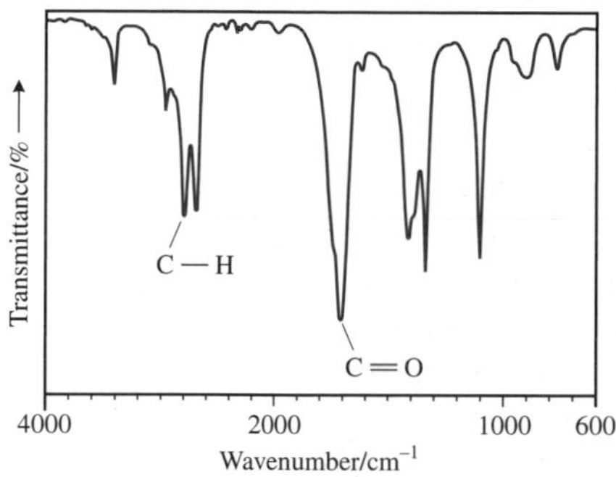

ethanoic acid

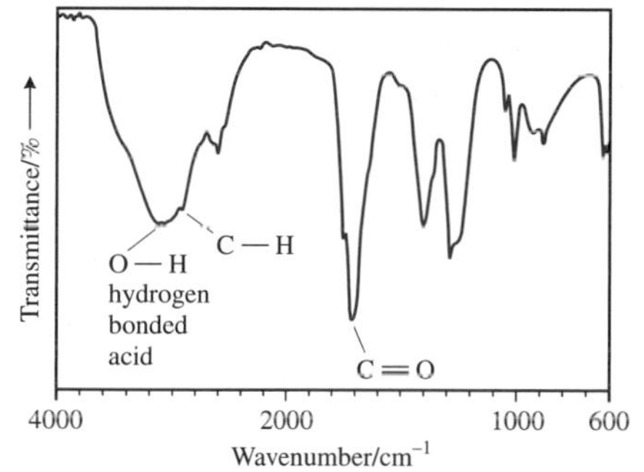

Figure 3.12.2 ◀ *The spectra of ethanol and two of its oxidation products: ethanal and ethanoic acid (see pages 215–216)*

3.13 Enthalpy changes

What makes things go? The short answer is that d
example: differences of temperature, differences o
pressure and differences of electrical potential.

In the natural environment the energy from the
stir the winds and vaporise water. The Sun also pr
photosynthesis in plants to make the concentrated
need to grow and move. In many mechanical syste
fuels which creates the high temperatures in engin
and vehicles moving. There is a close connection b
of state and chemical reactions.

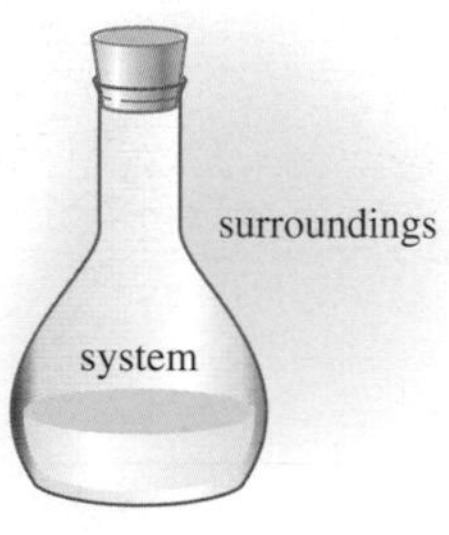

Figure 3.13.1 ▲
A system and its surroundings

Thermochemistry

Thermochemistry is the study of
Thermochemistry is important f
explain the stability of compoun
chemical change. With the help c
whether or not reactions are likel

Thermochemistry is a branch
quantitative science. Thermoche
use. This branch of chemistry ha
quantities such as enthalpy, entr
meaning.

System and surrounding

The term 'system' in thermocher
chemicals being studied. Everyth
such as the apparatus, the air in t
the universe.

An open system can exchang
Most chemical reactions in labor

A closed system cannot excha
energy.

Definition

The **enthalpy change** is the overall energy exchanged with the surroundings when changes happen at constant pressure and when the final temperature is the same as the starting temperature.

Enthalpy changes

When there is a change in a syste
transferred between the system a

The enthalpy change is the er
its surroundings when the chang

The symbol for an enthalpy c

Figure 3.13.6 ▶
A cold pack

Definition

The **enthalpy change of melting** is the enthalpy change when one mole of a solid turns to a liquid at its melting point. Another word for melting is 'fusion' so an alternative term is the **enthalpy change of fusion**, ΔH_{fus}.

Test yourself

1. Figure 3.13.7 shows the cooling curve for a test tube of water.

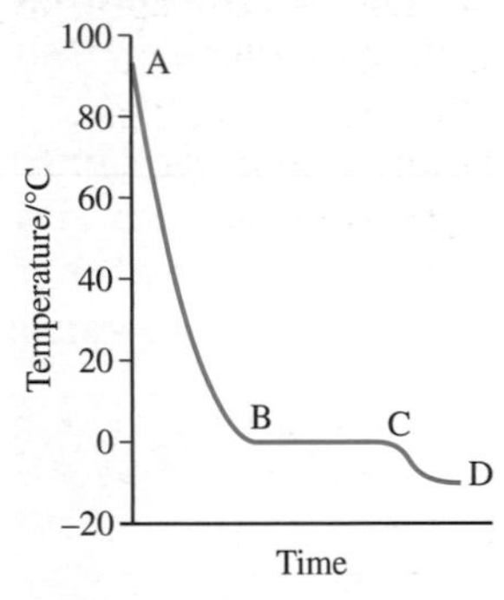

Figure 3.13.7 ▲

a) What happens to the water molecules as the water cools along the line AB?
b) Why is there no change in temperature between B and C?
c) What is in the tube after C?
d) Estimate the temperature in the freezing mixture surrounding the test tube

Enthalpy changes and changes of state

A cold pack (See Figure 3.13.6) helps to keep cool the contents of an insulated 'cool box'. The pack contains a liquid. When not in use the cold pack stays in a freezer where the liquid gives out energy and turns to a solid. In a cool box the chemical in the pack slowly melts, taking in energy from the surrounding air thus keeping the contents of the box cool.

Melting is an endothermic process. The energy makes the particles in a solid vibrate faster until they have enough energy to break free from their fixed positions (see Figure 2.2.5 on page 18).

A refrigerator takes advantage of the energy transfers when a liquid evaporates and condenses. A pump circulates a liquid with a low boiling point around a circuit of pipes. The liquid vaporises in the pipes inside the refrigerator. Evaporation is endothermic so the liquid takes in energy from the air inside the refrigerator keeping the food inside cool.

The pump compresses the vapour as it flows out at the bottom of the refrigerator. The compressed vapour is hot. As it flows through the pipes at the back of the refrigerator the fluid cools and condenses back to a liquid, giving out energy and heating up the air around the back of the cabinet.

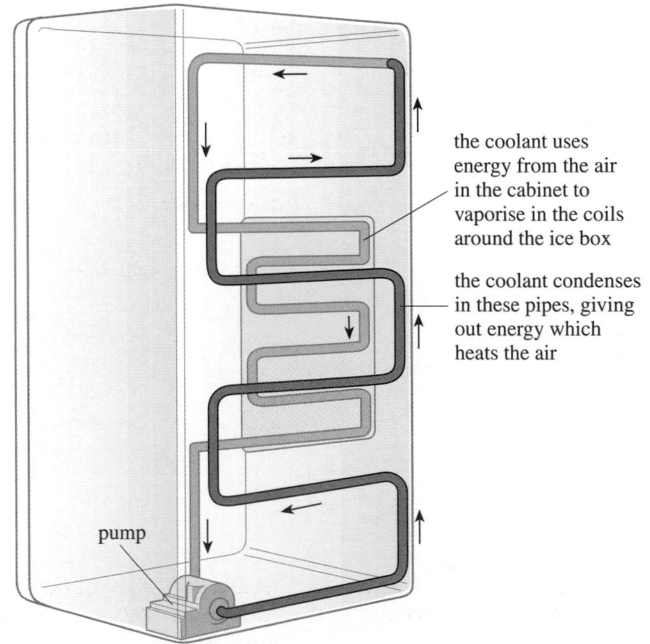

Figure 3.13.8 ▶
The coolant circuit in a refrigerator

Overall the circulating fluid transfers energy from inside of the refrigerator to the air in the room.

Evaporation separates the particles in a liquid so values for enthalpy of vaporisation give a measure of the strength of the bonding between particles in liquids.

The apparatus shown in Figure 3.13.9 contains ethanol. The immersion heater is connected to the electricity supply via a joulemeter to measure the energy transferred to the liquid as it boils.

Definition

The **enthalpy change of vaporisation**, ΔH_{vap}, for a liquid is the enthalpy change when one mole of the liquid turns to a vapour at its boiling point.

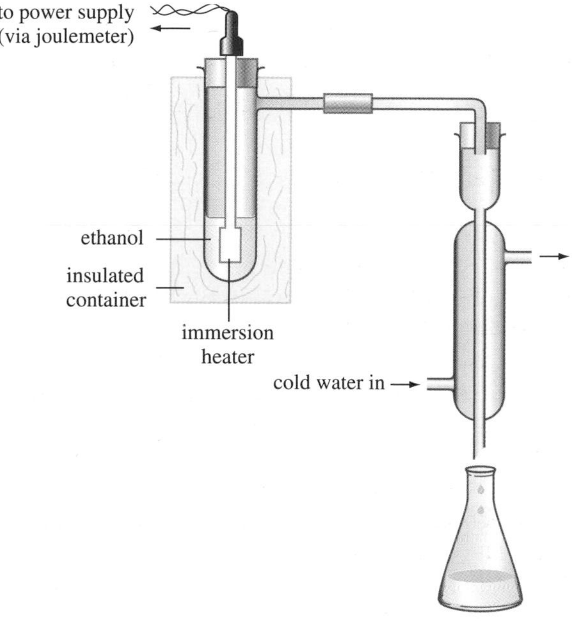

Figure 3.13.9 ◀ *Apparatus for measuring the enthalpy of vaporisation of a liquid*

Note

Physicists generally use 'specific latent heat' values when studying changes of state. Specific latent heat values are measured in joules per kilogram rather than joules per mole.

Substances with strong ionic or metallic bonding have much higher boiling points and enthalpies of vaporisation than substances consisting of molecules with weak intermolecular forces.

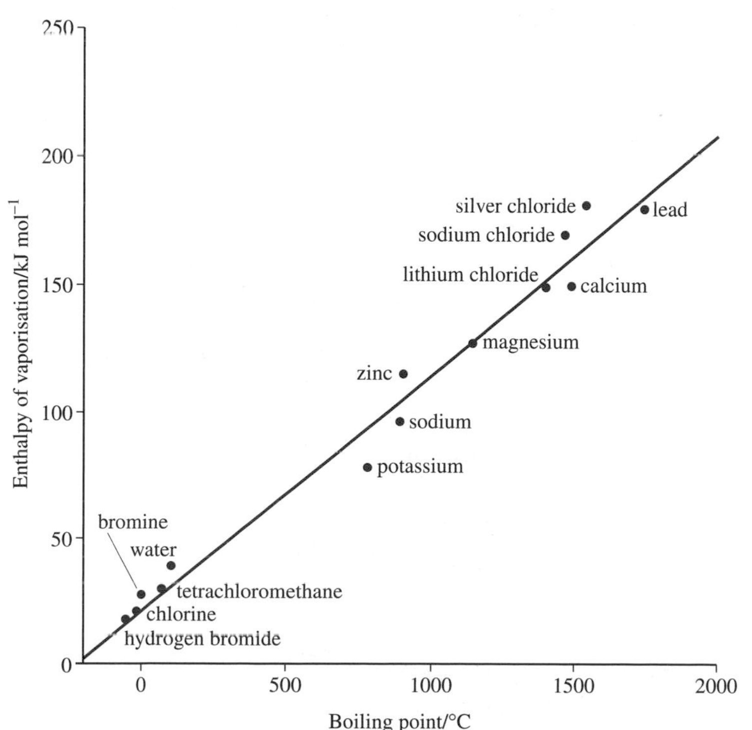

Figure 3.13.10 ◀ *Plot of enthalpy of vaporisation against boiling point*

Test yourself

2 The energy needed to vaporise 7.1 g ethanol, C_2H_5OH, in the apparatus shown in Figure 3.13.9 is 6050 J. What is the enthalpy of vaporisation of ethanol in kJ mol^{-1}?

3 Which of the elements or compounds in Figure 3.13.10:

a) have molecular structures
b) consist of giant structures of metal atoms
c) consist of giant structures of ions?

Enthalpy changes and reactions

Measuring enthalpies of combustion

All combustion reactions are exothermic. Burning fuels release energy to their surroundings. The easiest way to measure the enthalpy of combustion of a fuel is to burn a measured mass of fuel in such a way that the energy heats a container of water. Measuring the temperature rise of the water makes it possible to calculate the quantity of energy transferred from the flame to the water.

Enthalpy changes and chemical reactions

Enthalpies of combustion

A calorimeter is the name of the apparatus used to measure the energy change during a chemical reaction. Typically a calorimeter is insulated from its surroundings and contains water. The energy from the reaction heats up the water and the rest of the apparatus. An accurate thermometer measures the temperature rise.

Figure 3.13.11 shows an apparatus for measuring enthalpies of combustion.

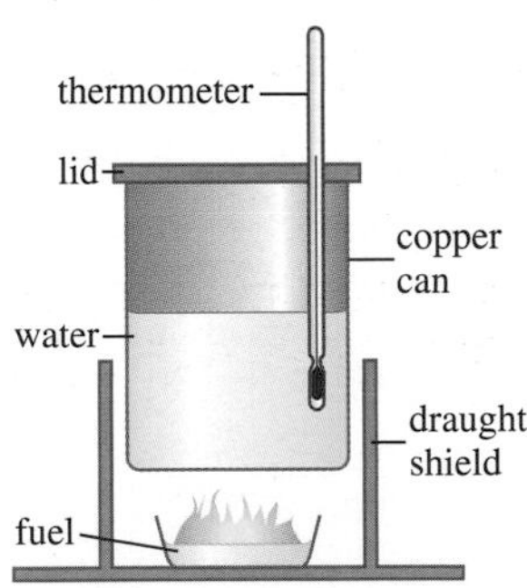

Figure 3.13.11 ▲
An approximate method for measuring the enthalpy of combustion of a fuel

Worked example

When 1.16 g propanone, CH_3COCH_3, was burnt in the apparatus shown in Figure 3.13.11, the temperature of 250 g water in the copper can rose from 19 °C to 41 °C. What is the enthalpy of combustion of propanone? Compare the answer to the value from a book of data and comment on any difference.

Notes on the method

Ignore the heat capacity of the copper can because it is small and other sources of error are larger.

Work with temperatures and temperature changes in Kelvin. The magnitude of the temperature rise is the same on the Celsius and Kelvin scales (see page 62).

$$\text{Energy transferred/J} = \text{mass of water/g} \times \text{temperature rise/K} \times \text{specific heat capacity/J mol}^{-1}\,\text{K}^{-1}$$

Answer

Temperature rise of the water = 22 K

Energy transferred to the water = 250 g × 22 K × 4.2 J mol^{-1} K^{-1}
= 23 100 J = 23.1 kJ

Molar mass of propanone (see pages 35–36) = 58 g mol^{-1}

Amount of propanone burnt = 1.16 g ÷ 58 g mol^{-1} = 0.02 mol

Energy from burning one mole propanone = 23.1 kJ ÷ 0.02 mol
= 1155 kJ mol^{-1}

Combustion is exothermic so the enthalpy change is negative.

$\Delta H_{combustion} = -1155\ \text{kJ mol}^{-1}$

This method of working wrongly assumes that all the energy from the flame heats the water. In practice a proportion of the energy heats the air and surrounding equipment. The same equipment can give much more accurate values if it is first calibrated to determine its overall heat capacity by measuring the temperature rise for a fuel with known enthalpy of combustion.

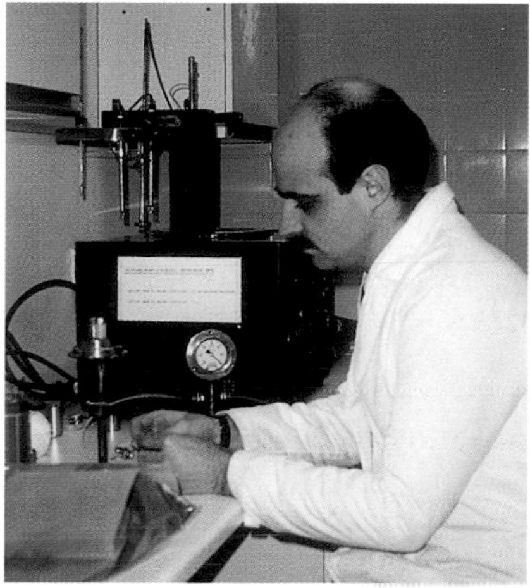

Figure 3.13.12 ▲
A scientist operating a bomb calorimeter

Accurate values for enthalpies of combustion are obtained using a bomb calorimeter.

A measured amount of a sample burns in oxygen under pressure. Excess oxygen makes sure that all the compound burns and that the elements are fully oxidised. There must be enough oxygen to make sure that any carbon in the compound is fully oxidised to carbon dioxide and that there is no carbon monoxide or soot.

The temperature rise of the whole apparatus is measured. The calorimeter can be calibrated using benzoic acid for which the standard enthalpy of combustion is known accurately. Alternatively the apparatus can be heated electrically to find the energy needed to bring about the same temperature rise as for the burning chemical.

A bomb calorimeter operates at constant volume so a correction is necessary to convert the results to enthalpy changes at constant pressure.

Bond breaking and bond forming

During combustion, the bonds in the fuel and oxygen break and new bonds form to make new molecules. Energy is needed to break bonds. Energy is released as bonds form.

Each type of bond in a molecule has its own 'bond enthalpy' (see pages 103–104). When hydrogen burns in oxygen:

$$2H_2(g) + O_2(g) \rightarrow 2H_2O(l)$$

The reaction is exothermic because more energy is released making four new O—H bonds in the two water molecules than in breaking all the bonds in the oxygen and hydrogen molecules.

Definition

The **specific heat capacity** of a material, c, is the energy needed to raise the temperature of 1 g of the material by 1 K. The specific heat capacity of water is given by $c = 4.2\ \mathrm{J\ g^{-1}\ K^{-1}}$. So it takes 4.2 joules of energy to raise the temperature of one gram of water by one degree. The energy, q, needed to raise the temperature of a mass of water, m, through a temperature change ΔT is given by: $q = mc\Delta T$.

Test yourself

4 Burning a butane, C_4H_{10}, lighter under a can of water raised the temperature of 200 g water from 18 °C to 28 °C. The lighter was weighed before and after and the loss in mass was 0.29 g. Estimate the molar enthalpy of combustion of butane.

Note

Bond breaking is endothermic.
Bond forming is exothermic.

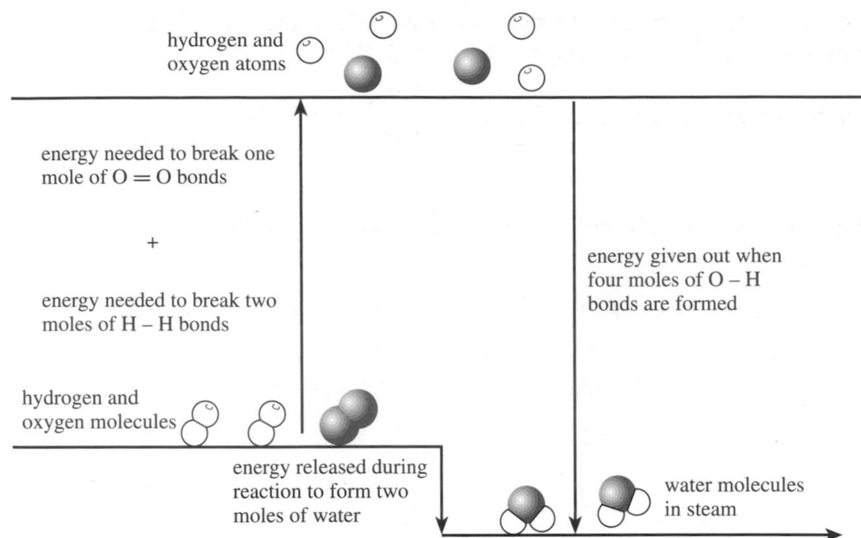

Figure 3.13.13 ▶
Energy level diagram for the combustion of hydrogen

Test yourself D

5 Look up the bond enthalpies for the H—H and Cl—Cl bonds. Calculate the overall energy change for the reaction

$H_2(g) + Cl_2(g) \rightarrow 2HCl(g)$

Draw an energy level diagram for the reaction (similar to Figure 3.13.13).

Enthalpy changes in solution

Enthalpy changes for reactions in solution can be compared quickly using an expanded polystyrene cup with a lid as the calorimeter. Expanded polystyrene is an excellent insulator and has a negligible specific heat capacity.

If the reaction is exothermic, the energy released cannot escape to the surroundings so it heats up the solution. If the reaction is endothermic, no energy can enter from the surroundings so the solution cools. If the solutions are dilute it is sufficiently accurate to calculate the energy changes, assuming that the density and specific heat capacity of the solutions are the same as water.

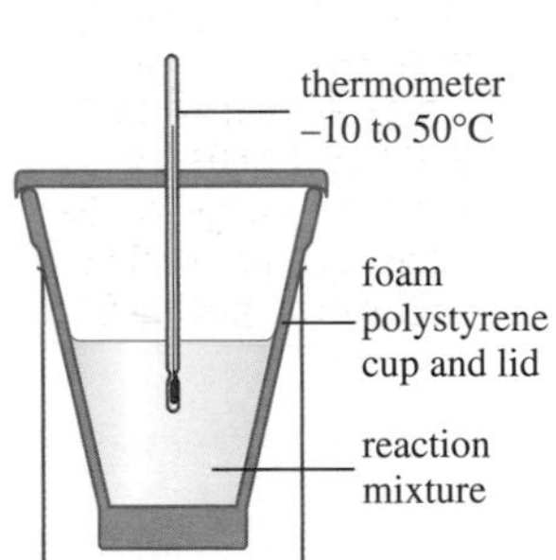

Figure 3.13.14 ▲
Measuring the enthalpy of a reaction in solution

Note

The enthalpy change is the energy exchanged with the surroundings when a reaction proceeds at a constant temperature and pressure. Here the energy from the exothermic reaction is kept in the system to heat up the water. The calculation shows the energy which would otherwise be lost to the surroundings during a constant temperature change.

Worked example

When 50 cm^3 of 2.0 mol dm^{-3} hydrochloric acid mixes with 50 cm^3 2.0 mol dm^{-3} sodium hydroxide in a polystyrene cup, the temperature rise is 13.3 °C. What is the enthalpy change for the neutralisation reaction?

Notes on the method

Assume that the density of the solutions is the same as water = 1 g cm^{-3} and that for both, the specific heat capacity, like water, is 4.2 J g^{-1} K^{-1}.

Note that the total volume of solution is 100 cm^3 so the mass of solution heating up can be taken as 100 g.

The enthalpy change for a reaction, ΔH, is the energy change for the amounts (in moles) shown in the chemical equation.

Answer

The amount of hydrochloric acid = the amount of sodium hydroxide

$$= \frac{50}{1000} \text{ dm}^3 \times 2.0 \text{ mol dm}^{-3}$$

$$= 0.1 \text{ mol}$$

Energy given out by the reaction and used to heat the water in the cup = $4.2\,\mathrm{J\,g^{-1}\,K^{-1}} \times 100\,\mathrm{g} \times 13.3\,\mathrm{K} = 5586\,\mathrm{J}$

$$\text{Energy given out per mole of acid} = \frac{5586\,\mathrm{J}}{0.1\,\mathrm{mol}} = 55\,860\,\mathrm{J\,mol^{-1}}$$

$$= 55.9\,\mathrm{kJ\,mol^{-1}}$$

The reaction is exothermic so the enthalpy change for the system is negative.

$$NaOH(aq) + HCl(aq) \rightarrow NaCl(aq) + H_2O(l) \quad \Delta H = -55.9\,\mathrm{kJ\,mol^{-1}}$$

The value is a little less than the accepted value because of energy losses. The accepted value for the enthalpy change for this reaction

$$\Delta H_{\text{neutralisation}} = -57.1\,\mathrm{kJ\,mol^{-1}}$$

The enthalpy of neutralisation for dilute solutions of strong acid with strong base is always close to $-57.1\,\mathrm{kJ\,mol^{-1}}$. The reason for this is that these acids and alkalis are fully ionised, so in every instance the reaction is the same (see page 31):

$$H^+(aq) + OH^-(aq) \rightarrow H_2O(l) \quad \Delta H = -57.1\,\mathrm{kJ\,mol^{-1}}$$

Standard enthalpy changes

Thermochemistry is a precise science and the values for enthalpy changes quoted in books of data are given for carefully defined conditions. Standard enthalpy changes, for example, are calculated for a temperature of 298 K (25 °C) and a pressure of 1 bar (= 10^5 Pa = 100 kPa). The symbol for a standard enthalpy change is $\Delta H^{\ominus}_{298}$.

In thermochemistry it is also important to specify the states of the chemicals. Equations should always include state symbols. The standard states of elements and compounds are their most stable state under standard conditions. The standard state of carbon, for example, is graphite which is energetically more stable than diamond.

Enthalpies of combustion

The standard enthalpy change of combustion of an element or compound is the enthalpy change when one mole of the substance burns completely in oxygen. The chemical and the products of burning must be in their normal stable (standard) states. For a carbon compound, complete combustion means that all the carbon burns to carbon dioxide and that there is no soot or carbon monoxide. When burning a compound containing hydrogen, the water formed must end up as a liquid, not a gas.

Values of enthalpies of combustion are much easier to measure than many other enthalpy changes. They can be calculated from measurements taken with a bomb calorimeter (see page 97). The importance of these values is that they can be used to calculate other enthalpies changes.

Chemists use two ways to summarise standard molar enthalpy changes of combustion. One way is to write the equation with the enthalpy change alongside it.

For the standard enthalpy of combustion of carbon:

$$C_{\text{graphite}} + O_2(g) \rightarrow CO_2(g) \quad \Delta H^{\ominus}_{c,298} = -393.5\,\mathrm{kJ\,mol^{-1}}$$

Test yourself

6 On adding 25 cm^3 of 1.0 mol dm^{-3} nitric acid to 25 cm^3 1.0 mol dm^{-3} potassium hydroxide in a plastic cup, the temperature rise is 6.5 °C. Calculate the enthalpy change for the neutralisation reaction.

7 When 4.0 g ammonium nitrate dissolves in 100 cm^3 water the temperature falls by 3.0 °C. Calculate the enthalpy change per mole when NH_4NO_3 dissolves in water under these conditions.

8 On adding excess powdered zinc to 25.0 cm^3 of 0.2 mol dm^{-3} copper(II) sulfate solution, the temperature rises by 9.5 °C. Calculate the enthalpy change for the displacement reaction of zinc with copper sulfate reaction.

Note

The superscript sign in $\Delta H^{\ominus}_{298}$ shows that the value quoted is for standard conditions. The symbol is pronounced 'delta H standard'.

Definition

The **standard enthalpy change of combustion** of a substance, $\Delta H^{\ominus}_{c,298}$, is the enthalpy change when one mole of the substance completely burns in oxygen under standard conditions with the reactants and products in their standard states.

Note

All combustion reactions are exothermic so $\Delta H^{\ominus}_{c}$ values are always negative.

Note

In thermochemistry the symbol O_2 represents one mole of oxygen molecules (that is 32 g oxygen) so $\frac{1}{2}O_2$ stands for half a mole of oxygen molecules (that is 16 g oxygen).

Definition

The **standard enthalpy change of formation** of a compound, $\Delta H^{\ominus}_{c,298}$, is the enthalpy change when one mole of the compound forms from its elements under standard conditions with the elements and the compound in their standard states.

Note

The definition of standard enthalpy of formation means that for an element, by definition, $\Delta H^{\ominus}_{f} = 0 \text{ kJ mol}^{-1}$. There is no change and so no enthalpy change when an element forms from itself. In other words, the standard enthalpy of formation of an element is zero.

The other way is to use a shorthand. For the standard enthalpy of combustion of methane:

$$\Delta H^{\ominus}_{c,298}[CH_4(g)] = -890 \text{ kJ mol}^{-1}$$

Like all thermochemical quantities, the precise definition of standard enthalpy of combustion is important.

Standard enthalpies of formation

The enthalpy change of formation of a compound is the enthalpy change when one mole of a compound forms from the elements. The elements and the compound formed must be in their stable standard states. The more stable state of an element is chosen where there are allotropes (see page 87).

As with standard enthalpies of combustion, there are two ways of representing standard molar enthalpies of formation.

One way is to write the equation with the enthalpy change alongside it. For the standard enthalpy of formation of water:

$$H_2(g) + \tfrac{1}{2}O_2(g) \rightarrow H_2O(l) \quad \Delta H^{\ominus}_{f,298} = -286 \text{ kJ mol}^{-1}$$

The other way is to use a shorthand. For the standard enthalpy of formation of ethanol:

$$\Delta H^{\ominus}_{f,298}[C_2H_5OH(l)] = -277 \text{ kJ mol}^{-1}$$

Like all thermochemical quantities the precise definition of standard enthalpy of formation is important.

Books of data tabulate values for standard molar enthalpies of formation. These tables are very useful because they make it possible to calculate the enthalpy changes for many reactions (see page 102).

Unfortunately it is usually difficult to measure enthalpy changes of formation directly. It is impossible, for example, to take carbon, hydrogen and oxygen and join them directly to make ethanol under any conditions. So chemists have had to find an indirect method for finding the standard enthalpy of formation of this and other compounds.

Hess's law

The energy change for a reaction is the same whether the reaction happens in one step or in a series of steps. So long as the reactants and the products are the same, the overall enthalpy change is the same whether the reactants are converted to products directly or through two or more intermediate reactions. This is Hess's law. In Figure 3.13.15 the enthalpy change for route 1 and the overall enthalpy change for route 2 are the same.

Hess's law makes it possible to calculate enthalpy changes which cannot be measured experimentally. Hess's law is used to calculate:

- standard enthalpies of formation from standard enthalpies of combustion,
- standard enthalpies of reaction from standard enthalpies of formation.

Hess's law is a chemical version of the law of conservation of energy. Suppose the enthalpy change for route 1 were more negative than the total enthalpy change for route 2 in Figure 3.13.15. It would be possible to go round a cycle from A to D direct and back to A via C and B ending up with the same starting chemical but with a net release of energy. This would contravene energy conservation.

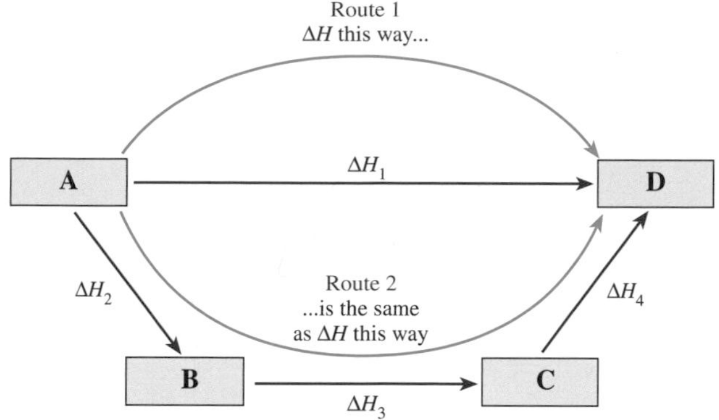

Figure 3.13.15 ◄ *A diagram to illustrate Hess's law.* $\Delta H_1 = \Delta H_2 + \Delta H_3 + \Delta H_4$

Note

Reversing the direction of a reaction reverses the sign of ΔH.

Enthalpies of formation from enthalpies of combustion

Figure 3.13.16 shows the form of energy cycles which chemists draw up to calculate enthalpies of formation from enthalpies of combustion.

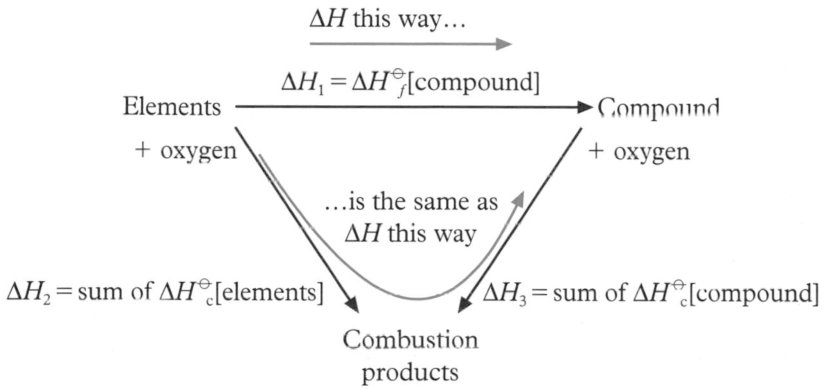

Figure 3.13.16 ◄ *Outline of a thermochemical cycle for calculating standard enthalpies of formation from standard enthalpies of combustion,* $\Delta H_1 = \Delta H_2 - \Delta H_3$

Standard enthalpies of combustion

propane
$\Delta H^{\ominus}_{c}\ [C_3H_8(g)] = -2220\ kJ\ mol^{-1}$

carbon
$\Delta H^{\ominus}_{c}\ [C_{graphite}] = -393\ kJ\ mol^{-1}$

hydrogen
$\Delta H^{\ominus}_{c}\ [H_2(g)] = -286\ kJ\ mol^{-1}$

Worked example

Calculate the enthalpy of formation of propane, C_3H_8, at 298 K given the standard enthalpies of combustion.

Notes on the method

Draw up a thermochemical cycle. Use Hess's law to produce an equation for the enthalpy changes. All the enthalpy changes are given except $\Delta H^{\ominus}_{f}[C_3H_8(g)]$. Pay careful attention to the signs. Put the value and sign for a quantity in brackets when multiplying, adding or subtracting enthalpy values.

Answer

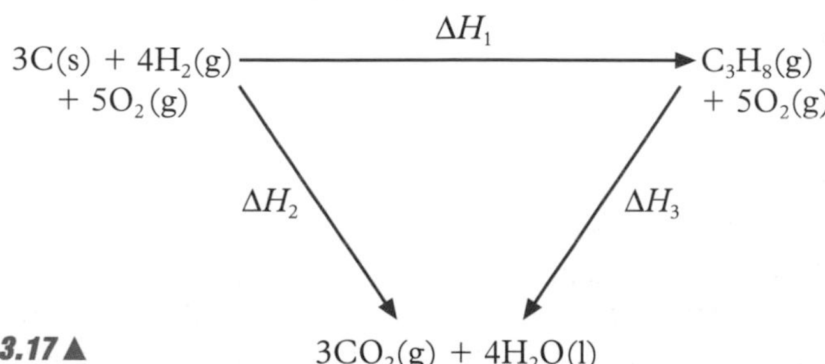

Figure 3.13.17 ▲

Test yourself D

9 By writing a balanced equation, show that the standard enthalpy of formation of carbon dioxide is the same as the standard enthalpy change of combustion of carbon (graphite).

10 Use the values for standard enthalpies of combustion to calculate the standard enthalpy of formation of methanol, CH_3OH.

According to Hess's law $\Delta H_1 = \Delta H_2 - \Delta H_3$

$\Delta H_1 = \Delta H^{\ominus}_f [C_3H_8(g)]$

$\Delta H_2 = 3 \times \Delta H^{\ominus}_c [C_{graphite}] + 4 \times \Delta H^{\ominus}_c [H_2(g)]$
$= 3 \times (-393\ kJ\ mol^{-1}) + 4 \times (-286\ kJ\ mol^{-1}) = -2323\ kJ\ mol^{-1}$

$\Delta H_3 = \Delta H^{\ominus}_c [C_3H_8(g)] = -2220\ kJ\ mol^{-1}$

Hence:

$\Delta H^{\ominus}_f [C_3H_8(g)] = (-2323\ kJ\ mol^{-1}) - (-2220\ kJ\ mol^{-1})$
$= -2323\ kJ\ mol^{-1} + 2220\ kJ\ mol^{-1}$
$= -103\ kJ\ mol^{-1}$

Test yourself D

11 The standard enthalpy of formation of methanol is -238.7 kJ mol^{-1}. Write the balanced equation for which the standard enthalpy of reaction is -238.7 kJ mol^{-1}.

12 When calculating standard enthalpies of reactions involving water why is it important to specify whether or not the H_2O is present as water or as steam?

13 Calculate the standard enthalpy of reaction of hydrazine, $N_2H_4(g)$, with oxygen to form nitrogen, $N_2(g)$, and steam, $H_2O(g)$, with the help of tables of standard enthalpies of formation.

Standard enthalpies of reaction from standard enthalpies of formation

Data books include tables of standard enthalpies of formation for inorganic and organic compounds. The great value of these tables is that they make it possible to calculate the standard enthalpy change for any reaction involving the substances listed in the tables.

The standard enthalpy change of a reaction is the enthalpy change when the amounts shown in the chemical equation react. Like other standard quantities in thermochemistry, the standard enthalpy change for reaction is defined at 298 K, 1 bar pressure with the reactants and products in their normal stable states under these conditions. The concentration of any solution is 1 mol dm^{-3}.

Thanks to Hess's law it is easy to calculate the enthalpy change for a reaction from tabulated values for standard enthalpies of formation.

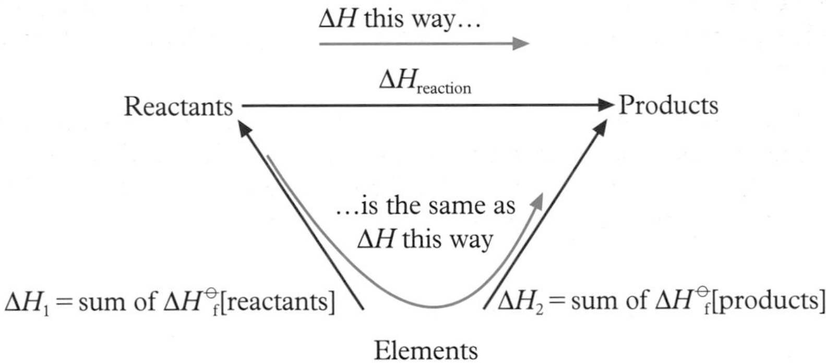

Figure 3.13.18 ▶ *Outline of a thermochemical cycle for calculating standard enthalpies of reaction from standard enthalpies of formation*

According to Hess's law: $\Delta H_{reaction} = -\Delta H_1 + \Delta H_2$

So $\Delta H_{reaction} = \{\text{sum of } \Delta H^{\ominus}_f [\text{products}]\} - \{\text{sum of } \Delta H^{\ominus}_f [\text{reactants}]\}$

Worked example

Calculate the enthalpy change for the reduction of iron(III) oxide by carbon monoxide.

$\Delta H^{\ominus}_f [Fe_2O_3] = -824\ kJ\ mol^{-1}$

$\Delta H^{\ominus}_f [CO] = -110\ kJ\ mol^{-1}$

$\Delta H^{\ominus}_f [CO_2] = -393\ kJ\ mol^{-1}$

Notes on the method

Write the balanced equation for the reaction.

Recall that by definition $\Delta H^{\ominus}_{f}$ [element] = 0 kJ mol^{-1}

Pay careful attention to the signs. Put the value and sign for a quantity in brackets when multiplying, adding or subtracting enthalpy values.

Answer

$Fe_2O_3(s) + 3CO(g) \rightarrow 2Fe(s) + 3CO_2(g)$

So $\Delta H^{\ominus}_{reaction} = \{\text{sum of } \Delta H^{\ominus}_{f}[\text{products}]\} - \{\text{sum of } \Delta H^{\ominus}_{f}[\text{reactants}]\}$

$$\Delta H^{\ominus}_{reaction} = \{2 \times \Delta H^{\ominus}_{f}[Fe] + 3 \times \Delta H^{\ominus}_{f}[CO_2]\} - \{\Delta H^{\ominus}_{f}[Fe_2O_3] + 3 \times \Delta H^{\ominus}_{f}[CO]\}$$

$$= \{0 + 3 \times (-393 \text{ kJ mol}^{-1})\} - \{(-824 \text{ kJ mol}^{-1}) + 3 \times (-110 \text{ kJ mol}^{-1})\}$$

$$= (-1179 \text{ kJ mol}^{-1}) - (-1154 \text{ kJ mol}^{-1})$$

$$\Delta H^{\ominus}_{reaction} = -25 \text{ kJ mol}^{-1}$$

Definition

Standard enthalpy of neutralisation
The enthalpy change of reaction when an acid and an alkali neutralise each other under standard conditions.

Enthalpy changes and the direction of change

Strike a match, it catches fire and burns. Put a spark to petrol and it burns furiously. These are two exothermic reactions which, once started, tend to go. They are examples of the many exothermic reactions which just keep going once that they have begun. In general, chemists expect that a reaction will go if it is exothermic.

What this means is that reactions which give out energy to their surroundings are the ones which happen. This ties in with the common experience that change happens in the direction in which energy is spread around and dissipated in the surroundings. So the sign of ΔH is a guide to the likely direction of change but it is not a totally reliable guide for three main reasons:

- The direction of change may depend on the conditions of temperature and pressure. One example is the condensation of a liquid. Steam condenses to water below 100 °C. This is an exothermic change. Above 100 °C the change goes in the opposite direction.

$$H_2O(g) \underset{\text{cool}}{\overset{\text{heat}}{\rightleftharpoons}} H_2O(l) \quad \Delta H = -44 \text{ kJ mol}^{-1}$$

- There are examples of endothermic reactions which go under normal conditions. So a reaction for which ΔH is positive can tend to happen. One example is the reaction of a solution of citric acid with sodium hydrogencarbonate. The mixture fizzes violently while cooling quickly.
- A reaction may be highly exothermic and tend to go, yet the rate of change may be so low that the mixture of chemicals is effectively inert. The change from diamond to graphite is exothermic, yet diamonds do not suddenly turn into black flakes.

Enthalpy changes and bonding

The overall enthalpy change for a reaction is the difference between the energy needed to break bonds in the reactants and the energy released as new bonds form in the products (see page 98).

When investigating bond breaking, chemists distinguish between:

- bond dissociation enthalpies – precise values for particular bonds
- average bond enthalpies – mean values which are a useful approximate guide.

The bond dissociation enthalpy is the enthalpy change on breaking one mole of a particular covalent bond in a gaseous molecule. In molecules with two or more bonds between similar atoms, the energies needed to break successive bonds are not the same. In water, for example, the energy needed to break the first OH bond in H—O—H(g) is 498 kJ mol^{-1} but the energy needed to break the second O—H bond in OH(g) is 428 kJ mol^{-1}.

Average bond enthalpies (or bond energies) are the average values of bond dissociation enthalpies used in approximate calculations to estimate enthalpy changes for reactions.

The mean values of bond enthalpies take into account the facts that:

- the successive bond dissociation enthalpies are not the same in compounds such as water or methane
- the bond dissociation enthalpy for a specific covalent bond varies slightly from one molecule to another.

Note

Energy is needed to break covalent bonds. Bond breaking is endothermic so bond enthalpies are positive.

Definition

Dissociation is a change in which a molecule splits into two or more smaller particles.

Worked example

Use average bond enthalpies to estimate the enthalpy of formation of hydrazine, N_2H_4.

Note on the method

Write out the equation showing all the atoms and bonds in the molecules to make it easier to count the numbers of bonds broken and formed.

Look up the mean bond energies in a book of data. The symbol E(N—H) stands for the average bond energy of a covalent bond between a nitrogen atom and a hydrogen atom.

Answer

The equation for the reaction:

$$\mathrm{N{\equiv}N + 2H{-}H \rightarrow \underset{H}{\overset{H}{>}}N{-}N\underset{H}{\overset{H}{<}}}$$

The energy needed to break the bonds in the reactants

$= E(\mathrm{N{\equiv}N})\ \mathrm{kJ\ mol^{-1}} + 2E(\mathrm{H{-}H})\ \mathrm{kJ\ mol^{-1}}$

$= 945\ \mathrm{kJ\ mol^{-1}} + 2 \times 436\ \mathrm{kJ\ mol^{-1}}$

$= 1817\ \mathrm{kJ\ mol^{-1}}$

The energy given out as new bonds form to make the product

$= E(\mathrm{N{-}N})\ \mathrm{kJ\ mol^{-1}} + 4E(\mathrm{N{-}H})\ \mathrm{kJ\ mol^{-1}}$

$= 158\ \mathrm{kJ\ mol^{-1}} + 4 \times 391\ \mathrm{kJ\ mol^{-1}}$

$= 1722\ \mathrm{kJ\ mol^{-1}}$

More energy is needed to break bonds than is given out when bonds are formed so the reaction is endothermic and the enthalpy change is positive.

$\Delta H = +\ 1817\ \mathrm{kJ\ mol^{-1}} - 1722\ \mathrm{kJ\ mol^{-1}} = +\ 95\ \mathrm{kJ\ mol^{-1}}$

Test yourself D

14 Calculate the average of the two bond dissociation energies of water. Compare your answer with the value for the average bond energy for the O—H bond.

15 Make a table to show the average bond enthalpies and bond lengths of the C—C, C=C and C≡C bonds. What two generalisations can you make based on your table?

16 Use average bond enthalpies to estimate the enthalpy change when ethene, $H_2C{=}CH_2(g)$, reacts with gaseous hydrogen to form ethane, $CH_3{-}CH_3(g)$.

17 Which is likely to give a more accurate answer – calculating the enthalpy change for a reaction
- from average bond energies
- from enthalpies of formation?

3.14 Reversible reactions

Some changes go in only one direction like baking bread. Once baked in an oven there is no way to split a loaf back into flour, water and yeast. Some chemical reactions are like this. Many other changes are reversible. Haemoglobin, for example, combines with oxygen as red blood cells flow through the lungs but then releases the oxygen for respiration as blood flows through the capillaries throughout the rest of the body. The study of reversible reactions helps chemists to answer the questions 'How far and in which direction?', answers they need when trying to make new chemicals in laboratories and in industry.

Burning fuels, such as methane, in air is an example of a one-way process. Once methane has burnt in air to make carbon dioxide and water, it is impossible to turn the products back to methane and oxygen. The combustion of methane is an irreversible process.

Many other chemical reactions are reversible. One example is the basis of a simple laboratory test for water. Hydrated cobalt(II) chloride is pink and so is a solution of the salt in water. Heating filter paper soaked in the solution in an oven makes the paper turn blue.

$$\underset{\text{pink}}{CoCl_2.6H_2O(s)} \rightarrow \underset{\text{blue}}{CoCl_2(s)} + 6H_2O(l)$$

The blue paper provides a sensitive test for water. The paper turns pink again if exposed to water or water vapour. At room temperature water rehydrates the blue salt.

$$\underset{\text{blue}}{CoCl_2(s)} + 6H_2O(l) \rightarrow \underset{\text{pink}}{CoCl_2.6H_2O(s)}$$

Figure 3.14.1 ▲
Using blue cobalt chloride paper to test for water

The reaction of ammonia with hydrogen is another reaction for which the direction of change depends on the temperature. At room temperature the two gases combine to make a white smoke of ammonium chloride.

$$NH_3(g) + HCl(g) \rightarrow NH_4Cl(s)$$

Heating makes this reversible reaction go the other way. Ammonium chloride decomposes at high temperatures to give hydrogen chloride and ammonia.

$$NH_4Cl(s) \rightarrow NH_3(g) + HCl(g)$$

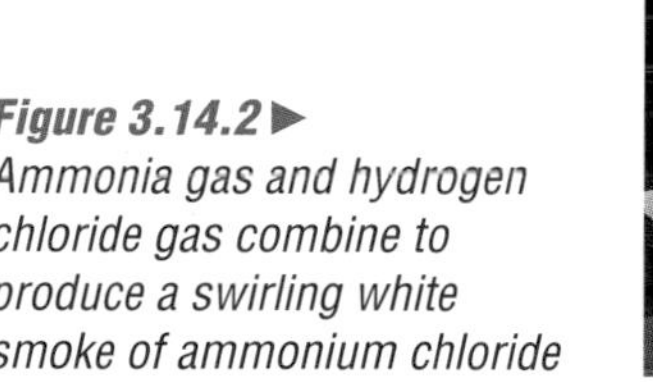

Figure 3.14.2 ►
Ammonia gas and hydrogen chloride gas combine to produce a swirling white smoke of ammonium chloride

Section four

Inorganic Chemistry

Contents

4.1 What is inorganic chemistry?

4.2 The periodic table

4.3 Oxidation numbers

4.4 Group 1

4.5 Group 2

4.6 Group 7

4.7 Water treatment

4.8 Inorganic chemistry in industry

4.9 Extraction of metals

4.10 Environmental issues

3.14 Reversible reactions

Some changes go in only one direction like baking bread. Once baked in an oven there is no way to split a loaf back into flour, water and yeast. Some chemical reactions are like this. Many other changes are reversible. Haemoglobin, for example, combines with oxygen as red blood cells flow through the lungs but then releases the oxygen for respiration as blood flows through the capillaries throughout the rest of the body. The study of reversible reactions helps chemists to answer the questions 'How far and in which direction?', answers they need when trying to make new chemicals in laboratories and in industry.

Burning fuels, such as methane, in air is an example of a one-way process. Once methane has burnt in air to make carbon dioxide and water, it is impossible to turn the products back to methane and oxygen. The combustion of methane is an irreversible process.

Many other chemical reactions are reversible. One example is the basis of a simple laboratory test for water. Hydrated cobalt(II) chloride is pink and so is a solution of the salt in water. Heating filter paper soaked in the solution in an oven makes the paper turn blue.

$$\underset{\text{pink}}{CoCl_2.6H_2O(s)} \rightarrow \underset{\text{blue}}{CoCl_2(s)} + 6H_2O(l)$$

Figure 3.14.1 ▲
Using blue cobalt chloride paper to test for water

The blue paper provides a sensitive test for water. The paper turns pink again if exposed to water or water vapour. At room temperature water rehydrates the blue salt.

$$\underset{\text{blue}}{CoCl_2(s)} + 6H_2O(l) \rightarrow \underset{\text{pink}}{CoCl_2.6H_2O(s)}$$

The reaction of ammonia with hydrogen is another reaction for which the direction of change depends on the temperature. At room temperature the two gases combine to make a white smoke of ammonium chloride.

$$NH_3(g) + HCl(g) \rightarrow NH_4Cl(s)$$

Heating makes this reversible reaction go the other way. Ammonium chloride decomposes at high temperatures to give hydrogen chloride and ammonia.

$$NH_4Cl(s) \rightarrow NH_3(g) + HCl(g)$$

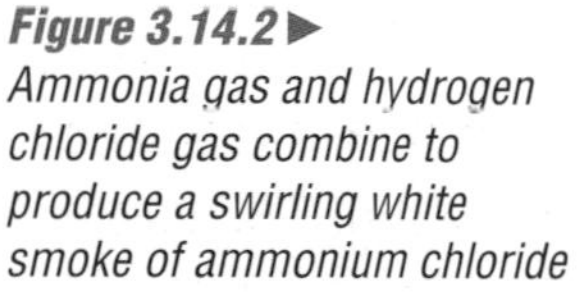

Figure 3.14.2 ►
Ammonia gas and hydrogen chloride gas combine to produce a swirling white smoke of ammonium chloride

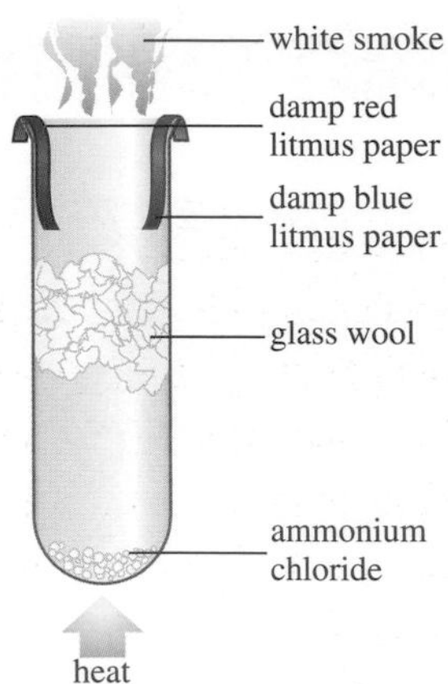

Figure 3.14.3 ► *Apparatus to show that ammonium chloride decomposes into two gases on heating. Ammonia gas diffuses through the glass wool faster than hydrogen chloride. After a short time the alkaline ammonia rises above the plug of glass wool and turns the red litmus blue. A while later both strips of litmus paper turn red as the acid hydrogen chloride arrives. A smoke of ammonium chloride appears above the tube when both gases meet and cool*

Changing the temperature is not the only way to alter the direction of change. Hot iron, for example, reacts with steam to make iron(III) oxide and hydrogen.

Supplying plenty of steam and sweeping away the hydrogen means that the reaction continues until all the iron changes to its oxide.

$$3Fe(s) + 4H_2O(g) \rightarrow Fe_3O_4(s) + H_2(g)$$

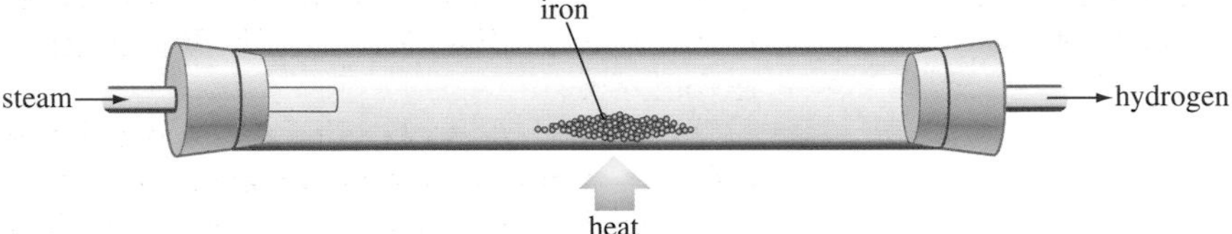

Figure 3.14.4 ► *The forward reaction goes when the concentration of steam is high and the hydrogen is swept away keeping its concentration low*

Altering the conditions brings about the reverse reaction. A stream of hydrogen reduces all the iron(III) oxide to iron so long as the flow of hydrogen sweeps away the steam formed.

$$Fe_3O_4(s) + H_2(g) \rightarrow 3Fe(s) + 4H_2O(g)$$

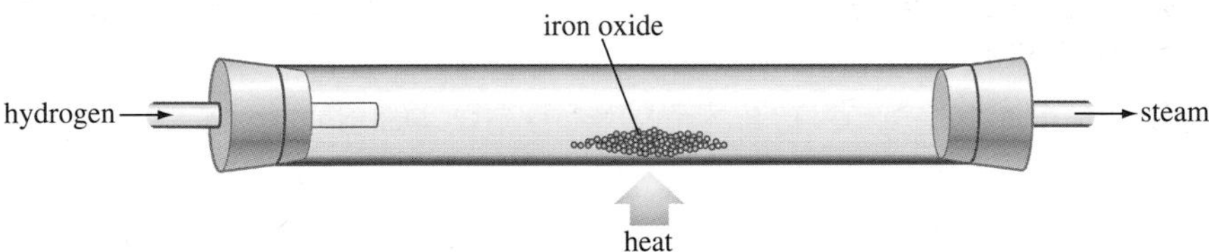

Figure 3.14.5 ► *The backward reaction goes when the concentration of hydrogen is high and the steam is swept away keeping its concentration low*

Note

In an equation, the chemicals on the left-hand side are the reactants. Those on the right are the products. The 'left-to-right' reaction is the 'forward' reaction and the 'right-to-left' reaction is the backward reaction.

Test yourself

1 How can these changes be reversed either by changing the temperature or by changing the concentration of a reactant or product?

a) freezing water to ice,
b) changing blue litmus to its red form
c) converting blue copper(II) sulfate to its white form.

2 Write a symbol equation to show the reversible change when iodine sublimes (see page 19).

3.15 Chemical equilibrium

Reversible changes often reach a state of balance, or equilibrium. What is special about chemical equilibrium is that nothing seems to be happening to the unaided eye, but at a molecular level there is ceaseless change. When chemists ask the question 'How far?' they want to know what the state of a reaction will be when it is at equilibrium. At equilibrium the reaction shown by an equation may be well to the right (mostly new products), well to the left (mostly unchanged reactants) or somewhere in between.

Note

The symbol $\rightleftharpoons$ represents a reversible reaction at equilibrium. In theory it is only possible to achieve a state of equilibrium in a closed system (see page 92).

Reaching an equilibrium state

Balance points exist in most reversible reactions when neither the forward nor the reverse reaction is complete. Reactants and products are present together and the reactions appear to have stopped. This is the state of chemical equilibrium.

One way to study the approach to equilibrium is to watch what happens on shaking a small crystal of iodine in a test tube with hexane and a solution of potassium iodide, KI(aq). The liquid hexane and the aqueous solution do not mix.

Iodine freely dissolves in hexane which is a non-polar solvent (see page 89). The non-polar iodine molecules mix with the hexane molecules. There is no reaction. The solution is a purply-violet colour – the same colour as iodine vapour. Iodine hardly dissolves in water but it does dissolve in a solution of potassium iodide. The solution is yellow, orange or brown depending on the concentration. In the solution iodine molecules, I_2, react with iodide ions, I^-, to form the tri-iodide ion, I_3^-.

Figure 3.15.1 is a study of changes which can be summed up by this equilibrium:

$$I_2(\text{in hexane}) + I^-(aq) \rightleftharpoons I_3^-(aq)$$

Figure 3.15.1 ▼
Two approaches to the same equilibrium state

CD-ROM

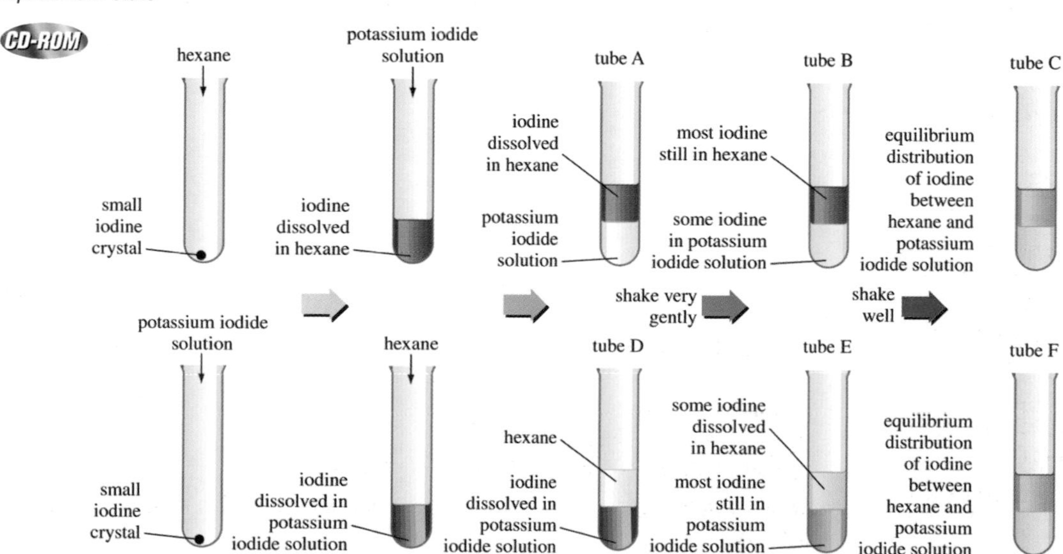

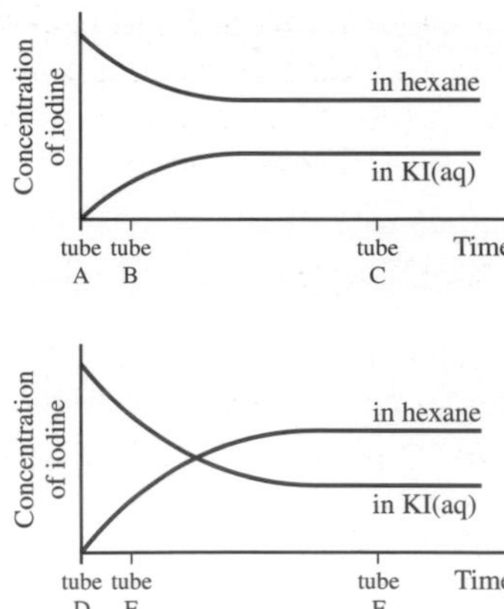

Figure 3.15.2 ▶ *Change of concentration of iodine with time in the mixtures shown in Figure 3.15.1*

Test yourself

1 Under what conditions are these in equilibrium:
  a) water and ice
  b) water and steam
  c) copper(II) sulfate crystals and copper(II) sulfate solution.

2 Draw a diagram to represent the movement of particles between a crystal and a saturated solution of the solid in a solvent.

The graphs in Figure 3.15.2 show how the iodine concentration in the two layers changes with shaking. After a little while no further change seems to take place. Tubes C and F look just the same. Both contain the same equilibrium system.

This demonstration shows two important features of equilibrium processes:

- at equilibrium the concentration of reactants and products does not change
- the same equilibrium state can be reached from either the 'reactant side' or the 'product side' of the equation.

Definition

In **dynamic equilibrium** the forward and backward reactions continue but at equal rates so the overall effect is no change. On a molecular scale there is continuous change. On the macroscopic scale nothing appears to be happening.

Dynamic equilibrium

Figure 3.15.3 shows what is happening at a molecular level not what the eye can see. Consider tube A in Figure 3.15.1. All the iodine molecules start in the upper hexane level. On shaking, some move into the aqueous layer. At first, molecules can only move in this direction (the forward reaction). The forward reaction begins to slow down as the concentration in the upper layer falls (Figure 3.15.3).

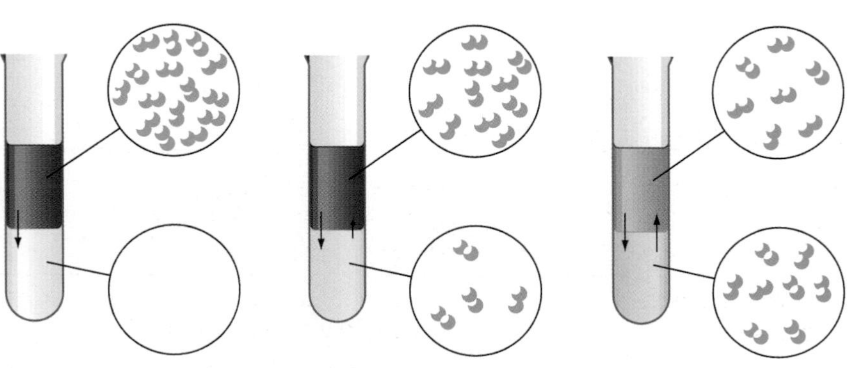

Figure 3.15.3 ▶ *Iodine molecules reaching dynamic equilibrium between hexane and a solution of potassium iodide*

Once there is some iodine in the aqueous layer, the reverse process can begin with iodine returning to the hexane layer. This backward reaction starts slowly but speeds up as the concentration of iodine in the aqueous layer increases.

There comes a time when both forward and backward reactions happen at the same rate. Movement of iodine between the two layers continues but overall there is no change. In tube C each layer is gaining and losing iodine molecules at the same rate. This is an example of dynamic equilibrium.

Note

Adding a catalyst does not affect the position of equilibrium. A catalyst speeds up the forward and backward reactions, so it shortens the time taken to reach equilibrium.

Factors affecting equilibria

Changing the conditions can disturb a system at equilibrium. At equilibrium the rate of the forward and backward reactions is the same. Anything which changes the rates can shift the balance.

Predicting the direction of change

Le Châtelier's principle is a qualitative guide to the effect of changes in concentration, pressure or temperature on a system at equilibrium. The principle was suggested as a general rule by the French physical chemist Henri Le Châtelier (1850–1936).

The principle states that when the conditions of a system at equilibrium change, the position of equilibrium shifts in the direction that tends to counteract the change.

Changing the concentration

Figure 3.15.4 shows the effects of changing concentration on this generalised equilibrium system:

$$A + B \rightleftharpoons C + D$$

The reaction of bromine with water provides examples of predictions based on Le Châtelier's principle. A solution of bromine in water is a yellow-orange colour because it contains bromine molecules in this equilibrium:

$$\underbrace{Br_2(aq)}_{\text{orange}} + \underbrace{H_2O(l) \rightleftharpoons OBr^-(aq) + Br^-(aq) + 2H^+(aq)}_{\text{colourless}}$$

Adding alkali turns the solution almost colourless. Hydroxide ions in the alkali react with hydrogen ions removing them from the equilibrium. As the hydrogen ion concentration falls, the equilibrium shifts to the right

Disturbance	How does the equilibrium mixture respond?	The result
Concentration of A increases	It moves to the right. Some A is used by reaction with B	More C and D are formed
Concentration of D increases	It moves to the left. Some of the added D is used up by reaction with C	More A and B are formed
Concentration of D decreases	It moves to the right to make up for the loss of D.	There is more C and less A and B in the new equilibrium

Figure 3.15.4 ▲
Le Châtelier's principle in action

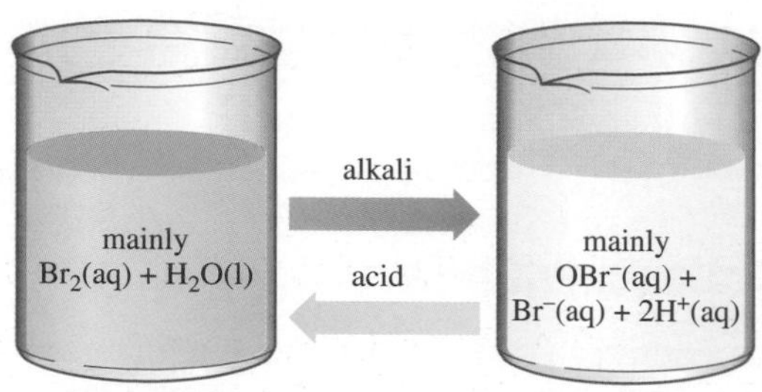

Figure 3.15.5 ► *The visible effects of adding alkali and acid to a solution of bromine in water*

converting orange bromine molecules to colourless ions. Lowering the hydrogen ion concentration slows down the backward reaction while the forward reaction goes on as before. The position of equilibrium shifts until once again the rates of the forward and backward reactions are the same.

Adding acid puts back the hydrogen ions, speeds up the backward reaction and makes the solution turn orange-yellow again. The equilibrium shifts to the left reducing the hydrogen ion concentration and increasing the bromine concentration until once again the forward and backward reactions are in balance.

Changing the pressure and temperature

Le Châtelier's principle helps to explain the conditions chosen for ammonia manufacture by the Haber process (see pages 162–163). The equilibrium system involved is:

$$N_2(g) + 3H_2(g) \rightleftharpoons 2NH_3(g) \quad \Delta H = -92.4 \text{ kJ mol}^{-1}$$

There are 4 mol of gases on the left-hand side of the equation but only 2 mol on the right. Le Châtelier's principle predicts that raising the pressure will make the equilibrium shift from left to right. This reduces the number of molecules and tends to reduce the pressure. So increasing the pressure increases the proportion of ammonia at equilibrium.

The reaction is exothermic from left to right and so endothermic from right to left. Le Châtelier's principle predicts that raising the temperature will make the position of equilibrium shift in the direction which takes in energy (tending to cool the mixture). So raising the temperature lowers the proportion of ammonia at equilibrium.

Definitions

A **phase** is one of the three states of matter – solid, liquid or gas. Chemical systems often have more than one phase. Each phase is distinct but need not be pure:

- a salt in equilibrium with its saturated solution in water is a two-phase system
- in the reactor for ammonia manufacture, the mixture of nitrogen, hydrogen and ammonia gases is one phase with the iron catalyst being a separate solid phase.

A **homogeneous equilibrium** is an equilibrium in which all the substances involved are in the same phase. An example is the equilibrium in a solution of bromine in water.

A **heterogeneous equilibrium** is an equilibrium system in which all the substances involved are not in the same phase. An example is the equilibrium state involving two solids and a gas formed on heating calcium carbonate in a closed container:

$$CaCO_3(s) \rightleftharpoons CaO(s) + CO_2(g)$$

Test yourself

3 Write an ionic equation for the reversible reaction of silver(I) ions with iron(II) ions to form silver atoms and iron(III) ions at equilibrium. Make a copy of the table in Figure 3.15.4 to show how Le Châtelier's principle applies to this equilibrium.

4 Yellow chromate(VI) ions, $CrO_4^{2-}(aq)$, react with aqueous hydrogen ions, $H^+(aq)$, to form orange dichromate(VI) ions, $Cr_2O_7^{2-}(aq)$ and water molecules. The reaction is reversible. Write an equation for the system at equilibrium. Predict how the colour of a solution of chromate(VI) ions will change:

a) on adding acid, followed by
b) hydroxide ions which neutralise hydrogen ions (see page 31).

5 Heating limestone, $CaCO_3$, in an closed furnace produces an equilibrium mixture of calcium carbonate with calcium oxide, CaO, and carbon dioxide gas. Heating the solid in an open furnace decomposes the solid completely into the oxide. How do you account for this difference?

3.16 Acid–base equilibria

Chemists have more than one way of explaining the behaviour of acids and bases. Each theory gives rise to a different definition of what counts as an acid and as a base. The theory that acids produce hydrogen ions in water (see page 30) has its limitations because it cannot cope with reactions which do not involve water.

The Brønsted–Lowry theory

The Danish chemist, Johannes Brønsted (1879–1947) published a new theory in 1923 at the same time as Thomas Lowry (1874–1936) in Cambridge came up with similar ideas.

The definition of an acid generally used today is based on their theory which defines an acid as a molecule or ion which can give away a hydrogen ion to something else. Acids are hydrogen ion donors.

According to this theory, hydrogen chloride molecules give hydrogen ions to water molecules when they dissolve in water producing hydrated hydrogen ions called oxonium ions. The water accepts the hydrogen ion and is here acting as a base. In this theory, a base is any molecule or ion which can accept a hydrogen ion from an acid.

$$\overset{\text{hydrogen chloride}}{HCl(g)} + H_2O(l) \xrightarrow{H^+} \rightleftharpoons \overset{\text{oxonium ion}}{H_3O^+(aq)} + Cl^-(aq)$$

(hydrogen chloride; oxonium ion — H⁺ transferred from HCl to H₂O)

Hydrogen ions, H^+, are hydrogen atoms which have lost an electron. Since a hydrogen atom consists of one proton and one electron this means that a hydrogen ion is just a proton. In water, hydrogen ions do not float around freely, they become attached to water molecules forming oxonium ions, H_3O^+. A lone pair of electrons on the oxygen atom forms a dative bond with the hydrogen atom (see pages 78–79).

An example of an acid–base reaction in the absence of water is the formation of a white smoke of ammonium chloride when ammonia gas mixes with hydrogen chloride gas.

$$NH_3(g) + HCl(g) \rightleftharpoons NH_4Cl(s)$$

(H⁺ transferred from HCl to NH₃)

Acid–base equilibria

Acid–base reactions are reversible. This is illustrated by a solution of an ammonia in water. There is a competition for protons (hydrogen ions).

On the left-hand side of the equation there is an acid and a base. During the forward reaction, ammonia molecules take protons from the water molecules. So here water is acting as an acid. During the reverse reaction. hydroxide ions take protons from ammonium ions.

$$\underset{\text{acid A}}{H_2O(l)} + \underset{\text{base B}}{NH_3(aq)} \rightleftharpoons \underset{\text{base A}}{OH^-(aq)} + \underset{\text{acid B}}{NH_4^+(aq)}$$

The system quickly reaches a state of dynamic equilibrium. The position of equilibrium depends on the relative strength of the acids and the bases. Ammonia is a weak base while the hydroxide ion is a strong base so the equilibrium is well over to the left-hand side.

Definitions

An **acid** is a proton *donor.*

A **base** is a proton *acceptor.*

Definitions

A **strong acid**, such as hydrogen chloride, readily gives away protons and is fully ionised in water.

A **weak** acid, such as ethanoic acid, is reluctant to give up protons and is only slightly ionised in water.

Definitions

An acid turns into its **conjugate base** when it loses a proton. A base turns into its **conjugate acid** when it gains a proton. Any acid–base equilibrium involves two conjugate acid–base pairs. In the example opposite they are:

- NH_4^+ and NH_3 and
- H_2O and OH^-.

3.17 Rates of chemical change

When people buy a pain killer or a cough medicine they expect it to work as well as it did the last time they bought the product. They may glance at the label and spot that it shows a 'sell by' date but they are unlikely to think about how that date is calculated. That is the job of the pharmacist who designed the medicine. The pharmacist has to know about the rate at which the chemicals slowly degrade in the bottle or pack. For many medicines, the shelf-life is the time for which they can be stored until the concentration of the active ingredient has dropped by no more than 10%.

CD-ROM

Test yourself

1 How is it possible to slow down or stop these reactions?

- **a)** iron corroding
- **b)** toast burning
- **c)** milk turning sour

2 How is it possible to speed up these reactions?

- **a)** fermentation in dough to make the bread rise
- **b)** solid fuel burning in a stove
- **c)** epoxy glues (adhesives) setting
- **d)** the conversion of chemicals in engine exhausts to harmless gases

3 In an experiment to study the reaction of magnesium with dilute hydrochloric acid, 48 cm^3 of hydrogen forms in 10 s at room temperature. Calculate the average rates of formation of:

- **a)** hydrogen in $cm^3\ s^{-1}$
- **b)** hydrogen in $mol\ s^{-1}$ (see page 41)
- **c)** the rates of appearance or disappearance of the other product and the reactants.

Chemical kinetics has practical importance to the chemical industry. Manufacturers aim to get the best possible yield in the shortest time. The development of new catalysts to speed up reactions is one of the frontier aspects of modern chemistry. The aim is to make manufacturing processes more efficient so that they use less energy and produce little or no harmful waste. The need for more efficient processes is now pressing as people have become more aware of the harm that waste chemicals can do to our health and to the environment.

Chemical reactions happen at a variety of speeds. Ionic precipitation reactions are very fast. Explosions are even faster. The rusting of iron, however, and other corrosion processes are slow and may continue for years.

Measuring reaction rates

Chemical equations say nothing about how quickly the changes occur. Chemists have to do experiments to measure the rates of reactions under various conditions.

The amounts or concentrations of chemicals change during any chemical reaction. Products form as reactants disappear. The rates at which these changes happen give a measure of the rate of reaction.

The rate of this reaction:

$$Mg(s) + 2HCl(aq) \rightarrow MgCl_2(aq) + H_2(g)$$

can be measured by:

- the rate of loss of magnesium
- the rate of loss of hydrochloric acid
- the rate of formation of magnesium chloride
- the rate of formation of hydrogen.

In this example it is probably easiest to measure the rate of formation of hydrogen by collecting the gas and recording its volume with time (Figure 3.17.1).

Chemists design their rate experiments to measure a property which changes with the amount or concentration of a reactant or product. Then:

$$\text{rate of reaction} = \frac{\text{change recorded in the property}}{\text{time for the change}}$$

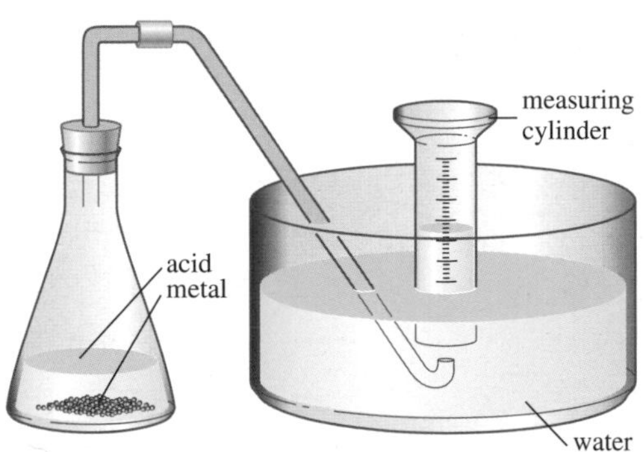

Figure 3.17.1 ◄
Collecting and measuring the gas produced when magnesium reacts with an acid

In most chemical reactions the rate changes with time. The graph in Figure 3.17.2 is a plot of results from a study of the reaction of magnesium with acid. The graph is steepest at the start when the reaction is at its fastest. As the reaction continues it slows down until it finally stops. This happens because one of the reactants is being used up.

The gradient at any point of a graph showing amount or concentration plotted against time measures the rate of reaction.

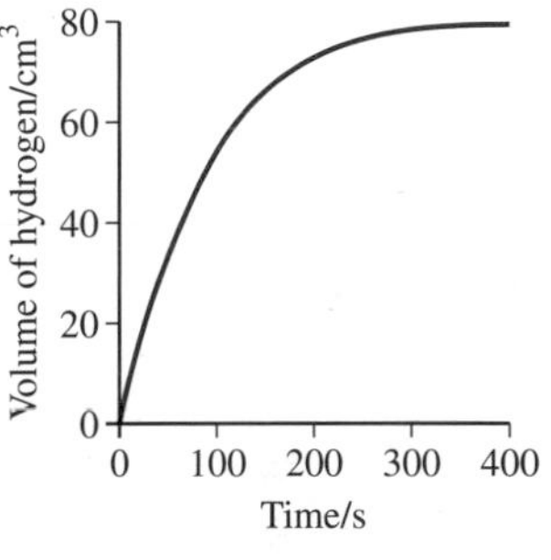

Figure 3.17.2 ▲
The volume of hydrogen plotted against time for the reaction of magnesium with hydrochloric acid

Concentration

In general the higher the concentration of the reactants, the faster the reaction. For gas reactions a change in pressure has the same effect as changing the concentration. A higher pressure compresses a mixture of gases and increases their concentration.

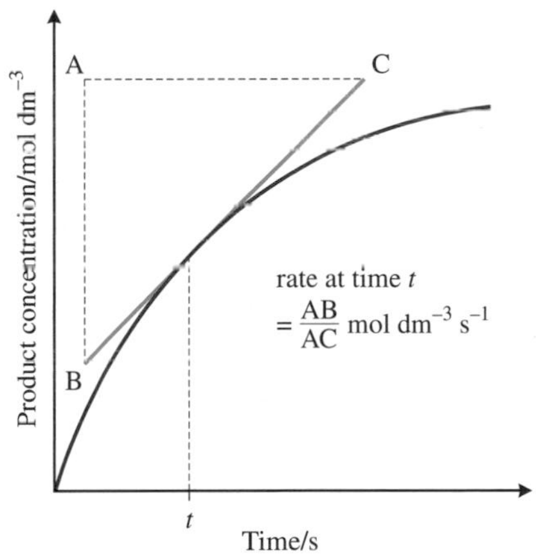
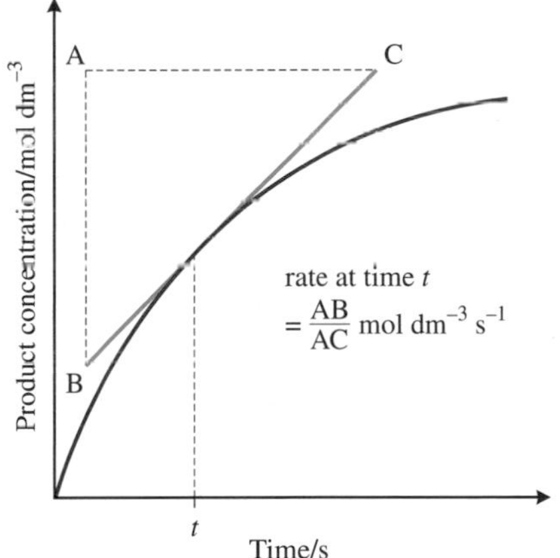

Figure 3.17.3 ◄
A graph showing quantity or concentration of a product plotted against time. The gradient at any point measures the rate of reaction

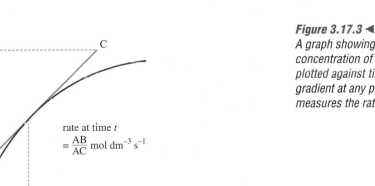

A useful way of studying the effect of changing the concentration on the rate is to find a way of measuring the rate just after mixing the reactants. Figure 3.17.4 is a graph for two different sets of conditions. One of the reactants was more concentrated to produce line A. Near the start it took t_A seconds to produce x mol of product. When the same reactant was less concentrated, the results gave line B. This time near the start it took t_B seconds to produce x mol of product.

Figure 3.17.5 illustrates an investigation of the effect of concentration on the rate at which thiosulfate ions in solution react with hydrogen ions to form a precipitate of sulfur.

$$S_2O_3^{2-}(aq) + 2H^+(aq) \rightarrow S(s) + SO_2(aq) + H_2O(l)$$

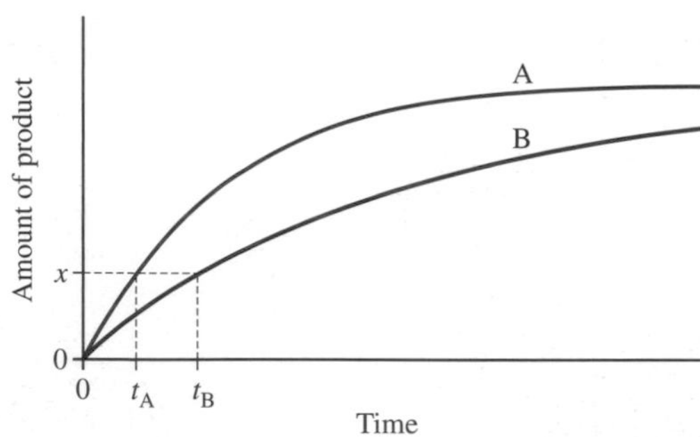

Figure 3.17.4 ▶
Formation of the same product starting with different concentrations of one of the reactants

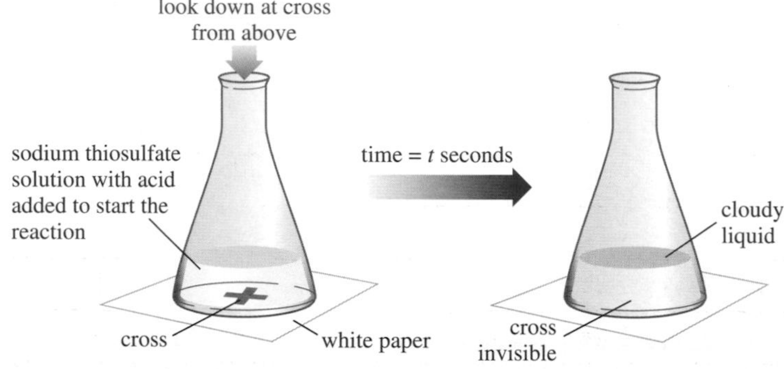

Figure 3.17.5 ▶
Investigating the effect of the concentration of thiosulfate ions on the rate of reaction in an acid solution. The hydrogen ion concentration is the same each time

Concentration sodium thiosulfate solution ($mol\ dm^{-3}$)	Time, *t*, for cross to be obscured (s)	Rate of reaction, 1/*t* (s^{-1})
0.15	43	0.023
0.12	55	0.018
0.09	66	0.015
0.06	105	0.0095
0.03	243	0.0041

In this example the quantity 'x' in Figure 3.17.4 is the amount of sulfur needed to hide the cross on the paper. This is the same each time. So the rate of reaction is proportional to 1/*t*.

Test yourself

Refer to Figure 3.17.5.

4 Plot a graph to show how the rate of reaction varies with the concentration of thiosulfate ions.

5 What is the relationship between rate and concentration for this reaction according to the graph?

Surface area of solids

Breaking a solid into smaller pieces increases the surface area in contact with a liquid or gas (see Figure 3.18.2 on page 117). This speeds up any reaction happening at the surface of the solid. This effect applies to any heterogeneous system (see page 100) including reactions between liquids which do not mix. Shaking breaks up one liquid into droplets which are dispersed in the other liquid, again increasing the surface area for reaction.

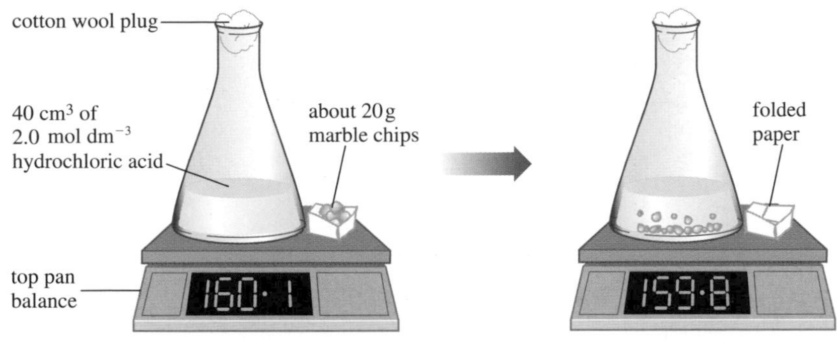

Time (s)	Mass of carbon dioxide formed (g) Small marble chips	Large marble chips
30	0.45	0.18
60	0.85	0.38
90	1.13	0.47
120	1.31	0.75
180	1.48	1.05
240	1.54	1.25
300	1.56	1.38
360	1.58	1.47
420	1.59	1.53
480	1.60	1.57
540	1.60	1.59
600	1.60	1.60

Figure 3.17.6 ▲
Comparing the rate of reaction of marble with acid using the same mass of larger and smaller marble chips CD-ROM

Figure 3.17.6 illustrates an investigation of the rate of reaction of calcium carbonate (as marble) with dilute nitric acid. Both sets of results were obtained using 20 g of marble chips and 40 cm^3 of 2 mol dm^{-3} nitric acid. The marble was in excess.

Test yourself

6 Refer to Figure 3.17.6.

a) Plot the two sets of results on the same axes.
b) Which reaction had the greater initial rate?
c) After what time did the reaction stop for each set of results? Why did the reaction stop?
d) For a given mass of marble, how is surface area related to particle size? What is the effect of changing the surface area of the solid on this reaction?

Temperature

Raising the temperature is a very effective way of increasing the rate of a reaction. Typically a 10 °C rise in temperature roughly doubles the rate of reaction (see Figure 3.17.7).

Bunsen burners, hot-plates and heating mantles are common in laboratories because chemists find it so convenient to speed up reactions by heating the mixtures of chemicals. For the same reason many industrial processes happen at high temperatures (see pages 161–165 and 167–171).

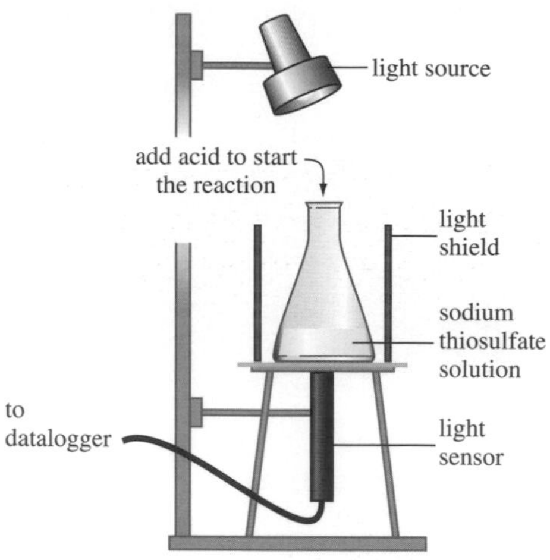

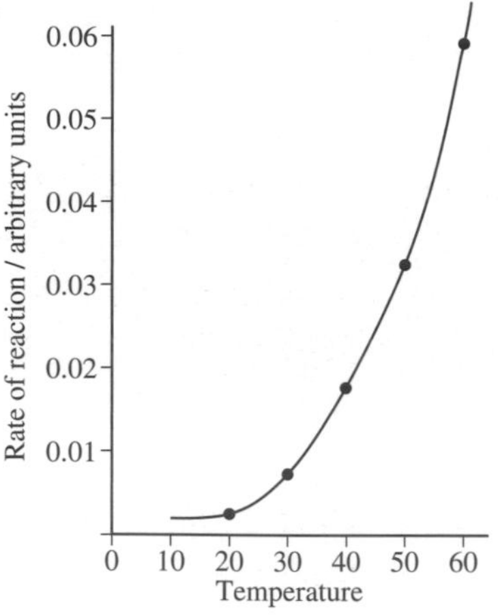

Figure 3.17.7 ▲
Effect of raising the temperature on the rate of decomposition of thiosulfate ions to sulfur

Test yourself

7 Figure 3.17.8 illustrates the effect of changing the conditions on the reaction of zinc metal with sulfuric acid. The red line shows the volume of hydrogen formed plotted against time using an excess of zinc turnings and 50 cm^3 of 2.0 mol dm^{-3} sulfuric acid at 20 °C.

a) Write a balanced equation for the reaction.

b) Draw the apparatus which could be used to obtain the results to plot the graph.

c) Identify which line of the graph shows the effect of carrying out the same reaction under the same conditions with the following changes:

i) adding a few drops of copper(II) sulfate solution to act as a catalyst

ii) raising the temperature to 30 °C

iii) using the same mass of zinc but in larger pieces

iv) using 50 cm^3 of 1.0 mol dm^{-3} sulfuric acid.

Catalysts

Catalysts speed up the rates of chemical reactions without themselves changing permanently. They can be recovered unchanged at the end of the reaction. Often a small amount of catalyst is effective.

Many catalysts are specific to a particular reaction. This is especially true of enzymes.

Catalysts speed up reactions but they do not change the position of equilibrium for a reversible reaction. If a catalyst is in the same phase as the reactants it is a **homogeneous catalyst**. If the catalyst is in a different phase it is a **heterogeneous catalyst**.

Here are some highlights in the development of industrial catalysts:

- 1908 Fritz Haber discovered how to make ammonia from nitrogen and hydrogen with a modified iron catalyst
- 1912 Paul Sabatier first used a nickel catalyst to hydrogenate unsaturated vegetable oils and turn them to solid fats for margarine
- 1930 Eugene Houdry developed catalytic cracking of oil fractions to make petrol
- 1942 Vladimir Ipatieff and Herman Pines found a catalytic method of alkylating hydrocarbons to produce branched hydrocarbons with high octane numbers to prevent knocking in petrol engines
- 1976 General Motors and the Ford Motor Corporation developed catalytic converters to cut pollution from motor vehicles.

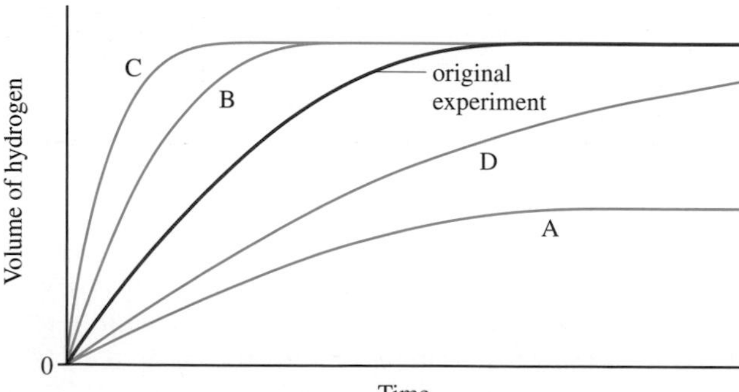

Figure 3.17.8 ▶

3.18 A collision model

Collision theory offers explanations for the effects of concentration, temperature and catalysts on reaction rates. The idea is that a chemical reaction happens when the molecules or ions of the reactants collide, making some bonds break and allowing new bonds to form.

Concentration, pressure and surface area

According to kinetic theory, the molecules in gases and liquids are in constant motion (see page 17). The molecules are forever bumping into each other. When they collide there is a chance that they will react.

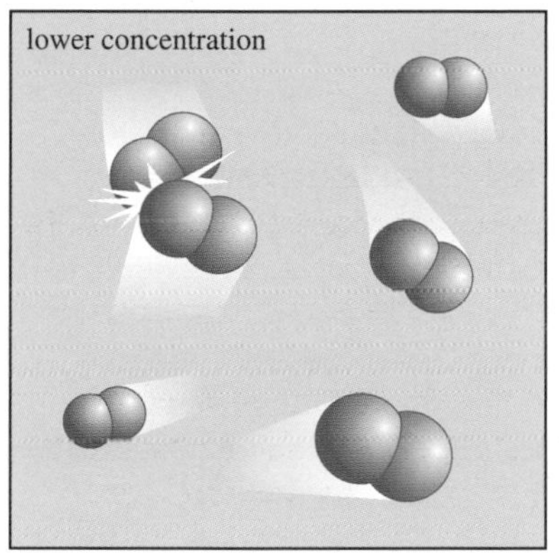

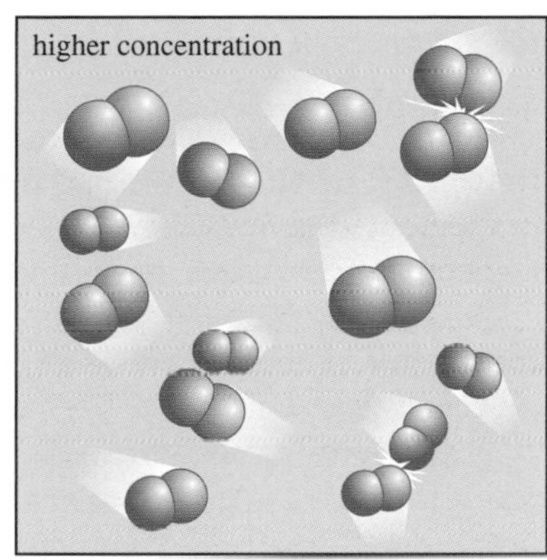

Figure 3.18.1 ◀
Raising the concentration means that the reacting particles are closer together. There are more collisions and reactions are faster

Figures 3.18.1 and 3.18.2 show how collision theory can account for the effects of changing concentration and surface area on reaction rates.

Increasing the pressure for a gas reaction squeezes the molecules closer together and so has the same effect as raising the concentration.

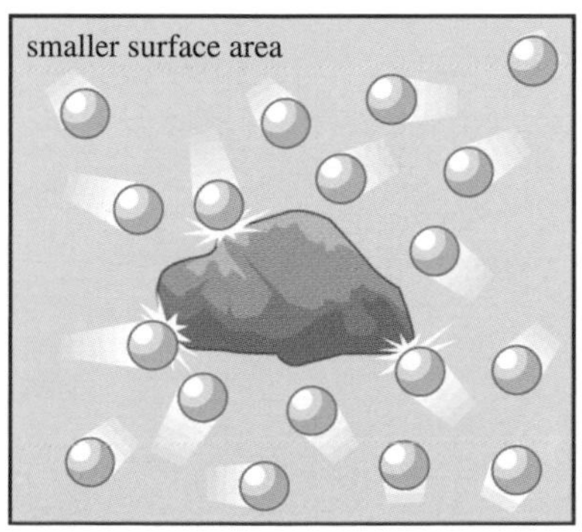

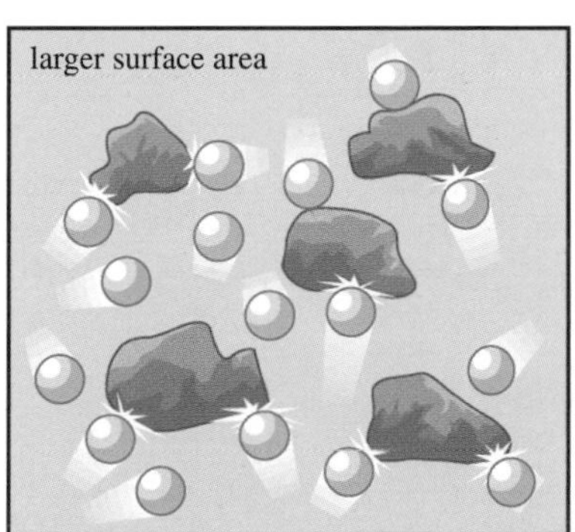

Figure 3.18.2 ◀
In a heterogeneous reaction of a solid with either a liquid or gas the reaction is faster if the solid is broken up into smaller pieces. Crushing the solid increases its surface area. Collisions can be more frequent and the rate of reaction is bigger

Definition

A **reaction profile** is a plot which shows how the total enthalpy of the atoms, molecules or ions changes during the progress of a change from reactants to products.

Temperature

It is not enough, however, for the molecules to collide. In soft collisions the molecules simply bounce off each other. Molecules are in such rapid motion that if every collision led to a reaction, all reactions would be explosively fast. Only pairs of molecules which collide with enough energy to stretch and break chemical bonds can lead to new products.

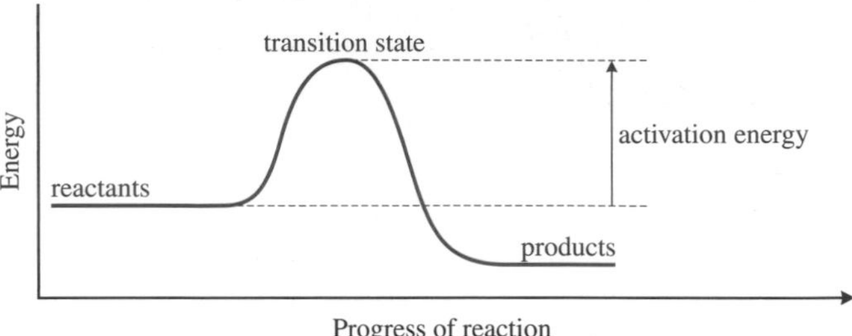

Figure 3.18.3 ▲
Reaction profile showing the activation energy for a reaction

Definition

Chemists use the term **activation energy** to describe the minimum energy needed in a collision between molecules if they are to react. The activation energy is the height of the energy barrier separating reactants and products during a chemical reaction.

Activation energies account for the fact that reactions go much more slowly than would be expected if every collision between atoms and molecules led to a reaction. Only a very small proportion of collisions bring about chemical change because molecules can only react if they collide with enough energy between them to overcome the energy barrier. So at around room temperature, only a small proportion of molecules have enough energy to react.

The Maxwell–Boltzmann curve describes the distribution of the kinetic energies of molecules (see page 65). As Figure 3.18.4 shows, the proportion of molecules with energies greater than the activation energy is small at around 300 K.

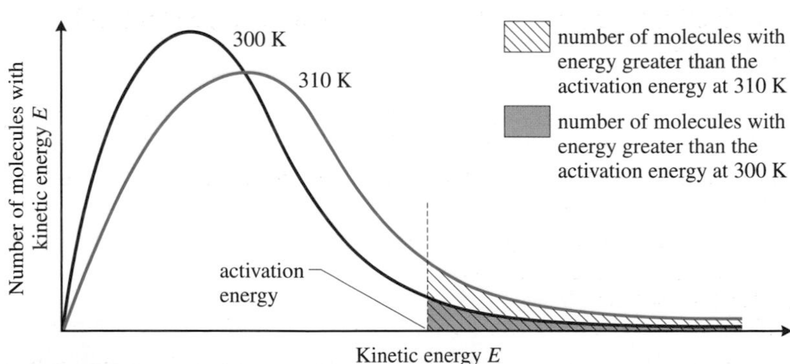

Figure 3.18.4 ▲
The Maxwell–Boltzmann distribution of molecular kinetic energies in a gas at two temperatures. The modal speed gets higher as the temperature rises. The area under the curve gives the total number of molecules. This does not change as the temperature rises so the peak height falls as the curve widens

The shaded areas in Figure 3.18.4 show the proportions of molecules having at least the activation energy for a reaction at two temperatures. This area is bigger at a higher temperature. So at a higher temperature there are more molecules with enough energy to react when they collide, and the reaction goes faster.

Catalysts

A catalyst works by providing for the reaction an alternative pathway with a lower activation energy. Lowering the activation energy increases the proportion of molecules with enough energy to react.

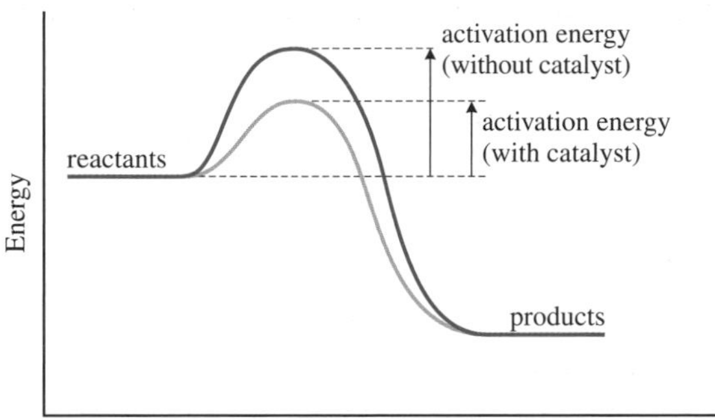

Figure 3.18.5 ◄
Reaction profile showing the effect of a catalyst on the activation energy of a reaction

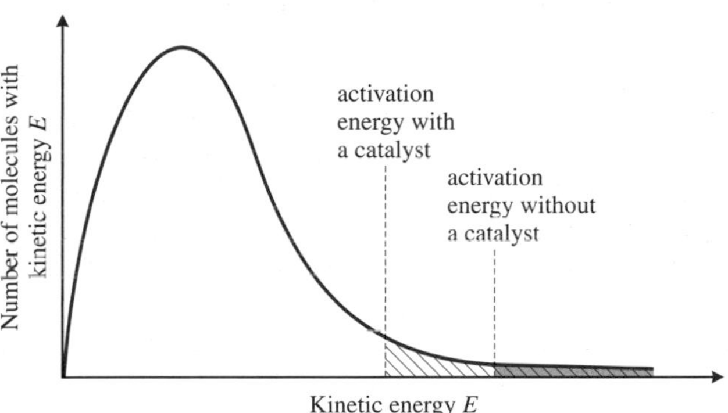

Figure 3.18.6 ◄
Distribution of molecular energies showing how the proportion of molecules able to react increases when a catalyst lowers the activation energy

Often a catalyst changes the mechanism of the reaction and make reactions more productive by increasing the yield of the desired product and reducing waste.

Test yourself

1 How and why do these changes affect the rate of reaction of zinc metal with sulfuric acid:

a) using zinc powder instead of zinc granules
b) adding some sodium carbonate crystals
c) adding some crushed ice
d) adding a few drops of copper(II) sulfate solution.

2 Explain why it takes a match or spark to light a Bunsen burner and why the gas keeps burning once it has been lit. Sketch a reaction profile to illustrate your explanation.

3.19 Stability

The study of energy changes (thermochemistry) and rates of reaction (kinetics) helps to explain why some chemicals or mixtures of chemicals are stable while others rapidly react.

Definitions

A chemical or mixture of chemicals is **thermodynamically stable** if there is no tendency for a reaction. Usually, but not always, a positive enthalpy change, ΔH, for the change indicates that the reaction does not tend to occur.

A chemical or mixture of chemicals is **kinetically stable** if nothing happens even when the enthalpy change for the reaction is negative. Kinetic stability is a consequence of a high activation energy for the change so that the reaction rate is very slow.

Compounds are stable if they do not tend to decompose into their elements or into other compounds. Sometimes there is no tendency for a reaction to go because the reactants are stable relative to the products. This is usually the case if the enthalpy change for the reaction is positive. Magnesium oxide, for example, has no tendency to split up into magnesium and oxygen. Sometimes a compound is stable at lower temperatures even though the enthalpy change suggests that the reaction should go. An example is the gas N_2O. This oxide is unstable relative to its elements. $\Delta H_f^\ominus$ (= +82 kJ mol^{-1}) is positive. The decomposition reaction from the compound into its elements is exothermic.

The compound tends to decompose into its elements but the rate is very slow under normal conditions because of the activation energy of the reaction. Chemists say that N_2O is kinetically 'stable' at room temperature.

Dinitrogen oxide does decompose on heating. It can even relight a glowing splint because the red hot wood increases the proportion of molecules with enough energy to overcome the activation energy for the decomposition reaction. When N_2O decomposes, it produces a mixture of gases with enough oxygen for a splint to burst into flames.

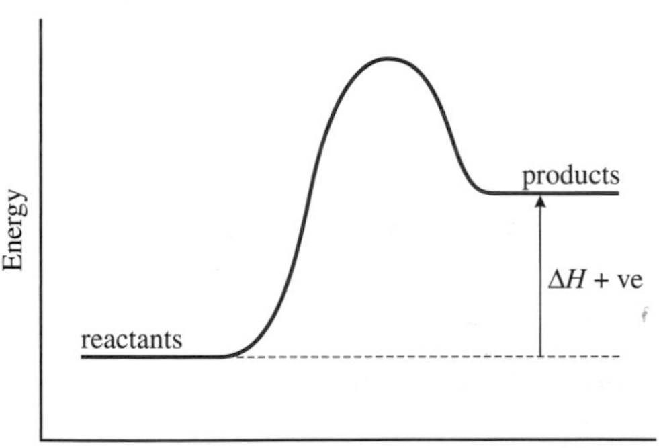

Figure 3.19.1 ▲
Reaction profile for a change which does not tend to go because the reactants are more stable than the products (thermodynamic stability)

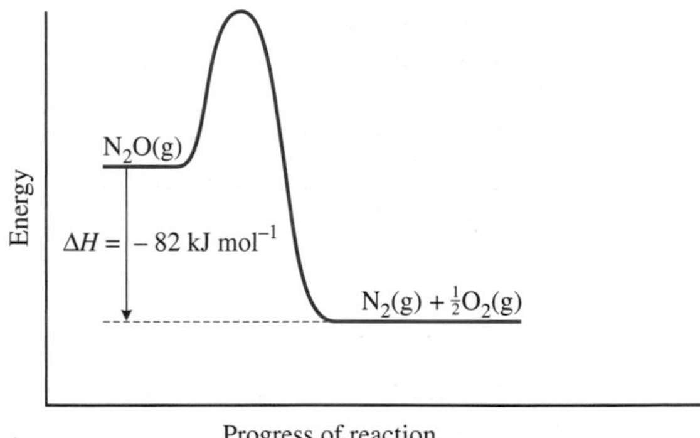

Figure 3.19.2 ▲
Reaction profile for the decomposition of dinitrogen oxide (kinetic stability)

Test yourself

1 Which of these are examples of thermodynamic stability and which are examples of kinetic stability?

a) Methane gas does not catch fire at room temperature when mixed with oxygen
b) Water does not decompose into hydrogen and oxygen when it boils and turns to steam
c) Diamonds do not change into graphite.

Review

This guidance will help you to organise your notes and revision. Check the terms and topics against the specification you are studying. You will find that some topics are not required for your course.

Key terms

Show that you know the meaning of these terms by giving examples. Consider writing the key term on one side of an index card and the meaning of the term with an example on the other side. Then you can easily test yourself when revising. Alternatively use a computer database with fields for the key term, the definition and the example. Test yourself with the help of reports which just show one field at a time.

- Fundamental particles
- Atomic number
- Mass number
- Isotopes
- Mass spectrometer
- Ionisation energies
- Energy levels
- Atomic orbitals
- Shielding
- Ideal gases
- Real gases
- Metallic bonding
- Covalent bonding
- Double bond
- Lone pair of electrons
- Dative covalent (co-ordinate) bonding
- Polar bonds
- Ionic radius
- Intermolecular forces
- Polar molecules
- Dipole-dipole interactions
- Van der Waals forces
- Hydrogen bonding
- Enthalpy change
- Dynamic equilibrium
- Strong acid
- Weak acid
- Activation energy
- Catalysts
- Heterogeneous
- Homogeneous

Symbols and conventions

Make sure that you understand the symbols and conventions which chemists use when describing atomic structures, working with physical quantities and carrying out thermochemical calculations. Illustrate your notes with examples.

- Symbols for isotopes showing the mass number and atomic number
- SI units
- Ways of representing the electron configurations of atoms
- Dot-and-cross diagrams
- Energy level diagrams for exothermic and endothermic reactions
- Standard conditions for enthalpy values

Section three study guide

Patterns and principles

Use tables, charts, concept maps or mind maps to summarise key ideas. Brighten your notes with colour to make them memorable.

- The building-up principle for electrons in atoms
- Patterns in the successive ionisation energies for an element
- Kinetic theory of gases and the ideal gas equation
- Physical properties of materials with: giant structures (metallic, ionic, covalent) and with molecular structures
- Crystal structures of metals, non-metals (diamond, graphite and iodine), ionic compounds (sodium chloride) and covalent compounds (ice, silicon dioxide)
- The octet rule and its limitations
- Patterns of solubility
- Hess' law and its applications
- Brønsted–Lowry theory of acids and bases
- Collision theory and the significance of the Maxwell–Boltzman distribution

Predictions

Use examples to show that you can apply chemical principles to make predictions.

- Predicting the shapes of molecules and ions
- Use of electronegativity values or Fajan's rules to predict the extent to which the bonding in a compound will be ionic or covalent
- Predicting the stability of compounds and the likely direction of chemical change using ΔH values
- Using Le Chatelier's principle to predict the effects of changes to pressure, concentration or temperature on the position of systems at equilibrium
- Predicting the effects of changing the conditions of temperature, concentration, pressure and particle size or the presence of catalysts on the rates of reactions

Laboratory techniques

Use labelled diagrams to illustrate and describe these practical procedures.

- Measurement of enthalpies of combustion
- Measurement of enthalpies of neutralisation
- Measurement of reaction rates

Calculations

Give your own worked examples, with the help of the Test Yourself questions to show that you can carry out calculations to work out the following from given data.

- Use of the ideal gas equation to calculate molar masses for gases and volatile liquids
- Determination of enthalpies of combustion and enthalpies of neutralisation from experimental results
- Use of Hess's law cycles to calculate
  - the standard enthalpy change of formation for a compound from standard enthalpy changes of combustion
  - the standard enthalpy change of reaction from standard enthalpy changes of formation
  - the standard enthalpy change of reaction from mean bond enthalpies.

Key skills

Problem solving

Planning, carrying out and interpreting a laboratory investigation calls for problem solving if you have to make your own decisions about the procedure, work independently and use your initiative. This could be an investigation of reaction rates, energy changes or reversible reactions.

Application of number

Quantitative practical work to investigate energy changes will give you opportunities to develop and practice the individual application of number skills you need. These include: selecting data from a large data set, working to the right number of significant figures, making measurements in appropriate units, using standard form, choosing the appropriate method of calculating the result, carrying out multi-stage calculations and using formulae plus checking to identify mistakes and experimental uncertainty.

Information technology

You can use chemical modelling software from a CD-ROM or the Internet to study the shapes of molecules. Compare the advantages and limitation of using software, physical models and textbook diagrams to study molecular shapes.

Section four

Inorganic Chemistry

Contents

4.1 What is inorganic chemistry?

4.2 The periodic table

4.3 Oxidation numbers

4.4 Group 1

4.5 Group 2

4.6 Group 7

4.7 Water treatment

4.8 Inorganic chemistry in industry

4.9 Extraction of metals

4.10 Environmental issues

4.1 What is inorganic chemistry?

Inorganic chemistry is the study of the hundred or so chemical elements and their compounds. The amount of information can be bewildering, hence the importance of the periodic table which offers a framework for giving a meaning to all the facts about properties and reactions.

Major chemicals from minerals

The scope of the chemical industry based on inorganic chemicals is illustrated by Figure 4.1.1. The chemical industry has flourished in the UK because many raw materials are available in these islands.

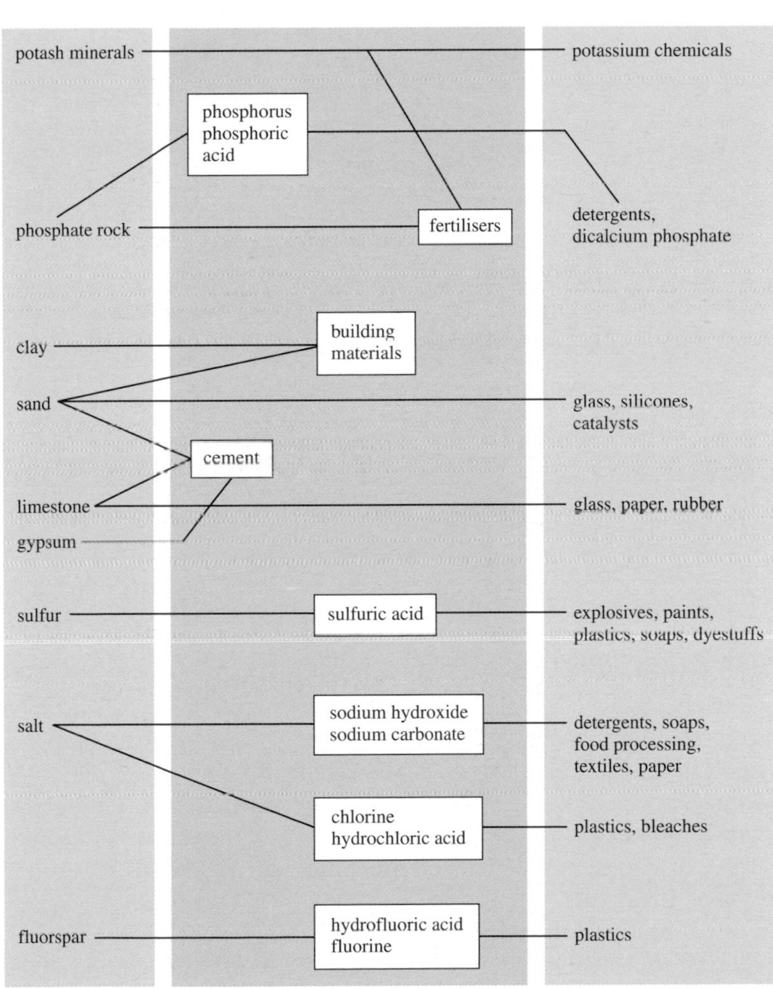

Figure 4.1.1 ◀ *Major chemicals from minerals and their uses*

Some of the finest china clay (kaolin) in the world is mined in Cornwall for use in the manufacture of high quality paper and pottery. Cheshire is famous for its huge underground salt deposits, while the Derbyshire Peak district is the source of high quality limestone and fluorspar. Potash minerals are mined in the north-east of England near Whitby.

Figure 4.1.2 ▲
A crystalline specimen of fluorspar, CaF_2

Themes in inorganic chemistry

Electron configuration

Studying inorganic chemistry means learning about the links between chemical behaviour and atomic structure. Especially important is the arrangement of electrons in atoms.

It is the outer electrons in the atoms which get involved when atoms meet and combine or react. Atoms with the same number and arrangement of outer electrons behave in similar ways.

Metals and non-metal characteristics

Metals are elements with atoms which have a weak hold on their outer electrons. Typically metal atoms lose their outer electrons and turn into positive ions. Atoms of reactive non-metals have a strong hold on their outer electrons. They tend to gain electrons and form negative ions.

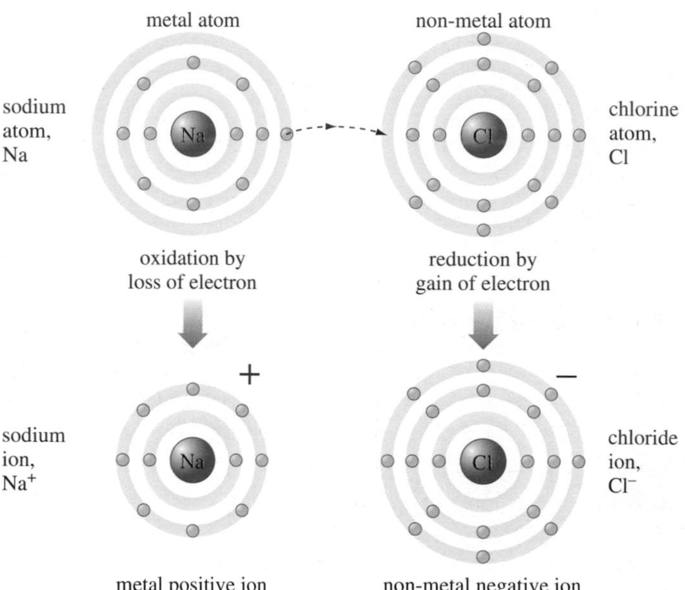

Figure 4.1.3 ►
Atoms into ions for metals and non-metals

Inorganic reactions

Inorganic chemicals take part in all the types of reaction described on pages 28–34. Explaining the changes depends on being able to interpret colour changes, gases forming, precipitates appearing and changes in pH.

Some test-tube tests depend on **ionic precipitation**. One method of distinguishing metal ions, for example, is based on the colour and solubility of metal hydroxides, many of which precipitate on adding alkalis to solutions of metal salts.

Plenty of inorganic reactions happen in water. Reactions between the molecules or ions with water often make the solution acidic or alkaline. Recognising **acid–base** reactions and their effects is also an important part of inorganic chemistry. Some inorganic compounds are destroyed in contact with water. The water molecules split them apart by **hydrolysis**.

Reactions which involve the loss or gain of electrons are **redox** reactions. Because of the importance of redox reactions in inorganic chemistry, it is quite common to organise the compounds of elements according to the degree to which they have been reduced or oxidised.

4.2 The periodic table

The periodic table helps chemists to bring order and patterns to the vast amount of information they have discovered about all the chemical elements and their compounds.

CD-ROM

Organising the chemical elements

The periodic table was a triumph for nineteenth century chemistry at time when chemists were discovering many new elements. In 1869 Dmitri Mendeléev published a version of the table on which all later versions have been based. He succeeded thanks to his insight that at that time there were still many undiscovered elements.

Mendeléev saw that when he arranged the elements in order of atomic mass, similar properties cropped up at intervals. He left gaps in his table for undiscovered elements and predicted their properties. His success with these predictions helped to persuade other scientists of the merits of his ideas when several of the missing elements were discovered in the next few years. His success was remarkable because at that time only 61 of the elements were known.

Note

The International Union of Pure and Applied Chemistry (IUPAC) now recommends that the groups should be numbered from 1 to 18. Groups 1 and 2 are the same as before. Groups 3 to 12 are the vertical families of d-block elements, the groups traditionally numbered 3 to 8 then become groups 13 to 18.

The periodic table today

Today chemists arrange the elements in order of atomic number (proton number). The horizontal rows in the table are periods. Each period ends with a noble gas. The vertical columns are groups arranged in blocks: the s-block, p-block, d-block and f-block based on the electron configurations of the elements (see pages 59–60). So the modern arrangement of elements in the table reflects the underlying electron configurations of the atoms.

The s-block elements are those in groups 1 and 2 to the left of the periodic table. For these elements the last electron added to the atomic structure goes into the s-orbital in the outer shell. All the elements in the s-block are reactive metals.

The p-block elements are those in groups 3, 4, 5, 6, 7 and 8 of the periodic table. For these elements the last electron added to the atomic structure goes into one of the three p-orbitals in the outer shell.

Definition

The **s-block elements** are the elements in groups 1 and 2 in the periodic table. For these elements the last electron added to the atomic structure goes into the s-orbital in the outer shell. All the elements in the s-block are reactive metals.

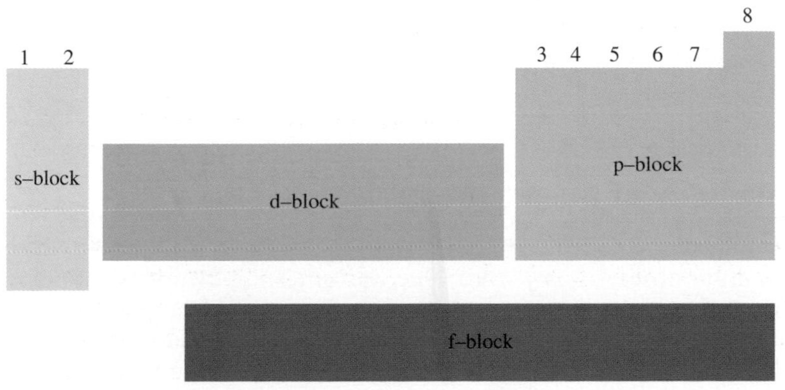

Figure 4.2.1 ◄ *Outline of a modern form of the periodic table showing the s-, p-, d- and f-blocks*

The d-block elements are those in the three horizontal rows of elements in periods 4, 5 and 6 for which the last electron added to the atomic structure goes into a d-orbital. In period 4, the d-block elements run from scandium ($1s^22s^22p^63s^23p^63d^14s^2$) to zinc ($1s^22s^22p^63s^23p^63d^{10}4s^2$).

Periodicity

When Mendeléev arranged the elements in order of atomic mass, he saw a repeating pattern. A repeating pattern is a periodic pattern. Hence the term 'periodicity'.

Perhaps the most obvious repeating pattern in the table is from metals on the left to non-metals on the right. Other patterns are revealed by plotting graphs of physical properties such as melting point against atomic number (see page 132). The formulae of simple compounds, such as chlorides, and the charges on simple ions also show periodic patterns when written into the periodic table.

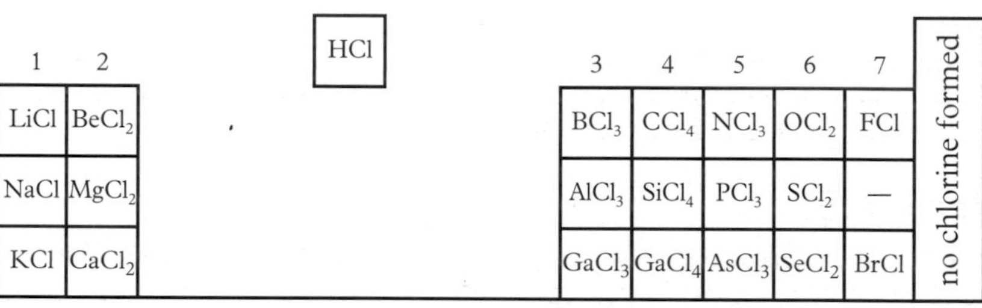

Figure 4.2.2 ▶ *Formulae of selected chlorides written into an outline periodic table*

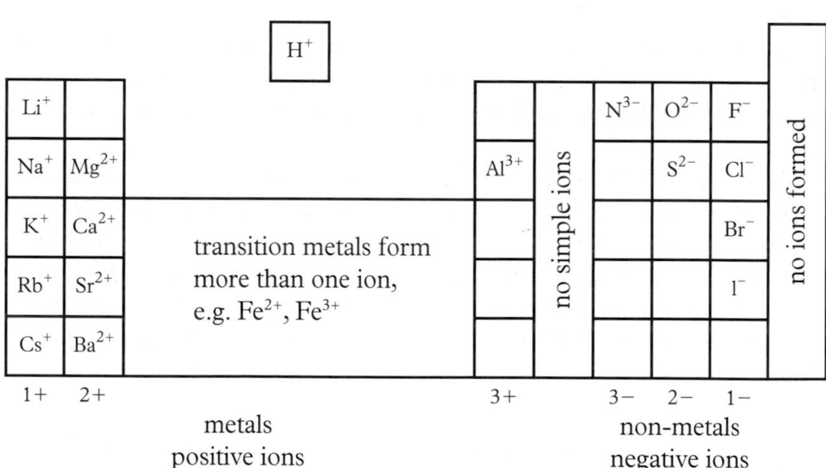

Figure 4.2.3 ▶ *Charges on selected simple ions written into the periodic table*

Periodicity of atomic properties

One of the tests of chemical theories and models is to see how well they account for observed properties. One way of testing theories of atomic structure (see pages 57–60) is to see how well they explain the periodic properties of the atoms of elements; properties such as atomic radius, ionisation energy and electron configuration.

Much of the chemistry of an atom is determined by the number and arrangement of its outer electrons. The reactivity of outer electrons is affected by the strength of the pull of the positive nucleus which depends both on the nuclear charge and the number of electrons in inner full shells.

Shielding

Shielding is the effect of electrons in inner shells which reduces the pull of the nucleus on the electrons in the outer shell of an atom. Thanks to shielding, the electrons in the outer shell are attracted by an 'effective nuclear charge' which is less than the full charge on the nucleus.

The shielding effect of an electron in an s orbital is more effective than the shielding effect of an inner electron in a p orbital. Also the effect of shielding by the inner electrons varies depending on the orbitals containing the outer electrons.

These minor differences account for some of the more subtle variations in properties.

Overall the effective nuclear charge 'felt' by an outer electron depends on:

- the charge on the nucleus
- the number and type (s, p or d) of inner shielding electrons
- the type of electron (s, p or d).

Atomic radii

Atomic radii measure the size of atoms in crystals and molecules. Chemists use X-ray diffraction and other techniques to measure the distance between the nuclei of atoms. The atomic radius of an atom cannot be defined precisely because it depends on the type of bonding and on the number of bonds.

Atomic radii for metals are calculated from the distances between atoms in metal crystals (metallic radii). The atomic radii for non-metals are calculated from the lengths of covalent bonds in crystals or molecules (covalent radii).

Atomic radii decrease from left to right across a period. Across the period Na to Ar, atomic radii fall from 0.191 nm for sodium to 0.099 nm for chlorine. From one element to the next across a period the charge on the nucleus increases by one as the number of electrons in the same outer shell increases by one. Shielding by electrons in the same shell is limited so the 'effective nuclear charge' increases and the electrons are drawn more tightly to the nucleus.

Test yourself

1 Take an outline of the periodic table.
   a) Write in the symbols of the elements mentioned in the specification you are studying. Use one colour for metals, another colour for non-metals and a third colour for 'in-between' elements.
   b) Shade in, number and name the groups of elements mentioned in the specification including the alkaline earth metals and the halogens.

2 Copy the style of Figure 4.2.2 to show the periodic patterns in the formulae of the oxides of the elements.

3 Describe in your own words the patterns in the charges of ions which you can find in Figure 4.2.3.

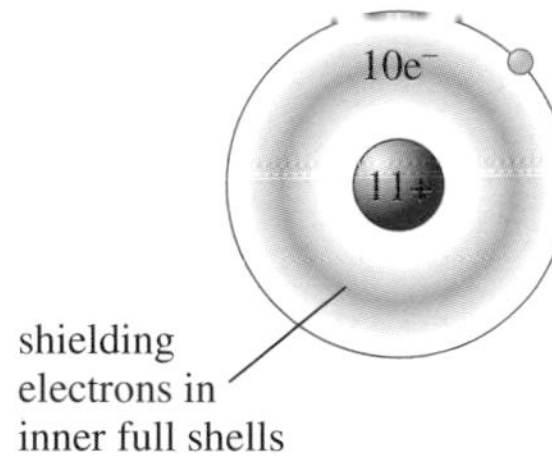

Figure 4.2.4 ▲ The shielding effect of electrons in inner shells reducing the pull of the nucleus on the electrons in the outer shell

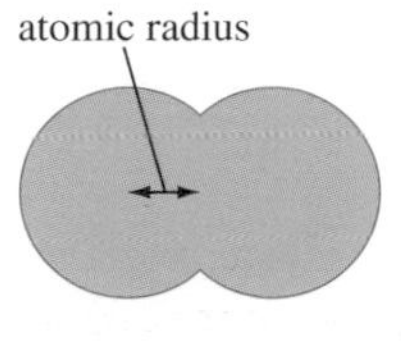

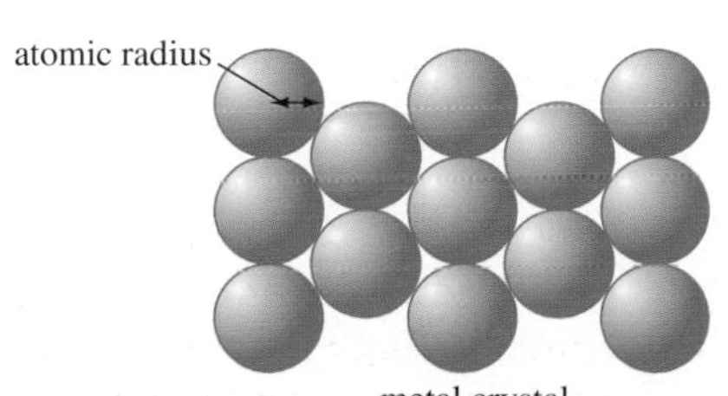

Figure 4.2.5 ◀ Atomic radii and the internuclear distances in crystals and molecules

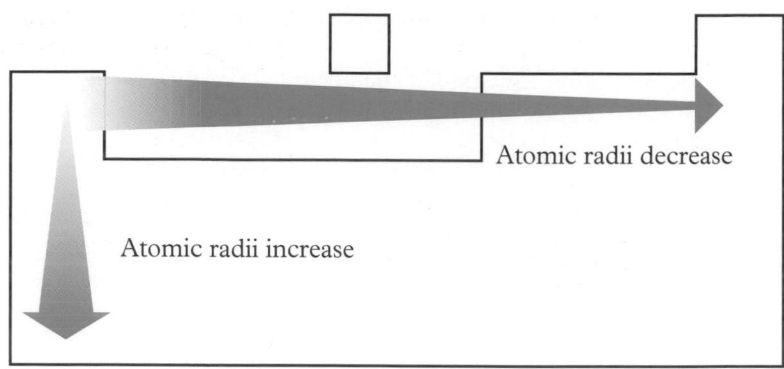

Figure 4.2.6 ▶ *Periodicity of atomic radii in the periodic table* CD-ROM

Test yourself

4 Arrange these elements in order of atomic radius: Al, B, C, K, Na.

5 Which atom or ion in each of these pairs has the larger radius?

a) Cl or Cl^-
b) Al or N

6 Write down the electron configurations for atoms of nitrogen and oxygen. Suggest an explanation for the fact that the first ionisation energy of oxygen is lower than the first ionisation energy of nitrogen.

Atomic radii increase down any group in the periodic table as the number of electron shells increases.

The overall effect of the two trends is a periodic pattern as shown in Figure 4.2.6.

Ionisation energies

There is a clear periodic trend in the first ionisation energies of the elements. The general trend is that first ionisation energies increase from left to right across a period. The nuclear charge increases across a period. The electrons are added progressively to the same outer shell so shielding by the full inner shells is roughly constant and the effective nuclear charge increases. As a result the outer electrons are more difficult to remove.

Figure 4.2.7 shows that the rising trend across a period is not smooth. There is a 2-3-3 pattern which reflects the way in which the electrons feed into s and p orbitals (see page 60).

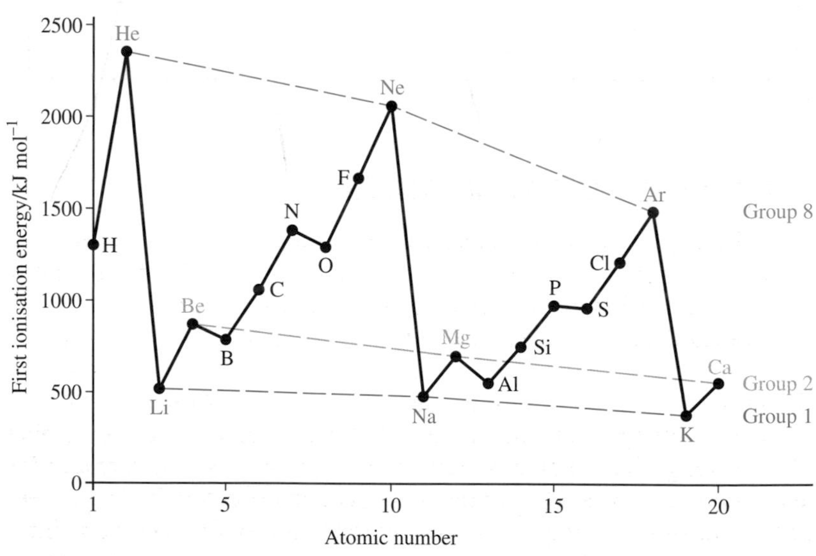

Figure 4.2.7 ▶ *Periodicity of ionisation energies* CD-ROM

Shielding accounts for the fall in the first ionisation energy for the elements down a group in the periodic table.

Electronegativity

Electronegativity measures the pull of an atom of an element on the electrons in a chemical bond. The stronger the pulling power of an atom, the higher its electronegativity (see page 81).

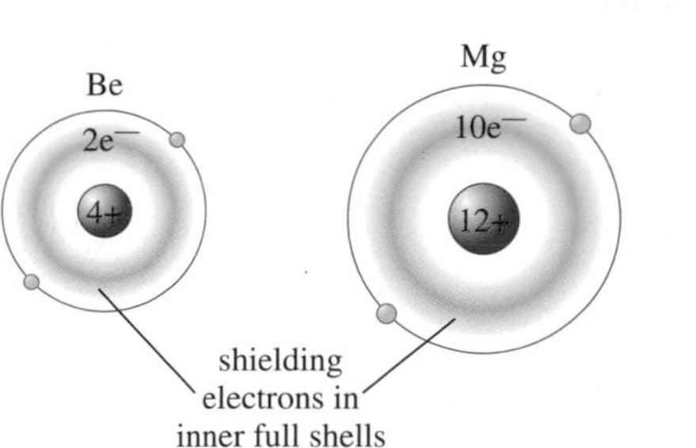

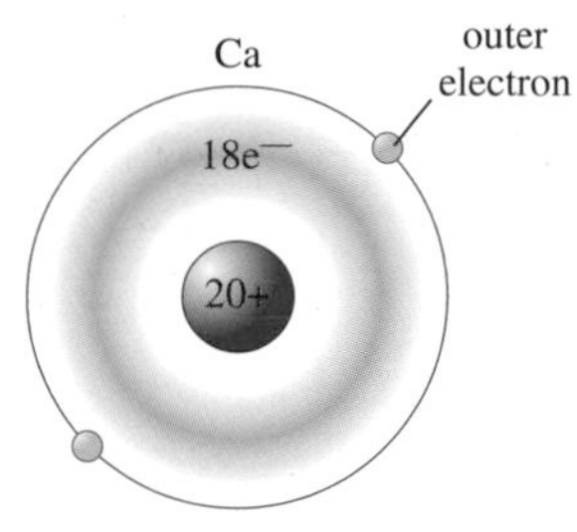

Figure 4.2.8 ◀
Shielding in the atoms of the elements in group 2 means that the 'effective nuclear charge' is 2+. Down the group the outer electron is held less strongly, being further from the same effective nuclear charge

Electronegativity increases from left to right across each period. The nuclear charge increases from one element to the next. The extra electrons to balance the increasing nuclear charge go into the same outer shell.

Electronegativity decreases down any group. Down a group the bonding electrons in the outer shell get further and further away from the same effective nuclear charge so the pull on these electrons gets less from one element to the next.

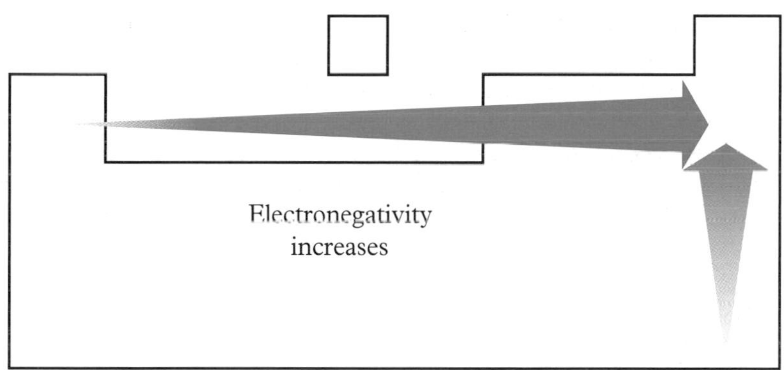

Figure 4.2.9 ◀
Periodicity of electronegativity

Periodicity, structure and bonding

Plots of the physical properties of the elements also show up repeating patterns. The change from metals on the left to non-metals on the right is reflected in the periodicity of electrical conductivity – from good conductors at the start of each period to insulators at the end of each period.

Figures 4.2.10 and 4.2.11 show the periodic patterns revealed by plotting the melting and boiling points of elements against atomic number.

The melting point of an element depends on both its structure and the type of bonding between the atoms. In a metal the bonding is strong (see page 68) but similar forces are present in the liquid so the melting point may not be very high. The more electrons each atom contributes to the shared delocalised electrons the stronger the bonding and the higher the melting point. Melting points rise from groups 1 to 2 to 3. In group 4 the elements carbon and silicon consist of covalent giant structures. These bonds are highly directional so very many of the bonds must break before the solid melts. The melting points of the Group 4 elements are at the peak of the graph.

The non-metal elements in groups 5, 6, 7 and 8 are molecular. The

Definition

Physical properties are the properties which describe how a substance behaves when it is chemically unchanged. Examples include:

- appearance (colour, transparency)
- mechanical properties (strength, hardness, ductility)
- electrical properties (electrical conductivity)
- thermal properties (melting and boiling points, thermal conductivity and enthalpies of melting and vaporisation).

Figure 4.2.10 ▶
Graph to show the periodicity of the melting point of the elements CD-ROM

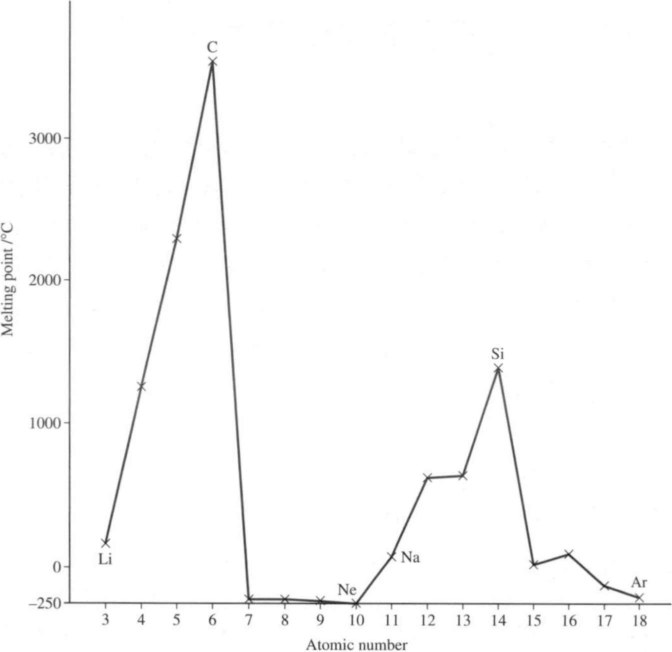

intermolecular forces between molecules are weak so these elements have low melting points.

The periodic trend for boiling points is similar to that for melting points. When liquids boil, the bonds between the particles must break completely. The boiling points of metals are typically much higher than the melting points because there is considerable metallic bonding between atoms in liquid metals. Melting group 4 elements such as carbon and silicon breaks up most of the bonding so that the boiling points are high but not so much higher than the melting points. The molecular non-metals evaporate easily so their boiling points, like their melting points, are low.

Figure 4.2.11 ▶
Graph to show the periodicity of the boiling point of the elements CD-ROM

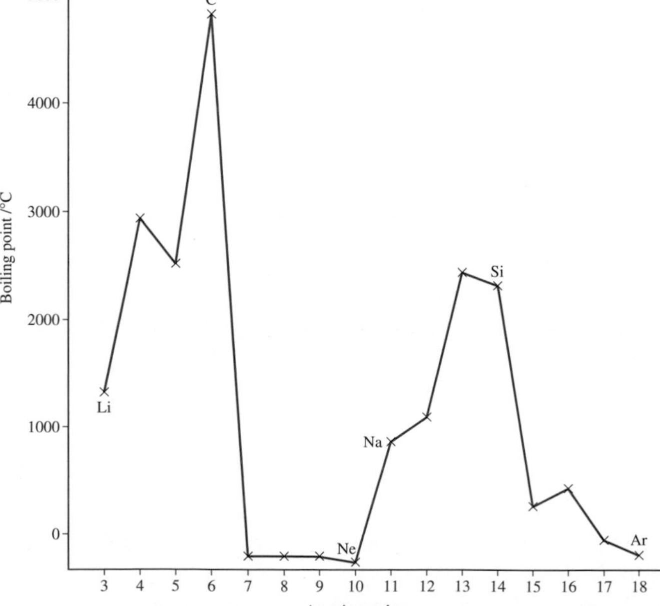

Test yourself D

7 Use a database or spreadsheet to explore the periodicity of the density of the elements helium to argon. Suggest an explanation for the periodic pattern in terms of structure and bonding.

4.3 Oxidation numbers

Chemists use oxidation numbers to keep track of the numbers of electrons transferred or shared when the atoms of elements combine forming ions or molecules. With the help of oxidation numbers it becomes much easier to recognise redox reactions. Oxidation numbers provide a useful way of organising the chemistry of elements such as chlorine which can be oxidised or reduced to varying degrees. Chemists have also chosen to base the names of inorganic compounds on oxidation numbers.

Oxidation numbers and ions

Oxidation numbers show how many electrons are gained or lost by an element when atoms turn into ions. They provide an alternative way of describing redox reactions involving electron transfer (see pages 28–30). In Figure 4.3.1 movement up the diagram involves the loss of electrons and a shift to more positive oxidation numbers – this is oxidation. Movement down the diagram involves the gain of electrons and a shift to less positive or more negative oxidation numbers. This is reduction.

The oxidation numbers of the elements are zero. In a simple ion, the oxidation number of the element is the charge on the ion.

Oxidation numbers distinguish between the compounds of elements such as iron which can exist in more than one oxidation state. In iron(II) chloride the Roman number II shows that iron is in oxidation state +2. Iron atoms lose two electrons when they react with chlorine to make iron(II) chloride.

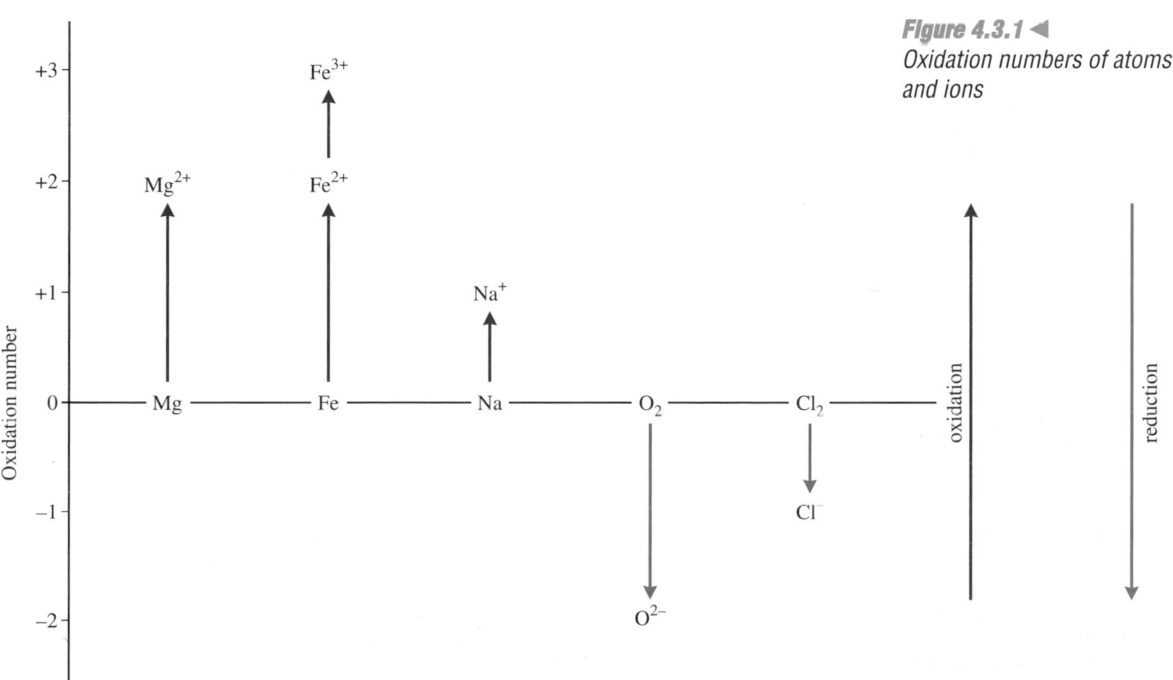

Figure 4.3.1 ◀ *Oxidation numbers of atoms and ions*

Figure 4.3.2 ►
Oxidation number rules

Oxidation number rules

1 The oxidation number of uncombined elements is zero.

2 In simple ions the oxidation number of the element is the charge on the ion.

3 The sum of the oxidation numbers in a neutral compound is zero.

4 The sum of the oxidation numbers for an ion is the charge on the ion.

5 Some elements have fixed oxidation numbers in all their compounds.

Metals		**Non-metals**	
group 1 metals (e.g. Li, Na, K)	+1	hydrogen (except in metal hydrides, H^-)	+1
group 2 metals (e.g. Mg, Ca, Ba)	+2	fluorine	−1
aluminium	+3	oxygen (except in peroxides, O_2^{2-}, and compounds with fluorine)	−2
		chlorine (except in compounds with oxygen and fluorine)	−1

Note

Note the use of 2− for the electric charge on a sulfate ion (number first for ionic charges) but the use of −2 to refer to the oxidation state of oxygen in the ion (charge first for oxidation states in ions and molecules).

NH_4^+	MnO_4^-
−3 +1	+7 −2
SO_4^{2-}	$Cr_2O_7^{2-}$
+6 −2	+6 −2

Figure 4.3.3 ▲
Oxidation numbers in ions

With the help of the rules in Figure 4.3.2 it is possible to extend the use of oxidation numbers to ions consisting of more than one atom. The charge on an ion such as the sulfate ion is the sum of the oxidation numbers of the atoms. The normal oxidation state of oxygen is −2. There are four oxygen atoms (four at −2) in the sulfate ion so the oxidation state of sulfur must be +6 to give an overall charge on the ion of 2−.

This is a helpful way of making sense of the chemistry of an element such as bromine (see Figure 4.3.4). A reaction turning bromine into BrO^- ions is oxidation. Further oxidation converts BrO^- ions to BrO_3^- ions.

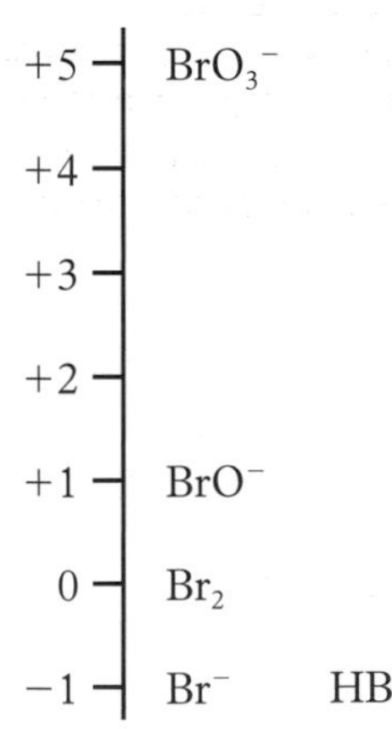

Figure 4.3.4 ▲
Oxidation numbers for bromine and some of its compounds

Oxidation numbers and molecules

The rules in Figure 4.3.2 make it possible to extend the definition of oxidation and reduction to molecules. In most molecules the oxidation state of an atom corresponds to the number of electrons from that atom which are shared in covalent bonds.

Where two atoms are linked by covalent bonds the more electronegative atom (see page 131) has the negative oxidation state. Fluorine always has a negative oxidation state of −1 because it is the most electronegative of all atoms. Oxygen normally has a negative oxidation state (−2) but it has a positive oxidation state (+1) when combined with fluorine.

The reason for writing oxidation states as +1, +2 and so on is to make quite clear that when dealing with molecules they do not refer to electric charges. Molecules are not charged and the sum of the oxidation states for all the atoms in a molecule is zero.

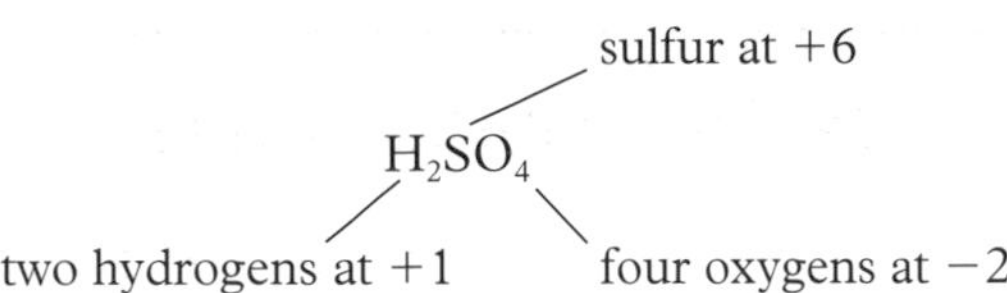

Figure 4.3.5 ◄
Oxidation states of elements in sulfuric acid

Periodicity of oxidation states

The oxidation states of elements lithium to chlorine in their oxides and hydrides reveal a periodic pattern when plotted against proton number (Figure 4.3.6). The most positive oxidation state for each element corresponds to the number of electrons in the outer shell of the atoms.

Typically the elements in the same group of the periodic table show a similar pattern in their oxidation states. In all the group 4 elements, for example, the main oxidation states are −4, +2 and +4, but the relative importance of these states varies down the group from carbon to lead.

Oxidation states and names of compounds

Names of inorganic compounds are becoming increasingly systematic but chemists still use a mixture of names. Most chemists prefer to call $CuSO_4.5H_2O$ hydrated copper(II) sulfate, or perhaps copper(II) sulfate-5-water but not the fully systematic name tetraaquocopper(II) tetraoxosulfate(VI)-1-water. The systematic name has much more to say about the arrangement of atoms, molecules and ions in the blue crystals but it is too cumbersome for normal use. The systematic name also shows the oxidation states of copper and sulfur in the compound.

These are some of the basic rules for common inorganic names:

- the ending '-ide' shows that a compound contains just the two elements mentioned in the name. The more electronegative element comes second, for example, sodium sulfide, Na_2S, carbon dioxide, CO_2, and phosphorus trichloride, PCl_3

Test yourself

1 What is the oxidation number of:

a) aluminium in Al_2O_3?
b) nitrogen in magnesium nitride, Mg_3N_2?
c) barium in barium nitrate, $Ba(NO_3)_2$?
d) nitrogen in the ammonium ion, NH_4^+?
e) phosphorus in PCl_5?

2 Are these elements oxidised or reduced in these conversions?

a) calcium to calcium bromide
b) chlorine to lithium chloride
c) chlorine to chlorine dioxide
d) sulfur to hydrogen sulfide
e) sulfur to sulfuric acid.

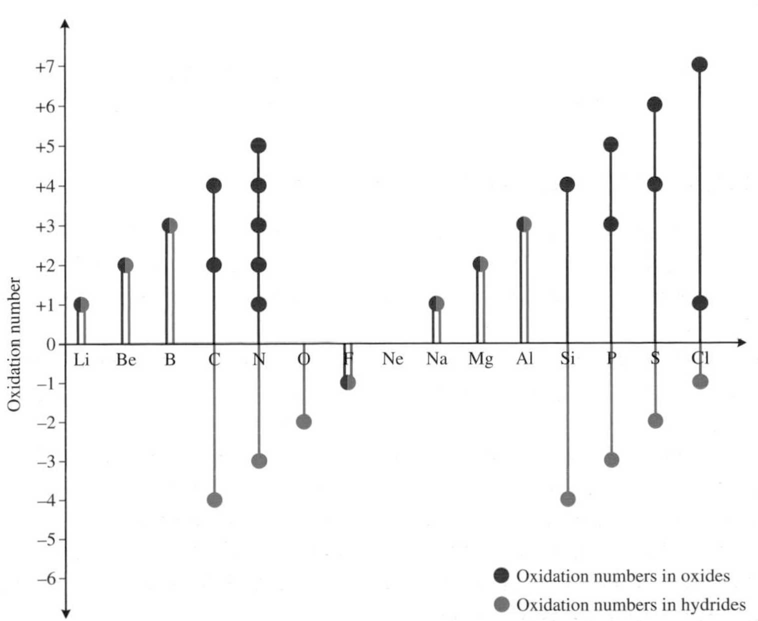

Figure 4.3.6 ◄
Oxidation numbers of elements in their oxides and hydrides. Note that p-block elements form oxides in a variety of oxidation states

Inorganic Chemistry
Section four

- the Roman numerals in names are the oxidation numbers of the elements, for example iron(II) sulfate, $FeSO_4$, and iron(III) sulfate, $Fe_2(SO_4)_3$
- the traditional names of oxoacids end '-ic' or '-ous' as in sulfuric (H_2SO_4) and sulfurous (H_2SO_3) acids and nitric (HNO_3) and nitrous (HNO_2) acids; where the '-ic' ending is for the acid in which the central atom has the higher oxidation number
- the corresponding traditional endings for the salts of oxoacids are '-ate' and '-ite' as in sulfate, SO_4^{2-}, and sulfite, SO_3^{2-}, and in nitrate, NO_3^-, and nitrite, NO_2^-.
- the more systematic names for oxoacids and oxo salts use oxidation numbers as in sulfate(VI) for sulfate, SO_4^{2-}, and sulfate(IV) for sulfite, SO_3^{2-}.

When in doubt chemists give the name and the formula. If necessary they may give two names: the systematic name and the traditional name.

Test yourself

3 Write the formulae of the compounds:

a) tin(II) oxide
b) tin(IV) oxide
c) sodium chlorate(III)
d) iron(III) nitrate(V)
e) potassium chromate(VI).

Balancing redox equations

Redox equations, like other balanced equations, show the amounts (in moles) of reactants and products involved in redox reactions. Oxidation numbers help to balance redox equations because the total decrease in oxidation number for the element reduced must equal the total increase in oxidation number for the element oxidised. This is illustrated here with the oxidation of hydrogen bromide by concentrated sulfuric acid. The main products are bromine, sulfur dioxide and water.

Step 1: Write down the formulae for the atoms, molecules and ions involved in the reaction

$$HBr + H_2SO_4 \rightarrow Br_2 + SO_2 + H_2O$$

Step 2: Identify the elements which change in oxidation number and the extent of change.

In this example only bromine and sulfur show changes of oxidation state.

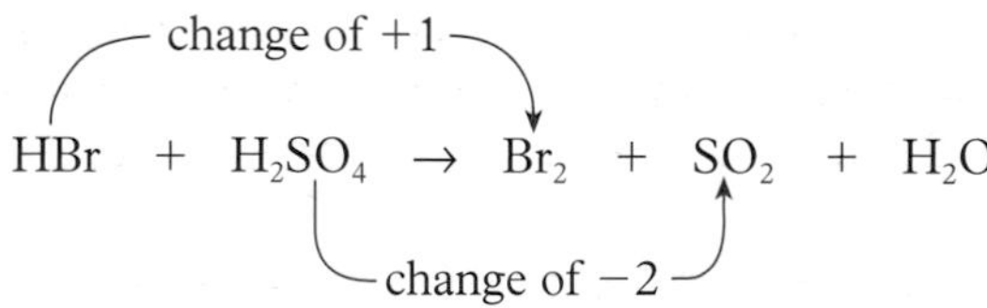

Step 3: Balance so that the total increase in oxidation number of one element equals the total decrease of the other element.

In this example the increase of +1 in the oxidation number of two bromine atoms balances the −2 decrease of one sulfur atom.

$$2HBr + H_2SO_4 \rightarrow Br_2 + SO_2 + H_2O$$

Step 4: Balance for oxygen and hydrogen.

In this example the four hydrogen atoms on the left of the equation join with the two remaining oxygen atoms to form water molecules.

$$2HBr + H_2SO_4 \rightarrow Br_2 + SO_2 + 2H_2O$$

Step 5: Add state symbols.

$$2HBr(g) + H_2SO_4(l) \rightarrow Br_2(l) + SO_2(g) + 2H_2O(l)$$

Test yourself

4 Write balanced equations for these redox reactions. State which element is oxidised and which is reduced in each example.

a) Fe with Br_2 to give $FeBr_3$
b) F_2 with H_2O to give HF and O_2
c) IO_3^- and H^+ with I^- to give I_2 and H_2O
d) $S_2O_3^{2-}$ and I_2 to give $S_4O_6^{2-}$ and I^-
e) Cl_2 with OH^- to give Cl^-, ClO^- and H_2O

4.4 Group 1

The group 1 elements are better known as the alkali metals. They have similar chemical properties because they all have one electron in an outer s orbital. These elements are more similar to each other than the elements in any other group. Even so, because of the increasing number of full, inner shells, there are trends in properties down the group from lithium to caesium. The element in period 7, francium, is very rare and all its isotopes are radioactive. CD-ROM

Inorganic Chemistry

Section four

The elements

The metals are soft and easily cut with a knife. They are shiny when freshly cut but quickly dull in air as they react with moisture and oxygen.

Lithium

- is a soft, shiny metal which turns dark grey in air
- is stored in oil
- floats on water and reacts, but quite slowly, forming hydrogen and LiOH which is soluble and strongly alkaline
- burns in air with a coloured flame (bright red) forming an oxide, Li_2O
- forms an ionic, colourless, crystalline chloride.

Figure 4.4.1 ▲
Lithium

Sodium

Sodium is a powerful reducing agent used for titanium extraction and the extraction of some other metals such as zirconium. Molten sodium is also the fluid which circulates through heat exchangers to transfer energy and raise steam in some nuclear power stations and other processes. Sodium is used in street lights. It has the following characteristics:

- soft, shiny metal which rapidly tarnishes in moist air
- stored in oil
- floats on water, melts and reacts violently forming hydrogen which catches fire and NaOH which is soluble and strongly alkaline
- produces a mixture of the oxide, Na_2O, and peroxide, Na_2O_2, when it burns in air
- forms an ionic, colourless, crystalline chloride Na^+Cl^-.

Figure 4.4.2 ▲
Sodium

Potassium

One of the main uses of potassium is to make the superoxide, KO_2, for use in emergency breathing apparatus (see page 140).

Potassium, as potassium ions, is an essential nutrient for plants and an ingredient of NPK fertilisers. Large, underground deposits of potassium chloride are mined as the mineral sylvinite just south of Teesside in the UK. Potassium has the following characteristics:

- very soft, shiny metal which rapidly tarnishes in moist air is stored in oil
- floats on water, melts and reacts violently forming hydrogen which catches fire and KOH which is soluble and strongly alkaline
- produces a superoxide when it burns in air, KO_2
- forms an ionic, colourless, crystalline chloride K^+Cl^-.

Figure 4.4.3 ▲
Potassium

lithium, Li	$[He]2s^1$
sodium, Na	$[Ne]3s^1$
potassium, K	$[Ar]4s^1$
rubidium, Rb	$[Kr]5s^1$
caesium, Cs	$[Xe]6s^1$

Figure 4.4.4 ▲ *The shortened forms of the electron configurations of group 1 metals (see page 60)*

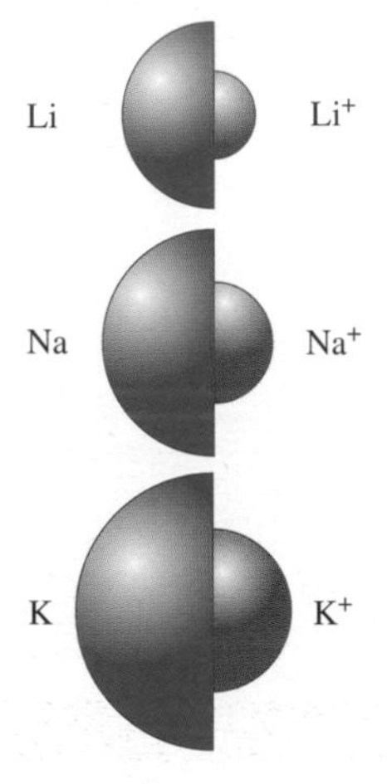

Figure 4.4.5 ▲ *Relative sizes of the atoms and ions of group 1 elements*

Atomic and ionic radii

Atomic radii and ionic radii increase down the group as the number of inner full shells of electrons increases. For each element the 1+ ion is smaller than the atom because of the loss of the outer shell of electrons. The tendency to react and form ions increases down the group.

Ionisation energies

The first ionisation energies decrease down the group as the increasing number of full shells means that the outer electrons get further away from the same effective nuclear charge (see page 129).

The atoms change in two ways down the group: the charge on the nucleus increases and the number of inner full shells also increases. The shielding effect of the inner electrons means that the effective nuclear charge attracting the outer electron is 1+. Down the group the outer electrons get further and further away from the same effective nuclear charge and so they are held less strongly and the ionisation energies decrease.

To remove a second electron to form a 2+ ion takes much more energy because the second electron has to be removed against the attraction of a much larger effective nuclear charge.

Figure 4.4.6 ► *Diagrams to represent the electron configurations of lithium and sodium*

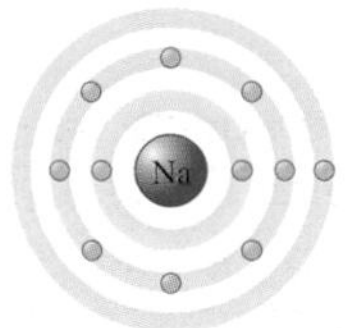

Flame colours

Flame tests help to detect some metal ions in salts. They are particularly useful in qualitative analysis to distinguish group 1 metal ions which are otherwise very similar.

Ionic compounds such as sodium chloride do not burn during a flame test. The energy from the flame excites electrons in the sodium atoms raising them to higher energy levels. The atoms then emit the characteristic yellow light as the electrons drop back to lower energy levels (see pages 56–58).

Metal ion	Colour
lithium	bright red
sodium	bright yellow
potassium	pale mauve

Figure 4.4.8 ▲ *Flame colours of group 1 metal compounds*

Figure 4.4.7 ▼ *Procedure for a flame test. Chlorides evaporate more easily and so colour flames more strongly than less volatile compounds. Concentrated hydrochloric acid converts involatile compounds such as carbonates to chlorides*

CD-ROM

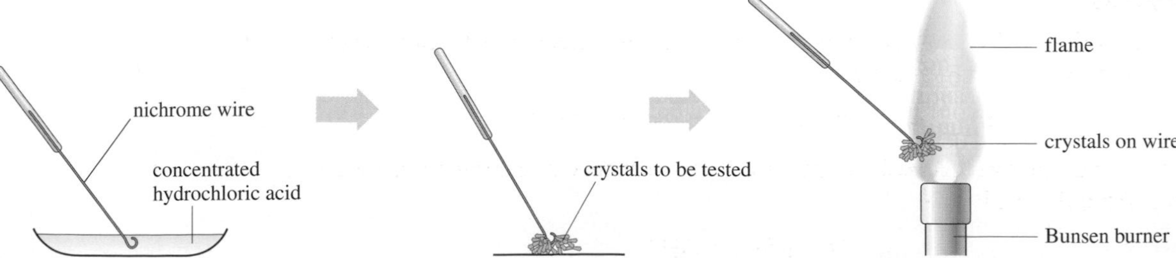

Oxidation states

When the atoms of alkali metals react, they lose their single s-electron from the outer shell turning into ions with a single positive charge: Li^+, Na^+, K^+ and so on. So the only oxidation state is +1 in the compounds of these metals.

The ions of alkali metals are colourless in crystals and in aqueous solution. Sodium or potassium compounds are only colourful as a result of the properties of the negative ions. Potassium chromate(VI) is yellow because CrO_4^- ions are yellow.

Reactions of the elements

The alkali metals are powerful reducing agents. They react by losing the outer s-electron to form M^+ ions. The ions of alkali metals are very unreactive. In aqueous solution they are often spectator ions (see page 34) taking no part in the chemical changes. This makes sodium and potassium compounds very useful as chemical reagents because the metal ions can generally be ignored.

Reaction with oxygen

All the metals burn brightly in oxygen on heating to form ionic oxides. Lithium forms a simple oxide, Li_2O.

$$4Li(s) + O_2(g) \rightarrow 2Li_2O(s)$$

Other products are possible with the other members of the group including peroxides containing the O_2^{2-} ion and superoxides with the O^2 ion.

Sodium forms mainly the peroxide, Na_2O_2, with some simple oxide, Na_2O, whilst the main product with potassium is the superoxide KO_2. The tendency to form a superoxide increases down the group as the size of the metal ion increases.

Reaction with water

All the metals react with water to form hydroxides, MOH, and hydrogen. The rate and violence of the reaction increases down the group. Lithium reacts steadily with cold water. Caesium reacts explosively.

Reaction with chlorine

All the metals react vigorously with chlorine to form colourless, ionic chlorides, M^+Cl^-. The chlorides are soluble in water. The crystal structures depend on the size of the metal ion (see page 73).

Properties of the compounds

Oxides

The oxides are basic. A basic oxide is an oxide of a metal which reacts with an acid to form a salt and water.

$$Li_2O(s) + 2HCl(aq) \rightarrow 2LiCl(aq) + H_2O(l)$$

It is the oxide ion in a basic oxide which acts as a base by taking a hydrogen ion from the acid.

$$O^{2-} + H^+ \rightarrow OH^-$$

Test yourself

1. Write out the full electron configurations of Li, Na and K atoms showing the numbers of s, p and d electrons in each shell (see page 60).
2. What is the electron configuration of a sodium ion?
3. Draw up a table to show both the ways in which the group 1 elements are the same and the ways in which their properties change down the group.

Note

The bonding is ionic in all the common compounds of group 1 metals. Do not draw lines to represent the bonding between alkali metal atoms and the atoms of other elements. Only use lines to represent covalent bonds. Represent sodium chloride as Na^+Cl^- and never as Na—Cl.

Basic oxides which dissolve in water are alkalis and this includes the simple oxides of group 1 metals. As the compound dissolves, the oxide ion, acting as a base, takes a hydrogen ion from water forming hydroxide ions.

$$2Na_2O(s) + H_2O(l) \rightarrow 2NaOH(aq)$$

The superoxide of potassium is an important ingredient of emergency breathing apparatus. The oxide removes carbon dioxide from moist exhaled air and replaces it with oxygen.

$$4KO_2(s) + 4CO_2(g) + 2H_2O(l) \rightarrow 4KHCO_3(s) + 3O_2(g)$$

Hydroxides

The hydroxides are all white solids commonly supplied as pellets or flakes.

The hydroxides of group 1 metals are:

- similar in that they all have the formula MOH, and are soluble in water forming alkaline solutions – they are strong bases
- different in that their solubility increases down the group.

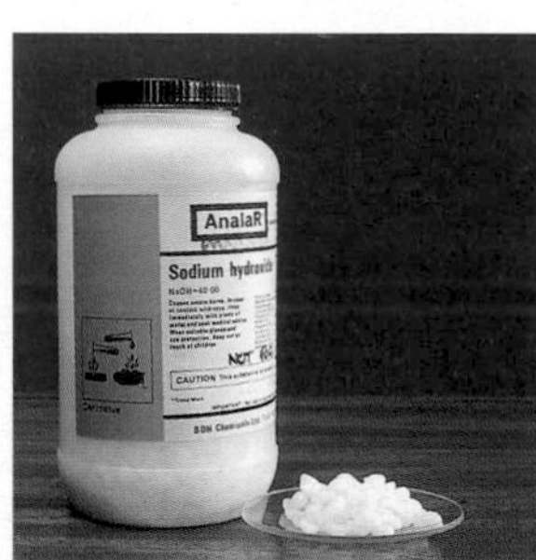

Figure 4.4.9 ▲
Sodium hydroxide, NaOH, is deliquescent which means that it picks up water from moist air and then dissolves in it. Sodium hydroxide is a strong base; it dissolves in water to form a highly alkaline solution. The traditional name for the alkali is caustic soda. Sodium hydroxide is highly corrosive and more hazardous to the skin and eyes than many acids

Carbonates

The carbonates are similar in that they all have the formula M_2CO_3 and, with the exception of lithium, do not decompose on heating. The carbonates of sodium and potassium are white powders which dissolve in water forming alkaline solutions because the carbonate ion is a base (see page 111). Carbonate ions react with water to form hydroxide ions which make the solution alkaline.

$$CO_3^{2-}(aq) + H_2O(l) \rightarrow HCO_3^-(aq) + OH^-(aq)$$

The carbonate acts as a base taking hydrogen ions from water molecules.

Nitrates

The nitrates are:

- similar in that they all have the formula MNO_3, are colourless crystalline solids, are very soluble in water and decompose on heating,
- different becoming more difficult to decompose down the group.

Lithium nitrate, like magnesium nitrate, decomposes on heating to the oxide, nitrogen dioxide and oxygen. The nitrates of sodium and potassium need strong heating to decompose and they form the nitrite:

$$2KNO_3(s) \rightarrow 2KNO_2(s) + O_2(g)$$

Note

It is the compounds of alkali metals which are alkaline not the elements themselves. It is the OH^- ions which make sodium hydroxide alkaline, not the sodium ions. If sodium ions were the cause of alkalinity then all sodium compounds would be alkaline including common salt, NaCl.

Test yourself

4 Write balanced equations for the reaction of:

a) sodium with water
b) potassium with chlorine
c) lithium oxide with water
d) sodium oxide with hydrochloric acid
e) potassium hydroxide with sulfuric acid.

4.5 Group 2

Group 2 elements belong to the family of alkaline earth metals. Many of the compounds of these elements occur as minerals in rocks, hence the name 'earth metals'. Chalk, marble and limestone are forms of calcium carbonate. Dolomite consists of a mixture of calcium and magnesium carbonates. Fluorspar is a form of calcium fluoride which is mined as the ornamental mineral Blue John in Derbyshire caves. These group 2 compounds are insoluble, unlike the equivalent group 1 compounds, so they do not dissolve in rain water. CD-ROM

The elements

The metals are harder and denser than group 1 metals and they have higher melting points. In air the surface of the metals is covered with a layer of oxide.

The first member of the group, beryllium, Be, is a strong metal with a high melting point but its density is much less than the density of iron. The element makes useful alloys with other metals.

The source of magnesium metal is electrolysis of molten magnesium chloride obtained either from sea water or from salt deposits. The low density of the metal helps to make light alloys, especially with aluminium. These alloys, which are strong for their weight, are especially valuable for cars and aircraft.

Barium is a soft, silvery-white metal. It is so reactive with air and moisture that it is generally stored under oil like the alkali metals.

Atomic and ionic radii

Atomic and ionic radii increase down the group. For each element the 2+ ion is smaller than the atom because of the loss of the outer shell of electrons. The tendency to react and form ions increases down the group.

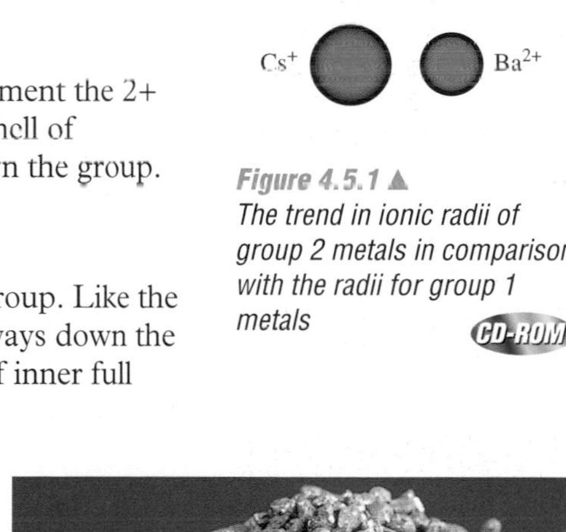

Figure 4.5.1 ▲ *The trend in ionic radii of group 2 metals in comparison with the radii for group 1 metals* CD-ROM

Ionisation energies

The first and second ionisation energies decrease down the group. Like the atoms of the alkali metals, the group 2 atoms change in two ways down the group: the charge on the nucleus increases and the number of inner full shells also increases.

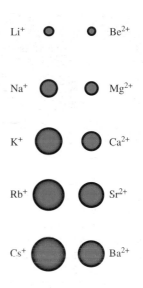

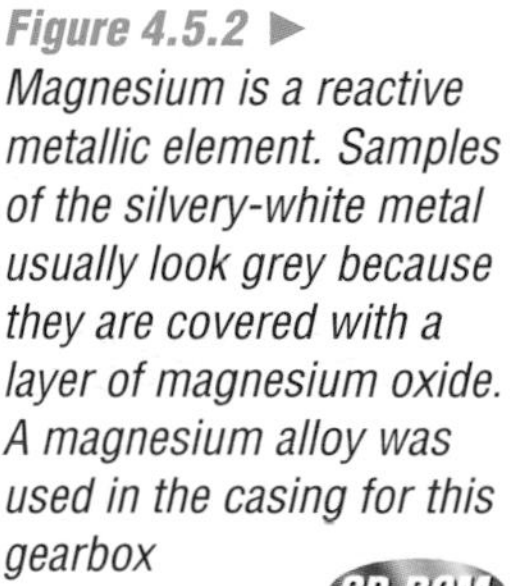

Figure 4.5.2 ► *Magnesium is a reactive metallic element. Samples of the silvery-white metal usually look grey because they are covered with a layer of magnesium oxide. A magnesium alloy was used in the casing for this gearbox* CD-ROM

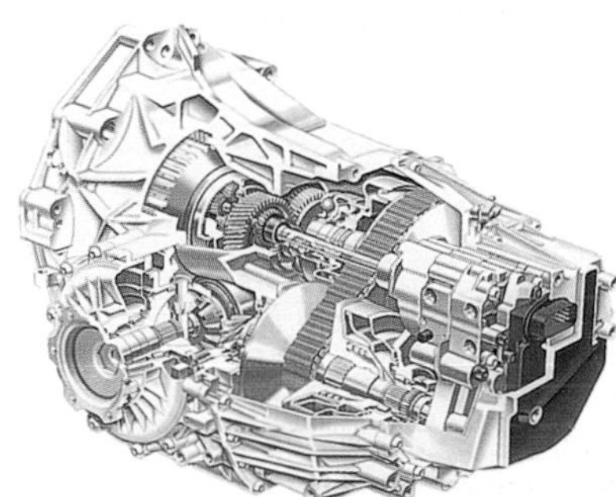

Figure 4.5.3 ▲ *Samples of the silvery metal calcium usually look grey because they are covered with a layer of calcium oxide*

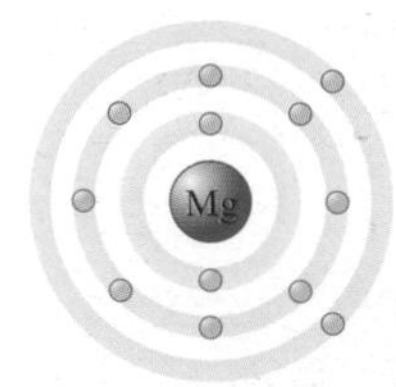

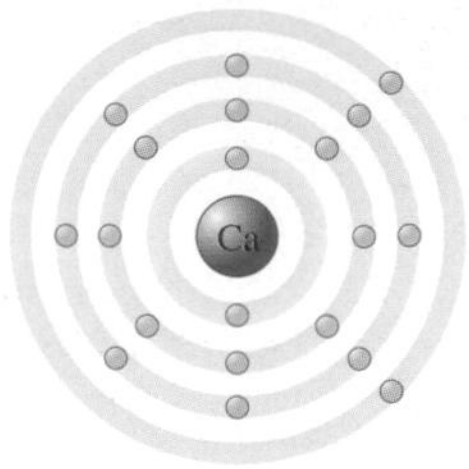

Figure 4.5.4 ▲
Diagrams to represent the electron configurations of magnesium and calcium atoms

The shielding effect of the inner electrons means that the effective nuclear charge attracting the outer electron is 2+. Down the group the outer s-electrons get further and further away from the same effective nuclear charge and so they are held less strongly and the ionisation energies decrease. This trend helps to account for the increasing reactivity of the elements down the group.

To remove a third electron to form a 3+ ion takes much more energy because the third electron has to be removed against the attraction of a much larger effective nuclear charge. This means that it is never energetically favourable for the metals to form M^{3+} ions.

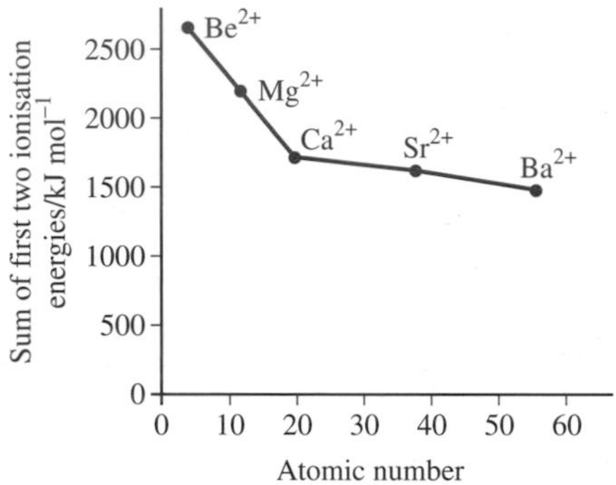

Figure 4.5.5 ▲
Plot to show the trend in the sum of the first two ionisation energies of group 2 metals: $M(g) \rightarrow M^{2+}(g) + 2e^-$

Test yourself

1 Write the full electron configurations of the atoms and ions of Mg, Ca and Ba showing the numbers of s-, p- and d-electrons (see page 60).

beryllium, Be	$[He]2s^2$
magnesium, Mg	$[Ne]3s^2$
calcium, Ca	$[Ar]4s^2$
strontium, Sr	$[Kr]5s^2$
barium, Ba	$[Xe]6s^2$

Figure 4.5.6 ▲
The shortened forms of the electron configurations of group 2 metals (see page 60)

Oxidation states

The group 2 metals all have similar chemical properties because they all have two electrons in an outer s orbital. When the metal atoms react to form ions, they lose the two outer electrons giving ions with a 2+ charge: Mg^{2+}, Ca^{2+}, Sr^{2+} and Ba^{2+}. So these elements exist in the +2 oxidation state in all their compounds.

Flame colours

Flame tests (see page 138) can help to identify compounds of calcium, strontium and barium. Beryllium and magnesium compounds do not colour a flame.

Metal ion	Colour
calcium	brick red
strontium	bright red
barium	pale green

Figure 4.5.7 ▲
Flame colours of group 2 metal compounds CD-ROM

Reactions of the elements

The metals are reducing agents. Apart from beryllium, they react by losing their two s-electrons to form M^{2+} ions (where M can represent Mg^{2+}, Ca^{2+}, Sr^{2+} or Ba^{2+}).

$$M \rightarrow M^{2+} + 2e^-$$

Beryllium is not a typical member of group 2. Beryllium chemistry is in many ways more like the chemistry of aluminium than of magnesium. This is because of the small size of the Be^{2+} ion which gives it the same tendency to polarise neighbouring negative ions as the Al^{3+} ion (see page 82). As a result, the bonding in beryllium compounds is polar covalent rather than ionic.

Reaction with oxygen

In the air, beryllium metal is coated with an inert layer of oxide which prevents further reaction unless the metal is heated to a high temperature.

The other group 2 metals burn brightly in oxygen on heating to form white, ionic oxides, $M^{2+}O^{2-}$.

Magnesium burns very brightly in air with an intense white flame forming the white solid magnesium oxide, MgO. For this reason magnesium powder is an ingredient of fireworks and flares.

Calcium also burns brightly in air but with a brilliant red flame forming the white solid calcium oxide, CaO.

Barium burns in excess air or oxygen to form a peroxide, BaO_2, which contains the peroxide ion O_2^{2-}.

Reaction with water and dilute acids

The oxide layer on beryllium prevents reaction with water and dilute acids but all the other metals in the group react. These reactions are not as vigorous as the reactions of the group 1 metals but, as in group 1, the rate of reaction increases down the group.

Magnesium reacts very slowly with cold water but much more rapidly on heating in steam.

$$Mg(s) + H_2O(g) \rightarrow MgO(s) + H_2(g)$$

Calcium reacts with cold water producing hydrogen and a white precipitate of calcium hydroxide.

$$Ca(s) + 2H_2O(l) \rightarrow Ca(OH)_2(s) + H_2(g)$$

Both metals dissolve rapidly in dilute acids such as hydrochloric acid.

$$Mg(s) + 2HCl(g) \rightarrow MgCl_2(s) + H_2(g)$$

Barium reacts even faster with cold water and dilute acids.

Reaction with chlorine

All the metals, including beryllium, react with chlorine on heating to form colourless chlorides, MCl_2.

$$Mg(s) + Cl_2(g) \rightarrow MgCl_2(s)$$

Properties of the compounds

The oxides

Apart from beryllium oxide, the oxides are basic. They react with acids forming salts.

$$BaO(s) + 2HNO_3(aq) \rightarrow Ba(NO_3)_2(aq) + H_2O(l)$$

Magnesium oxide is a white solid made by heating magnesium carbonate. In water it turns to magnesium hydroxide which is slightly soluble. Magnesium oxide has a high melting point and is used as a refractory ceramic to line furnaces.

Calcium oxide is a white solid made by heating calcium carbonate. It is a basic oxide. Calcium oxide reacts very vigorously with cold water, hence its traditional name 'quicklime'. The product is calcium hydroxide.

Beryllium oxide also has a high melting point and is a useful refractory. Chemically it is pretty inert and it does not react with water or dilute acids. Under extreme conditions it can be shown to be an amphoteric oxide.

Figure 4.5.8 ▲
Ornamental object made from Derbyshire Blue John which consists of calcium fluoride, CaF_2

Test yourself D

2 Write balanced equations for the reactions of:

a) barium with oxygen
b) calcium with dilute hydrochloric acid
c) strontium with chlorine
d) barium with water.

Definitions

A **basic oxide** is an oxide of a metal which reacts with an acid to form a salt and water. It is the oxide ion which acts as a base by taking a hydrogen ion from the acid. Basic oxides which dissolve in water are **alkalis**.

An **amphoteric oxide** can react both like a basic oxide and an acidic oxide. Many amphoteric oxides are, however, inert to aqueous reagents. It is usually easier to study amphoteric behaviour with the hydroxide.

Figure 4.5.9 ▲
Beryllium occurs naturally as beryl, an aluminosilicate mineral, which is colourless if pure but a brilliant green in emerald where chromium(III) ions replace some of the aluminium(III) ions

The hydroxides

Beryllium hydroxide is insoluble in water. Like the oxide it is amphoteric and this is shown by the fact that it will dissolve both in acids and in alkalis.

The hydroxides of the other elements in the group are:

- similar in that they all have the formula $M(OH)_2$, and are to some degree soluble in water forming alkaline solutions
- different in that the solubility increases down the group.

Magnesium hydroxide is the active ingredient in tablets of the antacid milk of magnesia. It is only very slightly soluble in water.

Calcium hydroxide, $Ca(OH)_2$, is only sparingly soluble in water forming an alkaline solution often called lime water.

The lime water test for carbon dioxide works because a solution of calcium hydroxide absorbs the gas forming a white, insoluble precipitate of calcium carbonate.

Test yourself D

3 Write balanced equations for the reactions of:
 a) magnesium oxide with dilute hydrochloric acid
 b) calcium oxide with water
 c) lime water with carbon dioxide.

4 With the help of a table of data show that the solubilities of the group 2 metal hydroxides increase down the group when measured in mol per 100 g water.

The chlorides

Like the group 1 chlorides, the chlorides of the elements magnesium to barium are ionic and they are soluble in water. These chlorides are usually hydrated. Calcium chloride, for example, crystallises from solution as a hydrate, $CaCl_2.6H_2O$.

Anhydrous calcium chloride, $CaCl_2$, is a cheap drying agent.

The bonding in beryllium chloride is covalent because the chloride ion is polarised by the very small, doubly-charged beryllium ions. Beryllium chloride vapour consists of $BeCl_2$ molecules which are linear (see page 79). The beryllium atom in $BeCl_2$ can accept two electron pairs. Solid beryllium chloride has an extended chain-like structure with chlorine atoms forming dative covalent bonds with beryllium atoms.

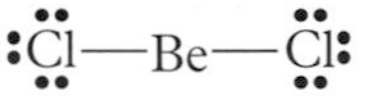

:Cl—Be—Cl:

$BeCl_2(g)$
molecules

Cl Cl Cl Cl Cl Cl
Be Be Be Be Be
Cl Cl Cl Cl Cl Cl

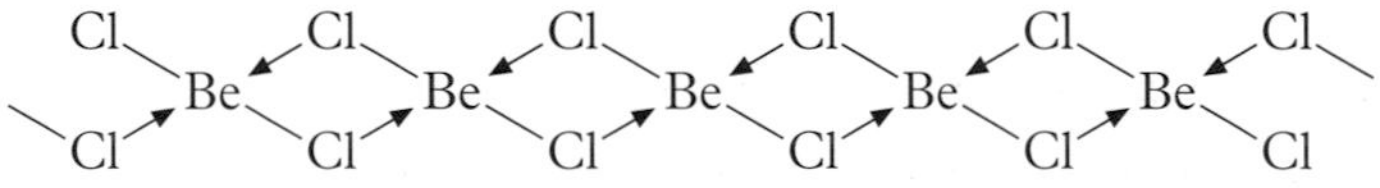

$BeCl_2(s)$

Figure 4.5.11 ►
The structure of beryllium chloride CD-ROM

The carbonates

The carbonates of group 2 metals (Mg to Ba) are:

- similar in that they all have the formula MCO_3, are insoluble in water, react with dilute acids and decompose on heating to give the oxide and carbon dioxide:

$$CaCO_3(s) \rightarrow CaO(s) + CO_2(g)$$

- different in that they become more difficult to decompose down the group (they become more thermally stable).

Calcium carbonate occurs naturally as limestone, chalk and marble. Limestone is an important mineral. Some of the rock is quarried for road building and construction.

Figure 4.5.10 ▲
Barium occurs naturally as barytes, $BaSO_4$, (above) and also as witherite, $BaCO_3$

Pure limestone is also used in the chemical industry. Heating limestone in a furnace at 1200 K converts it to calcium oxide (quicklime). The reaction of quicklime with water produces calcium hydroxide (slaked lime).

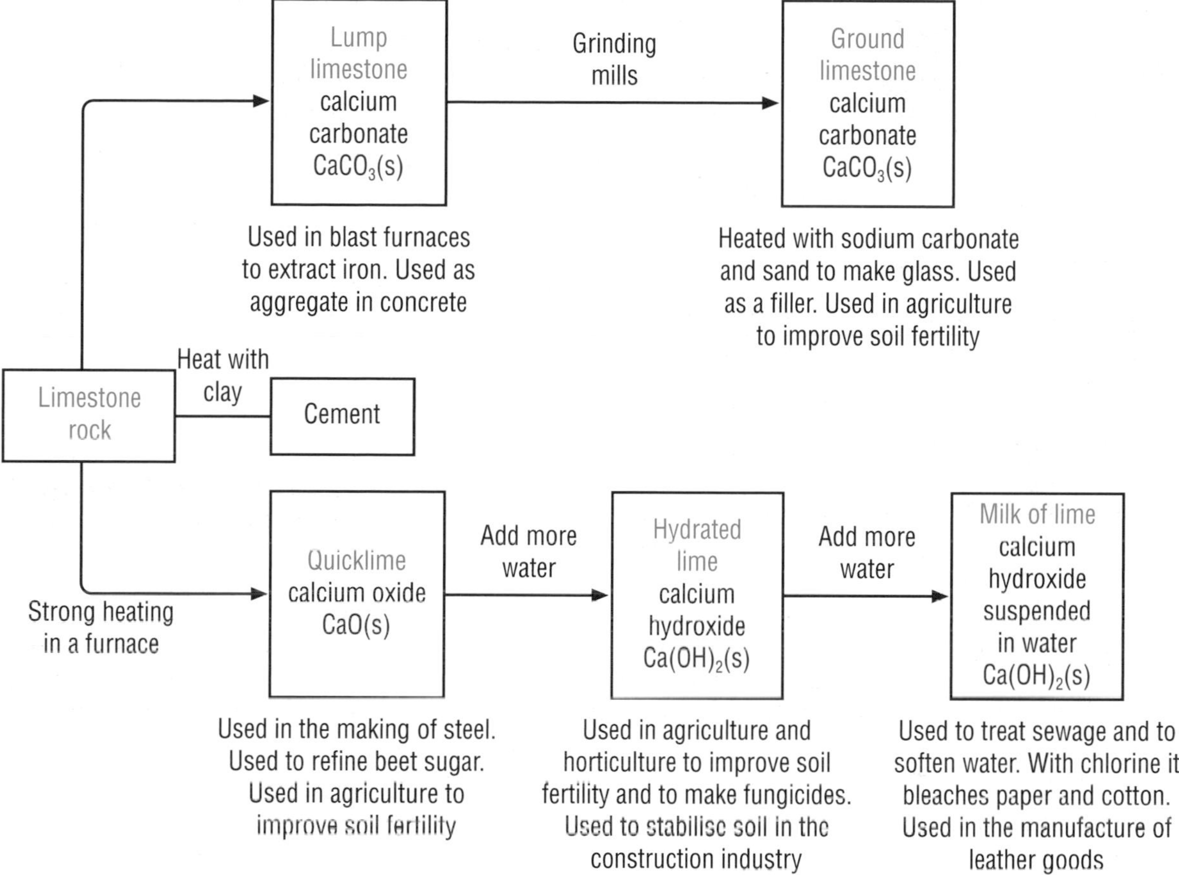

Figure 4.5.12 ▼ *Products from limestone and their uses*

The nitrates

The nitrates of group 2 metals (Mg to Ba) are:

- similar in that they all have the formula $M(NO_3)_2$, are colourless crystalline solids, are very soluble in water and decompose to the oxide on heating:

$$2Mg(NO_3)_2 \rightarrow 2MgO(s) + 4NO_2(g) + O_2(g)$$

- different in that they become more difficult to decompose down the group.

The sulfates

The sulfates are:

- similar in that they are all colourless solids with the formula MSO_4
- different in that they become less soluble down the group.

Epsom salts consist of hydrated magnesium sulfate, $MgSO_4.7H_2O$, which is a laxative.

Plaster of Paris is the main ingredient of building plasters and much is

Figure 4.5.13 ▲
Crystals of gypsum – an hydrated form of calcium sulfate

used to make plasterboard. The white powder is made by heating the mineral gypsum in kilns to remove most of the water of crystallisation.

$$CaSO_4.2H_2O(s) \rightarrow CaSO_4.\tfrac{1}{2}H_2O(s) + \tfrac{3}{2}H_2O(g)$$

Stirring plaster of Paris with water produces a paste which soon sets as it turns back into interlocking grains of gypsum. Plaster makes good moulds because it expands slightly as it sets so that it fills every crevice.

Barium sulfate absorbs X-rays strongly so it is the main ingredient of 'barium meals' used to diagnose disorders of the stomach or intestines. Soluble barium compounds are toxic but barium sulfate is very insoluble and so cannot be absorbed into the bloodstream from the gut. X-rays cannot pass through the 'barium meal' which therefore creates a shadow on the X-ray film.

A soluble barium salt can be used to test for sulfate ions because barium sulfate is insoluble even when the solution is acidic. Adding a solution of barium nitrate or barium chloride to an acid solution produces a white precipitate only if sulfate ions are present.

$$Ba^{2+}(aq) + SO_4^{2-}(aq) \rightarrow \underset{\text{white precipitate}}{BaSO_4(s)}$$

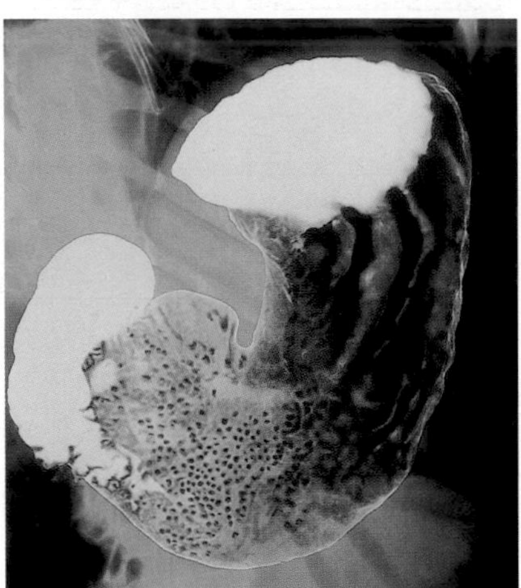

Figure 4.5.14 ►
X-ray photograph of the digestive system of a patient who has taken a barium meal

Thermal stability of carbonates and nitrates

Most of the compounds of group 1 and 2 elements are ionic. So chemists try to explain differences in the properties of the compounds of these elements in terms of two factors:

- the charge on the metal ions
- the size of the metal ions.

Group 2 carbonates and nitrates are generally less stable than the corresponding group 1 compounds. This suggests that the larger the charge on the metal ion, the less stable the compounds.

Down either group 1 or group 2 the carbonates become more stable. This suggests that the larger the metal ion, the more stable the compounds.

Figure 4.5.15 shows the temperatures at which the carbonates of group 2 metals begin to decompose. The figures confirm that magnesium carbonate is the least stable – it decomposes easily when heated with a Bunsen flame. Barium carbonate is the most stable. (Beryllium carbonate is so unstable that it does not exist.)

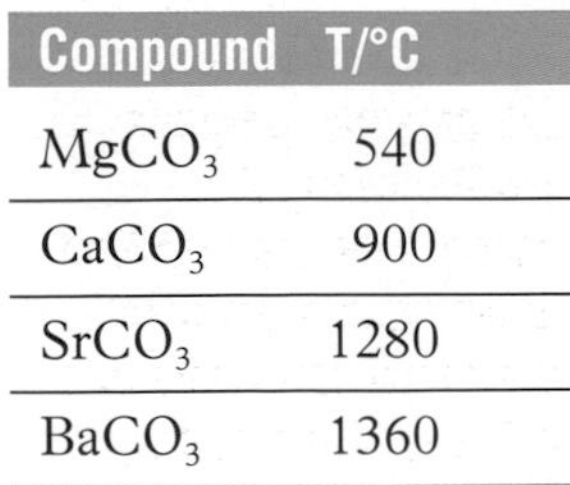

Compound	T/°C
$MgCO_3$	540
$CaCO_3$	900
$SrCO_3$	1280
$BaCO_3$	1360

Figure 4.5.15 ▲
Temperature at which group 2 carbonates begin to decompose

CD-ROM

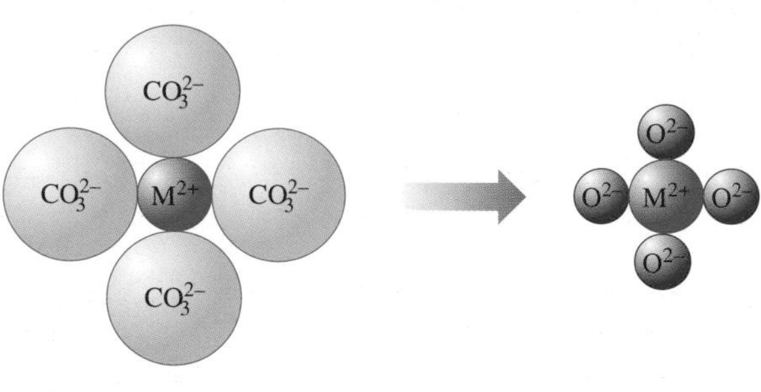

Figure 4.5.16 ◄ *Decomposition of a group 2 carbonate to a group 2 oxide. The smaller the metal ion, the less stable the carbonate*

Chemists explain the trend in thermal stability by analysing the energy changes. Two of the energy quantities they take into account are:

- the energy needed to break up the carbonate ion into an oxide ion and carbon dioxide
- the energy given out as the 2+ and 2− charges get closer together when the larger carbonate ions break up into smaller oxide ions and carbon dioxide.

It turns out that all the carbonates are thermally stable at room temperature but become unstable as the temperature rises (as shown by Figure 4.5.15). The key factor is the energy released as the ions get closer together. This is greater when the metal ion is small than when the metal ion is large.

Detailed analysis of the energy changes takes time, so chemists find it convenient to correlate the stability of compounds such as carbonates and nitrates with the polarising power of the metal ions (see page 82). Generally the greater the polarising power of the metal ion, the less stable the carbonates and nitrates and the more easily they decompose to the oxide.

Test yourself D

5 Draw and label a diagram of a simple apparatus to show that magnesium carbonate decomposes on heating.

6 Write equations for:

a) the thermal decomposition of magnesium carbonate
b) the reaction of magnesium carbonate with hydrochloric acid
c) the thermal decomposition of calcium nitrate
d) the reaction of barium nitrate solution with zinc sulfate solution.

7 With the help of a table of data show that the solubilities of the group 2 metal sulfates decrease down the group from Mg to Ba when measured in mol per 100 g water.

Definitions

Refractories are materials with very high melting points used to line furnaces and to make crucibles.

Ceramics are materials such as pottery, glasses, cement, concrete, graphite. Ceramics also include a wide range of crystalline compounds such as magnesium oxide, aluminium oxide and silicon nitride. These materials are all inorganic materials which are heated to a high temperature in a furnace at some stage during their manufacture.

Antacids are ingredients of indigestion tablets taken to neutralise acid in the stomach.

Note

Whenever chemists use the term 'stability' they are making comparisons. For the group 2 carbonates the question is which is more stable, the metal carbonate or a mixture of the metal oxide and carbon dioxide?

4.6 Group 7

Fluorine, chlorine, bromine and iodine belong to the family of halogens. All four are very reactive non-metals. They are interesting chemically because they do much vigorous chemistry. The elements are hazardous because they are so reactive. For the same reason they are not found free in nature – they exist as compounds with metals. Many of the compounds of group 7 elements are salts – hence the name 'halo-gen' meaning 'salt-former'. The halogens are important economically as the ingredients of plastics, pharmaceuticals, photographic chemicals, anaesthetics and dyestuffs.

CD-ROM

Note

Chemists sometimes use the symbol X to represent any halogen.

The rarest halogen, astatine, is not described here. All of its 20 known isotopes are highly radioactive. The longest-lived isotope, astatine-210, has a half-life of 8.3 hours.

The elements

The halogens all consist of diatomic molecules, X_2, linked by a single covalent bond. They are all volatile. Intermolecular forces increase down the group as the numbers of electrons in the atoms increase (see page 84). The larger molecules are more polarisable than the smaller molecules, so melting points and boiling points rise down the group. Under laboratory conditions, chlorine is a yellow-green gas, bromine is a dark red liquid, while iodine is a greyish solid. Fluorine is a pale yellow gas but it is much too dangerously reactive to be used in normal laboratories.

The halogens have similar chemical properties because they all have seven electrons in the outer shell – one less than the next noble gas in group 8.

Figure 4.6.1 ▼ *Shortened form of the electron configurations of the halogens*

fluorine, F	$[He]2s^22p^5$
chlorine, Cl	$[Ne]3s^23p^5$
bromine, Br	$[Ar]3d^{10}4s^24p^5$
iodine, I	$[Kr]4d^{10}5s^25p^5$

Figure 4.6.2 ► *Iodine is a lustrous grey-black solid at room temperature which sublimes when gently warmed to give a purple vapour*

Fluorine is the most electronegative of all elements. Its oxidation state is −1 in all its compounds. Uses of fluorine include the manufacture of a wide range of compounds consisting of only carbon and fluorine (fluorocarbonds). The most familiar of these is the very slippery, non-stick polymer, poly(tetrafluorethene).

Chlorine is a powerful oxidising agent which reacts directly with most elements. In its compounds, chlorine is usually present in the −1 oxidation states but chlorine can be oxidised to positive oxidation state by oxygen and fluorine. Most chlorine is used in the production of polymers such as pvc. Water companies use chlorine to kill bacteria in drinking water. Another important use of the element is to bleach paper and textiles.

Test yourself

1 Write down the full electron configurations of:

a) a chlorine atom
b) a chloride ion
c) a bromine atom
d) a bromide ion.

2 Write balanced equations for the reactions of:

a) bromine with magnesium
b) chlorine with iron
c) iodine with iron.

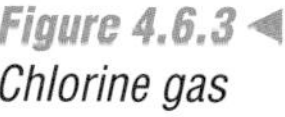

Figure 4.6.3 ◀ *Chlorine gas*

Bromine, like the other halogens, is an oxidising element. It is a less powerful oxidising agent than chlorine. Until recently an important use of bromine was for the manufacture of petrol additives to prevent lead deposits building up in engines running on leaded fuel. Bromine compounds, such as bromomethane, are used to kill soil pests under valuable crops such as strawberries. Silver bromide is used to make photographic film and paper.

Iodine, like the other halogens, is an oxidising agent but a less powerful oxidising agent than bromine. Iodine and its compounds are used to make pharmaceuticals, photographic chemicals and dyes. In many regions sodium iodide is added to table salt to supplement the iodine in the diet and drinking water and so prevent goitre – a swelling of the thyroid gland in the neck. Iodine is needed in the diet so that the thyroid gland in the neck can make the hormone thyroxine which regulates growth and metabolism.

Note

Iodine is only very slightly soluble in water. It is much more soluble in a solution of potassium iodide because of the formation of the tri-iodide ion, $I_3^-(aq)$. Iodine solution in aqueous potassium iodide is a brownish-yellow colour. Iodine dissolves freely in non-polar solvents such as hexane forming a solution with the same violet colour as iodine vapour.

Reactions of the elements

The halogens are powerful oxidising agents. Apart from fluorine, chlorine is the strongest oxidising agent and iodine the weakest in the group.

Halogen atoms are highly electronegative (see page 81). They form

Figure 4.6.4 ▶ *Bromine is a dark red liquid at room temperature but is very volatile and gives off a choking orange vapour*

ionic compounds or compounds with polar bonding. Electronegativity decreases down the group.

Reactions with metal elements

Chlorine and bromine reacts with s-block metals to form ionic halides in which the halogen atoms gain one electron to fill the 4p energy levels.

Iodine also reacts with metals to form iodides but because of the polarisability of the large iodide ion, the iodides formed with small cations or highly charged cations are essentially covalent (see page 82). Examples are lithium iodide, magnesium iodide and aluminium iodide.

Hot iron burns brightly in a stream of chlorine gas forming iron(III) chloride. The reaction with bromine is similar but much less exothermic.

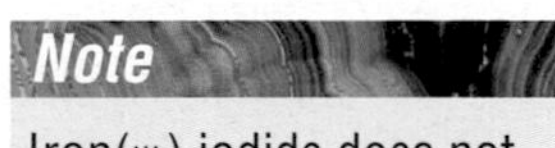

Note

Iron(III) iodide does not exist because iodide ions reduce iron(III) ions to their lower oxidation state. Heating iron in iodine vapour produces iron(II) iodide.

Reactions with non-metal elements

Chlorine reacts with most non-metals to form molecular chlorides. Hot silicon, for example, reacts to form silicon tetrachloride, $SiCl_4(l)$, and

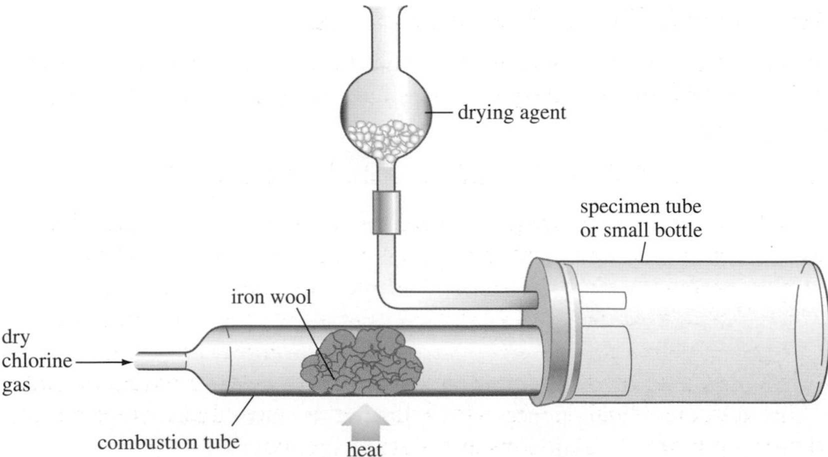

Figure 4.6.5 ▶ *Laboratory apparatus for making anhydrous iron(III) chloride*

phosphorus produces phosphorus trichloride, $PCl_3(l)$. Chlorine does not, however, react directly with carbon, oxygen or nitrogen.

Hydrogen burns in chlorine to produce the colourless, acidic gas hydrogen chloride, HCl. Igniting a mixture of chlorine and hydrogen gases leads to a violent explosion.

Bromine also oxidises non-metals such as sulfur and hydrogen on heating forming molecular bromides. A mixture of bromine vapour and hydrogen gas reacts smoothly with a pale bluish flame.

$$H_2(g) + Br_2(g) \rightarrow 2HBr(g)$$

Iodine oxidises hydrogen on heating forming hydrogen iodide. Unlike the reactions of chlorine and bromine, this is a reversible reaction.

$$H_2(g) + I_2(g) \rightleftharpoons 2HI(g)$$

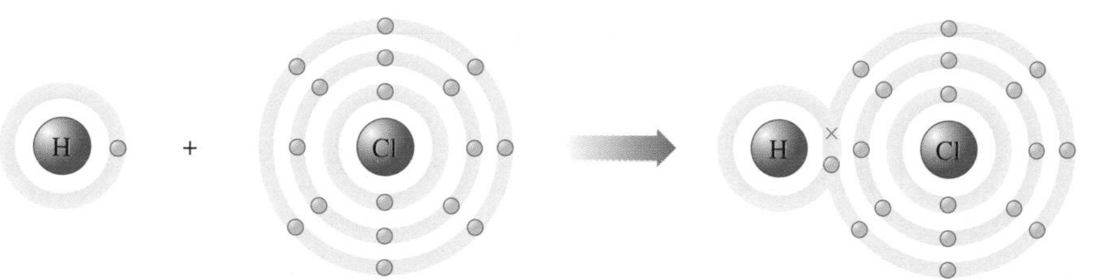

Figure 4.6.6 ◀
Covalent bonding in a hydrogen chloride molecule

Oxidation state −1

Halide ions are the ions of the halogen elements in oxidation state −1. They include the fluoride, F^-, chloride, Cl^-, bromide, Br^- and iodide, I^-, ions.

Displacement reactions

In group 7, a more reactive halogen takes the place of (or displaces) a less reactive halogen from a halide. So bromine reacts with a solution of an iodide to produce iodine and a bromide. Bromine has a stronger tendency to gain electrons and turn into ions than iodine. The order of reactivity for the halogens is chlorine > bromine > iodine. The more reactive halogen oxidises the ions of a less reactive halogen.

$$Br_2(aq) + 2I^-(aq) \rightarrow 2Br^-(aq) + I_2(s)$$

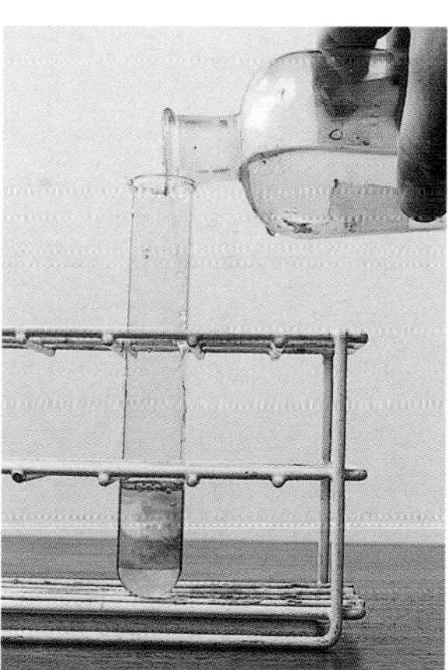

Figure 4.6.7 ▲
A test tube containing a mixture formed by mixing a solution of chlorine in water with aqueous potassium iodide. Shaking with a hydrocarbon solvent produces a violet colour in the organic solvent, showing that iodine is present in the mixture. Only non-polar molecules dissolve in the non-polar solvent. All the ions stay in aqueous solution

Reactions of halides with concentrated sulfuric acid

Warming solid sodium chloride with concentrated sulfuric acid produces clouds of hydrogen chloride gas. This acid–base reaction can be used to make hydrogen chloride.

$$NaCl(s) + H_2SO_4(l) \rightarrow HCl(g) + NaHSO_4(s)$$

Both sulfuric acid and hydrogen chloride are strong acids. The reaction goes from left to right because the hydrogen chloride is a gas and escapes from the reaction mixture. So the reverse reaction cannot happen.

This type of reaction cannot be used to make hydrogen bromide or hydrogen iodide because bromide and iodide ions are strong enough reducing agents to reduce sulfur from the +6 state to lower oxidation states.

The reactions of halide ions with sulfuric acid show that there is a trend in the strength of the halide ions as reducing agents:

Note

Acidify the unknown halide solution with dilute nitric acid before testing with silver nitrate solution. This prevents the precipitation of other silver compounds.

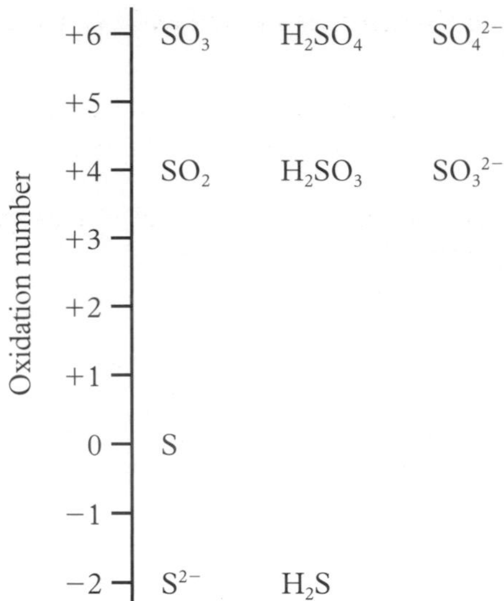

Figure 4.6.8 ▶ *Oxidation states of sulfur compounds*

Test yourself

3 Explain why silicon tetrachloride is a liquid in terms of structure and bonding.

4 Write equations for the reactions and use oxidation numbers to show that:

a) phosphorus is oxidised when it reacts with chlorine
b) chlorine is reduced when it displaces iodine from a solution of potassium iodide.

5 Describe the colour changes expected on adding: CD-ROM

a) a solution of chlorine in water to aqueous sodium bromide
b) a solution of bromine in water to aqueous potassium iodide.

6 Draw the laboratory apparatus for collecting several large test tubes full of hydrogen chloride gas from the reaction of sodium chloride with concentrated sulfuric acid.

7 With the help of oxidation numbers, write a balanced equation for the reaction of sodium bromide with concentrated sulfuric acid.

- with sodium chloride the product is hydrogen chloride gas
- sodium bromide turns to orange bromine molecules as the bromide ions reduce H_2SO_4 to SO_2, together with some hydrogen bromide gas
- iodide ions are the strongest reducing agents, so with sodium iodide little or no hydrogen iodide forms; instead the products are iodine molecules mixed with sulfur, S and hydrogen sulfide, H_2S. Sulfur is reduced from +6 to −2.

So the trend as reducing agents is: $I^- > Br^- > Cl^-$.

Chlorine is the strongest oxidising agent so it has the greatest tendency to form negative ions. Conversely the chloride ions are reluctant to give up their electrons and turn back into chlorine molecules. So chloride ions are the weakest reducing agents.

Iodine is the weakest oxidising agent so it has the least tendency to form negative ions. Conversely iodide ions are the strongest reducing agents being most ready to give up electrons and turn back into iodine molecules.

Hydrogen halides

The hydrogen halides are compounds of hydrogen with the halogens. They are all colourless, molecular compounds with the formula HX where X stands for Cl, Br or I. The bonds between hydrogen and the halogens are polar.

Hydrogen chloride, hydrogen bromide and hydrogen iodide are similar in that they are:

- colourless gases at room temperature which fume in moist air
- very soluble in water forming acid solutions (hydrochloric, hydrobromic and hydriodic acids)
- strong acids so they ionise completely in water.

H^+ transferred

$$HCl + H_2O \longrightarrow Cl^-(aq) + H_3O^+(aq)$$

oxonium ion

Figure 4.6.9 ▲ *The reaction of hydrogen chloride with water. Hydrogen chloride is a strong acid – it is fully ionised in aqueous solution*

Hydrogen chloride, hydrogen bromide and hydrogen iodide show some trends down group 7:

- they become less thermally stable – heating does not decompose hydrogen chloride, but a hot wire will decompose hydrogen iodide into hydrogen and iodine
- they become easier to oxidise to the halogen – hydrogen iodide is a strong reducing agent.

Which halide?
Silver nitrate solution distinguishes between halides. Silver fluoride is soluble so there is no precipitate on adding silver nitrate to a solution of fluoride ions. The other silver halides are insoluble; adding silver nitrate to a solution of one of these halide ions produces a precipitate.

$$Ag^{+}(aq) + Cl^{-}(aq) \rightarrow AgCl(s)$$

Silver chloride is white and quickly turns purply-grey in sunlight. This distinguishes it from silver bromide which is a creamy colour and silver iodide which is a brighter yellow.

The colour changes are not very distinct but a further test can help to distinguish the precipitates. Silver chloride easily dissolves in dilute ammonia solution. Silver bromide will redissolve in concentrated ammonia solution. Silver iodide will not redissolve in ammonia solution at all.

Oxidation states +1, +3 and +5

Chlorine oxoanions form when chlorine reacts with water and alkalis.

Reactions with water

Chlorine dissolves in water. It reacts reversibly with water forming a mixture of weak chloric(I) acid and strong hydrochloric acid. This is an example of a disproportionation reaction.

$$Cl_2(aq) + H_2O(l) \rightleftharpoons HOCl(aq) + Cl^{-}(aq) + H^{+}(aq)$$

(Cl in Cl_2 → +1 in HOCl; Cl in Cl_2 → −1 in Cl^{-})

Bromine reacts in a similar way but to a much lesser extent. Iodine is almost insoluble in water and hardly reacts at all.

Reactions with alkali

When chlorine dissolves in potassium (or sodium) hydroxide solution at room temperature it produces chlorate(I) and chloride ions.

$$Cl_2(aq) + 2OH^{-}(aq) \rightarrow ClO^{-}(aq) + Cl^{-}(aq) + H_2O(l)$$

The active ingredient in household bleach is sodium chlorate(I) made by dissolving chlorine in sodium hydroxide solution – both products of the electrolysis of brine (see page 159).

On heating, the chlorate(I) ions disproportionate to chlorate(V) and chloride ions:

$$\underset{+1}{3ClO^{-}(aq)} \rightarrow \underset{+5}{ClO_3^{-}(aq)} + \underset{-1}{2Cl^{-}(aq)} \quad \text{oxidation states}$$

Test yourself

8 Write ionic equations for the reactions of silver nitrate solution with:

a) potassium iodide solution

b) sodium bromide solution. CD-ROM

Definition

A **disproportionation reaction** is a change in which the same element both increases and decreases its oxidation number. So one element is both oxidised and reduced.

Note

Bromine and iodine react with alkalis in a similar way to chlorine. The BrO^{-} and IO^{-} ions are less stable, however, and so they disproportionate to the +5 state at a lower temperature.

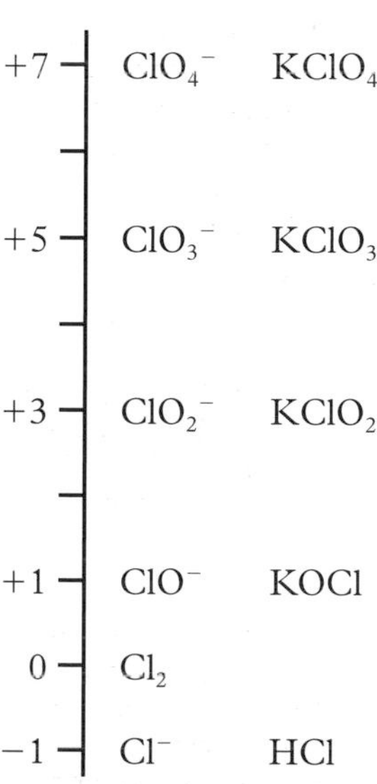

Figure 4.6.10 ▲
Oxidation states of chlorine

4.7 Water treatment

A life-saving application of chlorine chemistry is water treatment. Either on its own or as sodium chlorate(I) (bleach), chlorine is a powerful disinfectant which quickly kills the bacteria and other micro-organisms which cause disease. Since the nineteenth century, the treatment of drinking water with chlorine has helped to control the spread of diseases such as typhoid and cholera. In Europe today it is chlorine which makes almost all drinking water safe.

Disinfection

Chlorine disinfects tap water by forming chloric(I) acid, HOCl.

$$Cl_2(aq) + H_2O(l) \rightleftharpoons HOCl(aq) + H^+(aq) + Cl^-(aq)$$

Chloric(I) acid is a powerful oxidising agent and a weak acid. It is an effective disinfectant because the molecule can pass through the cell walls of bacteria, unlike ClO^- ions. Once inside the cells the HOCl molecules break them open and kill the organism by oxidising and chlorinating molecules which make up the structure of its cells.

Safe homes and swimming pools

Chlorine gas is very hazardous, so for household cleaning it is dissolved in sodium hydroxide to make domestic bleach, sodium chlorate(I). Sodium chlorate ionises fully in water.

$$NaOCl(aq) \rightarrow Na^+(aq) + OCl^-(aq)$$

Chloric(I) acid is a weak acid so in bleach solution some of the chloric(I) ions take hydrogen ions from water molecules and turn into the unionised acid.

$$OCl^-(aq) + H^+(aq) \rightleftharpoons HOCl(aq)$$

The position of this equilibrium can be controlled by altering the pH of the solution. When the pH is low, the hydrogen ion concentration is high and, as Le Chatelier's principle predicts, the equilibrium shifts to the right giving mainly HOCl.

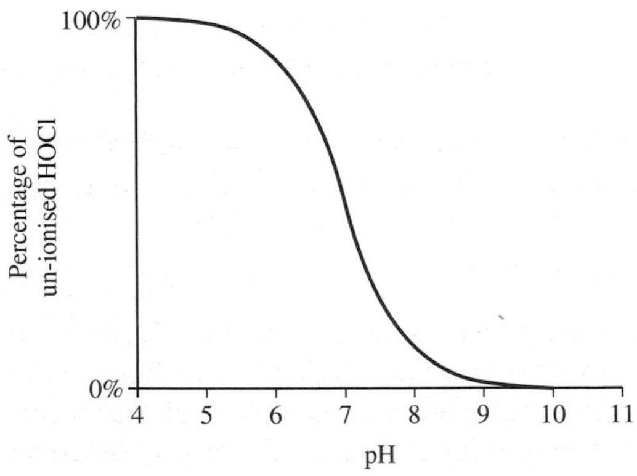

Figure 4.7.1 ▶ *Graph to show how the concentration of chloric(I) acid varies over a range of pH values at 20 °C*

Definition

The pH of a solution is a measure of the hydrogen ion concentration. The definition of pH means that the hydrogen ion concentration rises as the pH falls. The hydroxide ion, OH^-, concentration rises as the pH rises.

pH	1	4	7	9	14
H^+/mol dm^{-3}	1×10^{-1}	1×10^{-4}	1×10^{-7}	1×10^{-9}	1×10^{-13}
OH^-/mol dm^{-3}	1×10^{-13}	1×10^{-10}	1×10^{-7}	1×10^{-5}	1×10^{-1}

Conversely, when the pH is high, the hydrogen ion concentration is low and the equilibrium shifts to the left.

Swimming pools are sterilised with chlorine compounds which produce chloric(I) acid when they dissolve in water. Swimming pool managers have to check the pH of the water carefully. They aim to keep the pH in the range 7.2–7.8. If the pool water is too alkaline (higher pH) the concentration of HOCl is not high enough to kill bacteria. If the water is too acidic (lower pH) the water may be uncomfortable to bathers; it may also start to corrode the pipes through which the circulating water flows and start to etch the concrete surfaces.

Test yourself

1 Write an equation to show the reaction of chlorine with sodium hydroxide solution to make domestic bleach.

2 Why do domestic bleach bottles carry the warning that the bleach should never be used together with acidic descalers?

3 Show that Le Chatelier's principle can help to account for the shape of the graph in Figure 4.7.1.

Analysing bleach

Chlorine bleach solutions slowly decompose and in time they become ineffective. It is important to be able to check on the concentration of bleach to ensure that it is sufficiently concentrated to destroy bacteria and viruses. The guidelines recommend that diluted bleach used for killing micro-organisms should contain at least 1% by mass of available chlorine.

The general technique which chemists can use to measure the concentration of oxidising agents, including bleach, is an iodine–thiosulfate titration. The method is based on the fact that oxidising agents such as bleach convert iodide ions to iodine quantitatively in acid conditions.

Adding acid reverses the reaction to form bleach and converts the chlorate and chloride ions to chlorine.

$$OCl^-(aq) + Cl^-(aq) + 2H^+(aq) \rightleftharpoons Cl_2(aq) + H_2O(aq)$$

The chlorine then oxidises iodide ions to iodine. So this technique measures the total 'available chlorine' in a solution of bleach.

$$Cl_2(aq) + 2e^- \longrightarrow 2Cl^-(aq)$$

$$2I^-(aq) \longrightarrow I_2(aq) + 2e^-$$

The iodine stays in solution in excess potassium iodide forming a yellow-brown colour.

The iodine produced is then titrated with a standard solution of sodium thiosulfate which reduces iodine molecules back to iodide ions. This also happens quantitatively, exactly as in the equation:

$$I_2(aq) + 2S_2O_3^{2-}(aq) \rightarrow 2I^-(aq) + S_4O_6^{2-}(aq)$$

The greater the amount of oxidising agent added, the more iodine is formed and so the more thiosulfate needed from a burette to react with it. On adding thiosulfate from a burette, the colour of the iodine gets paler. Near the end point the solution is a very pale yellow. Adding a little soluble

Note

Iodine is only very slightly soluble in water. It dissolves in potassium iodide solution because it forms $I_3^-(aq)$. A reagent labelled 'iodine solution' is normally $I_2(s)$ in KI(aq).

The $I_3^-(aq)$ ion is a yellow-brown colour which explains why aqueous iodine looks quite different from a violet solution of iodine in a non-polar solvent such as hexane.

starch solution as an indicator near the end point gives a sharp colour change from blue-black to colourless. Starch solution gives an intense blue-black colour with iodine.

Worked example

Calculate the concentration in grams per litre of the available chlorine in a solution of diluted bleach standardised by this method. A 10.0 cm^3 sample of bleach was run into a flask from a burette. Excess potassium iodide was dissolved in the solution which was then acidified with dilute ethanoic acid. The iodine formed was titrated with a 0.10 mol dm^{-3} solution of sodium thiosulfate from a burette. The volume of sodium thiosulfate solution needed to decolourise the blue iodine–starch colour at the end point was 27.6 cm^3.

Notes on the method

From the equations (see page 155) work out the amount in moles of $S_2O_3^{2-}$ equivalent to 1 mol available Cl_2.

There is then no need to consider the amounts of iodine in the calculations.

Look up the molar mass of chlorine: $M_r(Cl_2) = 71.0$ g mol^{-1}.

Only use your calculator in the final stages of the calculation to avoid repeated rounding errors.

Answer

From the equations, 1 mol Cl_2 produces 1 mol I_2 which then reacts with 2 mol $S_2O_3^{2-}$.

So 2 mol $S_2O_3^{2-}$ is equivalent to 1 mol Cl_2.

The amount of thiosulfate in 27.6 cm^3 (= 0.0276 dm^3) solution = 0.0245 dm^3 × 0.10 mol dm^{-3}

So the amount of Cl_2 in the flask was 0.5 × 0.0276 dm^3 × 0.10 mol dm^{-3}

The available chlorine came from 10.0 cm^3 (= 0.010 dm^3) of the diluted bleach.

So the concentration of the bleach
= (0.5 × 0.0276 dm^3 × 0.10 mol dm^{-3}) ÷ 0.010 dm^3
= 0.138 mol dm^{-3}

The mass concentration = 0.138 mol dm^{-3} × 71.0 g mol^{-1} = 9.80 g dm^{-3}

This is slightly below the recommended concentration for diluted bleach which should be at least 1% (10 g dm^{-3}).

Test yourself

4 Excess potassium iodide was dissolved in a 20 cm^3 sample of bleach which was then acidified and titrated with a 0.20 mol dm^{-3} solution of sodium thiosulfate. The average titre was 20.6 cm^3. Calculate the concentration of the bleach solution.

4.8 Inorganic chemistry in industry

The inorganic chemical industry converts raw materials, such as minerals, natural gas, water and air, into useful products such as fertilisers, paints and pigments. This industry continues to be important in the UK because of large deposits of minerals, such as salt, potash, limestone and gypsum, with easy access to sea water and large reserves of natural gas.

A chemical plant consists not only of the reaction vessels and equipment for separating and purifying products, but also the storage vessels, pumps and pipes, sources of energy and heat exchangers together with the control room.

Bulk chemicals are manufactured on a scale of thousands or even millions of tonnes per year. Examples are sulfuric acid, ammonia and chlorine. They are mainly used as the starting point for making other substances. Fine chemicals are made on a much smaller scale – a few tonnes or hundreds of tonnes – as pesticides or pharmaceuticals, for example.

Speciality chemicals are manufactured for their particular properties as thickeners, stabilisers, flame retardants and so on.

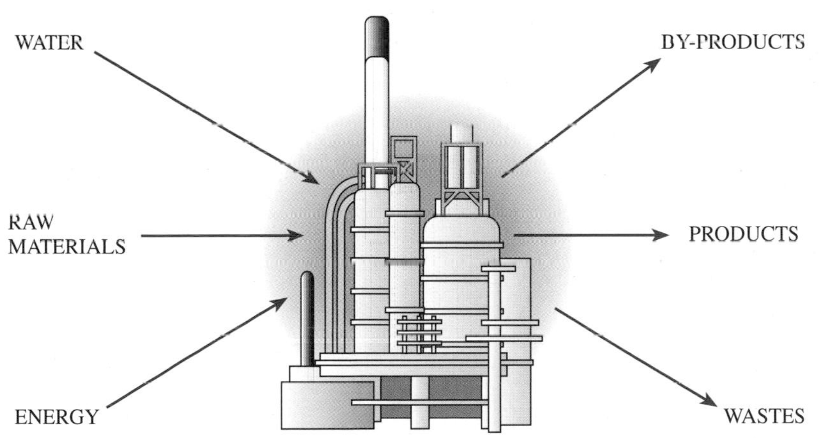

Figure 4.8.1 ◀ A chemical plant with its inputs and outputs

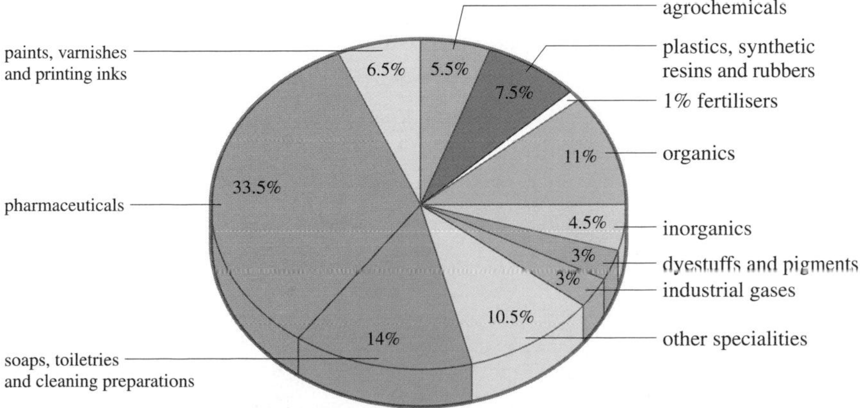

Figure 4.8.2 ◀ Sectors of the UK chemical industry showing the proportions contributed to the total value added in 1993

Main raw materials	Large scale processes and products	Uses of the products
Salt (sodium chloride) and limestone for the chlor-alkali industry	Electrolysis of brine for making chlorine, sodium hydroxide and hydrogen Solvay process to make sodium carbonate	Manufacture of bleaches, disinfectants, solvents, some polymers, glass and paper
Sulfur from underground deposits of the element or from the purification of oil and gas, oxygen from the air	Contact process for sulfuric acid manufacture	Manufacture of paints, pigments, fertilisers, detergents, plastics and many uses in the chemical, metallurgical and petrochemical industries
Nitrogen from the air and natural gas oil fractions	Haber process for ammonia manufacture. Catalytic oxidation of ammonia for nitric acid manufacture	Manufacture of fertilisers, dyes, pigments, detergents, explosives, plastics and fibres
Calcium phosphate rock	Treatment of phosphate rock with concentrated sulfuric acid to make phosphoric(V) acid and phosphates	Fertiliser industry. Manufacture of washing powders, toothpaste, food industry, enamels and glazes
Fluorite (calcium fluoride)	Action of concentrated sulfuric acid on fluorite to make hydrogen fluoride. Electrolysis of fluorides in hydrogen fluoride to make fluorine	Etching and polishing glass and integrated circuits. Manufacture of fluorocarbons and hydrofluorocarbons (to replace CFCs). Pharmaceuticals. The polymer PTFE

Figure 4.8.3 ▲
Major sectors of the inorganic chemical industry. Note that this table excludes metal extraction from ores

Bromine manufacture

In the UK the chemical industry extracts bromine from sea water. A four stage process concentrates and separates the bromine. Sea water contains about 65 ppm of bromide ions and so the plant has to process 20 000 tonnes of water to produce 1 tonne of bromine.

Stage 1 Oxidation of bromide ions to bromine

Sea water is filtered and acidified to pH 3.5 to prevent chlorine and bromine reacting with water (see page 153). Then chlorine displaces bromine.

$$Cl_2 + 2Br^- \rightarrow 2Cl^- + Br_2$$

Stage 2 Separation of bromine vapour

A blast of air through the reaction mixture carries away the displaced bromine and helps to concentrate it.

Stage 3 Formation of hydrobromic acid

The air containing the bromine vapour meets sulfur dioxide gas and a fine mist of water producing hydrobromic acid, which is ionised in solution. After this stage the concentration of bromine in the solution is 1500 times greater than in sea water.

$$Br_2(g) + SO_2(g) + 2H_2O(l) \rightarrow 4H^+(aq) + 2Br^-(aq) + SO_4^{2-}(aq)$$

Stage 4 Displacement and purification of bromine

The solution from stage 3 now trickles down a tower against an upward flow of chlorine gas and steam. The chlorine oxidises bromide ions to bromine which evaporates in the steam. The mixture of steam and bromine is cooled and condensed, producing a dense lower bromine layer under a layer of water.

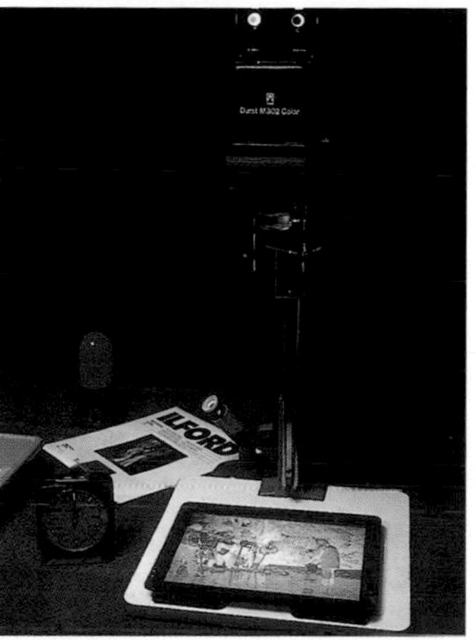

Figure 4.8.4 ◄ *Bromine is used to make flame retardants, agricultural chemicals, synthetic rubber for the inner lining of tubeless tyres, dyes and a range of chemical intermediates. Silver bromide is a light-sensitive chemical used in photography*

Chlor-alkali industry

Electrolysis of brine is the basis of the chlor-alkali industry which manufactures chlorine, hydrogen and sodium hydroxide.

Brine is a solution of sodium chloride in water. During electrolysis chlorine forms at the positive electrode (anode). Hydrogen bubbles off at the negative electrode (cathode) while the solution turns into sodium hydroxide.

The cell used for the electrolysis of brine has to be carefully designed because chlorine reacts with sodium hydroxide. The cell therefore has to keep the chlorine and sodium hydroxide apart.

The two main types of cell used for the process in the UK are the flowing mercury cell and the membrane cell. The mercury cell is being phased out as mercury is expensive and hazardous. Traces of mercury escaping into the environment cause pollution.

Test yourself

1. Draw a flow chart to summarise the stages in the manufacture of bromine from sea water.
2. Identify the oxidising and reducing agents involved in the manufacture of bromine.
3. Write the equation for the reaction of chlorine with water. Use the equation to explain why acidifying the solution minimises the extent to which chlorine reacts with water.
4. Why has the demand for chlorine increased by a factor of 30 since 1950?
5. Why does the chlor-alkali industry have to balance the demand for chlorine and the demand for sodium hydroxide?

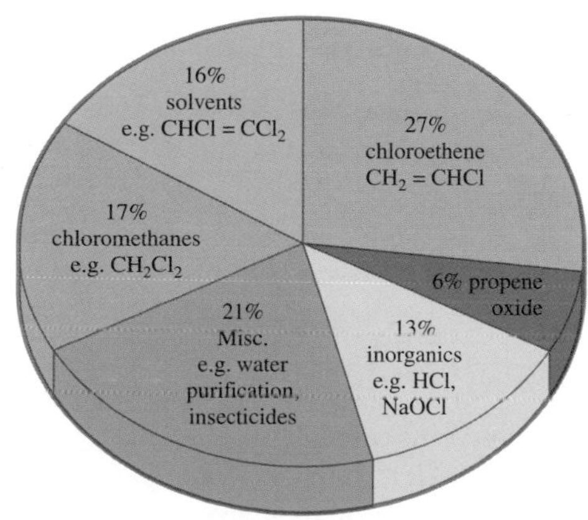

Figure 4.8.5 ◄ *Major uses of chlorine to manufacture chlorine compounds*

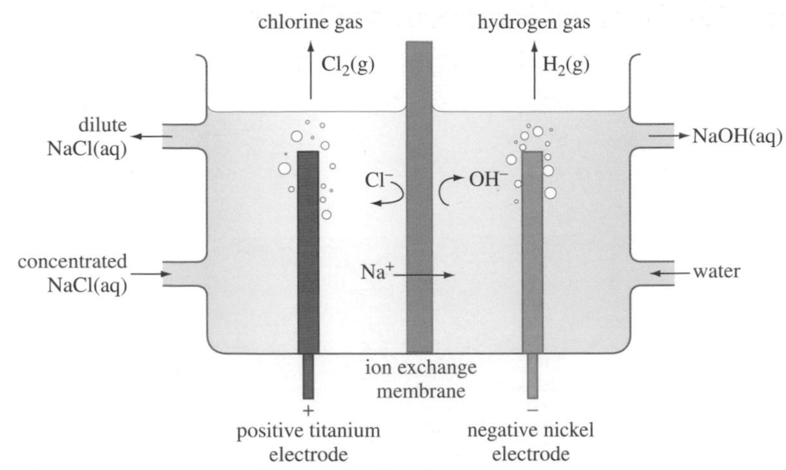

Figure 4.8.6 ▶ *Main inputs and outputs of a membrane cell. The membrane allows the solution to pass through but stops the chlorine mixing with the alkali. The membrane has ion exchange properties. It lets positive ions through but not negative ions*

Figure 4.8.7 ▶ *Sodium hydroxide is a white, translucent solid supplied as flakes or pellets. UK industry produces over a million tonnes of the alkali each year. Sodium hydroxide is widely used for manufacturing other chemicals including soaps and detergents, rayon fibres as well as aluminium, sodium cyanide and sodium peroxide*

Figure 4.8.8 ▶ *Hydrogen is an important industrial chemical. It is one of the products of the electrolysis of brine. Hydrogen is used to hydrogenate vegetable oils for margarine and similar spreads. (See page 162 for details of the large scale manufacture of hydrogen for making ammonia). In addition, hydrogen–oxygen fuel cells supply electricity to the Shuttle*

Sulfuric acid manufacture

The Contact process which makes sulfuric acid from sulfur produces over 150 million tonnes of the acid in the world each year.

The process operates in three main stages:

Stage 1 Burning sulfur to make sulfur dioxide

$$S(s) + O_2(g) \rightarrow SO_2(g)$$

This is a highly exothermic reaction. The hot gas is cooled in heat exchangers which produce steam used to generate electricity. Sulfuric acid plants do not have electricity bills. Electricity and steam exported from the plant help to make the process economic.

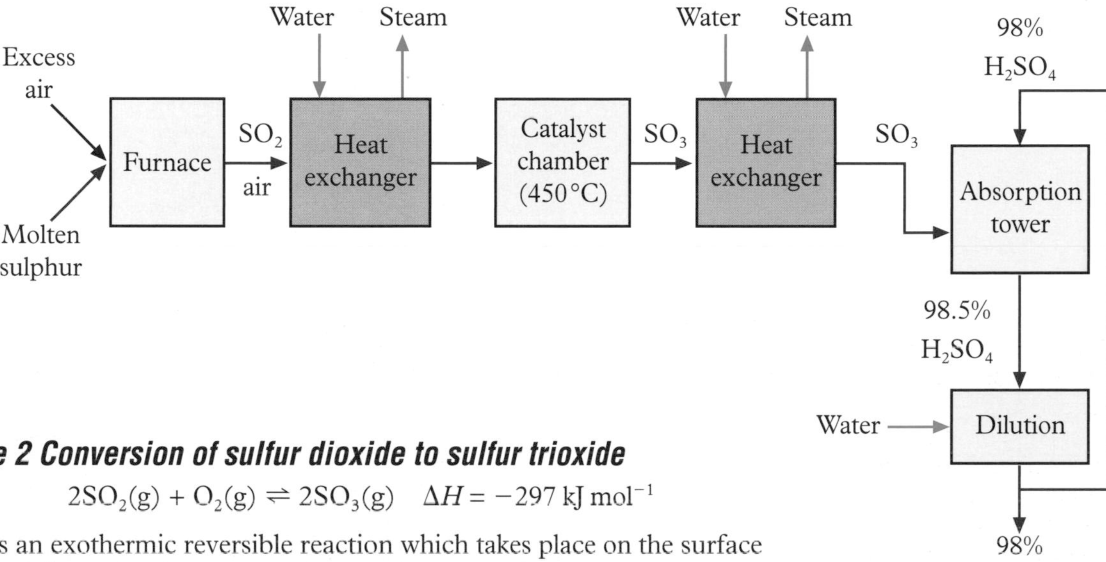

Figure 4.8.9 ▼ *Outline flow diagram for the manufacture of sulfuric acid*

Stage 2 Conversion of sulfur dioxide to sulfur trioxide

$$2SO_2(g) + O_2(g) \rightleftharpoons 2SO_3(g) \quad \Delta H = -297\ \text{kJ mol}^{-1}$$

This is an exothermic reversible reaction which takes place on the surface of a vanadium(v) oxide catalyst. The temperature effect on equilibrium means that raising the temperature lowers the percentage conversion to SO_3 but the temperature must be high enough to make the reaction go fast enough. The catalyst is not active below 380 °C and works best at a higher temperature.

Increasing the pressure would increase the conversion to sulfur dioxide but the extra cost is not usually justified.

Typically the gas mixture passes through four catalyst beds. Between each bed the gas mixture is cooled in a heat exchanger and cold air with more oxygen is added to the mixture. After the third bed the level of conversion is 98%. To ensure conversion of the remaining SO_2, sulfur trioxide is absorbed from the gas stream and more air is added before the gases flow through the fourth bed of catalyst. So three factors contribute to converting as much sulfur dioxide as possible to sulfur trioxide: cooling, adding more of one of the reactants and removing the product.

Stage 3 Absorption

$$H_2O(l) + SO_3(g) \rightarrow H_2SO_4(l)$$

Sulfur dioxide cannot be dissolved in water directly because the violent reaction produces a hazardous mist of acid. Instead the gas is absorbed in 98% sulfuric acid. Sulfur trioxide passes up a tower packed with pieces of ceramic down which the concentrated acid trickles. The circulating acid is kept at the same strength by drawing off product while adding water.

Environmental legislation requires very low emissions of acid gases into the air. A modern plant converts 99.7% of the sulfur dioxide into sulfuric acid in stages 2 and 3.

Test yourself

6 Which of the conditions used to manufacture sulfuric acid help to increase:

a) the proportion of a required product at equilibrium (see page 109)

b) the rate at which products are formed?

Figure 4.8.10 ▶
Sulfuric acid is needed on a large scale to make many other chemicals including phosphate fertilisers, paints and pigments, detergents, plastics, fibres and dyes

Ammonia manufacture

The Haber process synthesises ammonia by combining nitrogen with hydrogen in the presence of an iron catalyst.

$$N_2(g) + 3H_2(g) \rightleftharpoons 2NH_3(g) \quad \Delta H = -92.4 \text{ kJ mol}^{-1}$$

The hydrogen comes from a process called steam reforming which converts natural gas and steam to hydrogen and carbon monoxide. The process happens in the presence of a nickel oxide catalyst at 800 °C under pressure.

$$CH_4(g) + H_2O(g) \rightleftharpoons 3H_2(g) + CO(g) \quad \Delta H = +210 \text{ kJ mol}^{-1}$$

Injecting air into the gas mixture adds nitrogen. The quantity of air is controlled so that the final gas composition will be N_2:$3H_2$. Oxygen in the injected air reacts with some of the hydrogen forming steam, which in turn reacts with any remaining methane in the secondary reformer.

A further reaction with excess steam converts the CO into CO_2 in the shift reactor. The reaction happens in the presence of an iron(III) oxide catalyst at 400 °C.

$$H_2O(g) + CO(g) \rightleftharpoons H_2(g) + CO_2(g) \quad \Delta H = -42 \text{ kJ mol}^{-1}$$

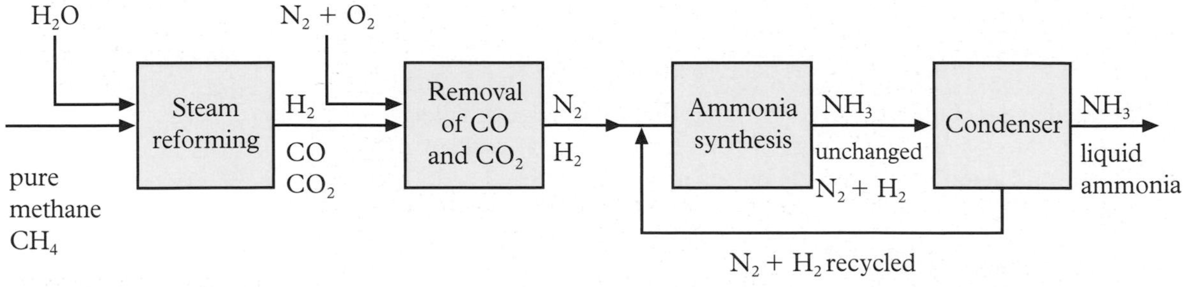

Figure 4.8.11 ▲
Flow diagram for the synthesis of ammonia

Hot potassium carbonate removes the carbon dioxide from the gas mixture.

The mixture of nitrogen and hydrogen is now ready for ammonia synthesis. The reaction between the two gases is very slow at room temperature. Raising the temperature increases the rate of reaction but the reversible reaction is exothermic so, according to Le Chatelier's principle (see pages 109–110), the higher the temperature the lower the yield of ammonia at equilibrium. A catalyst makes it possible for the reaction to go fast enough without the temperature being so high that the yield is too low.

Also according to Le Chatelier's principle, increasing the pressure raises the percentage of ammonia at equilibrium. Economics limits the pressure used. The higher the pressure, the higher the capital cost of the pipes and reaction vessels to contain the gases. Also a raised pressure increases the running costs because of the power used to run compressors.

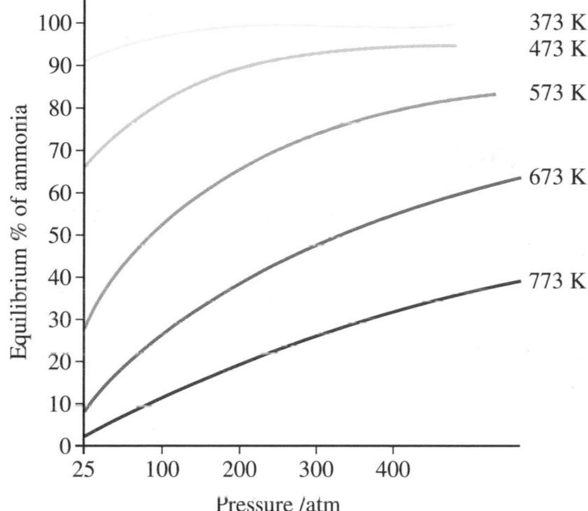

Figure 4.8.12 ◀ *Graph showing how the equilibrium yield of ammonia varies with pressure and temperature*

The Haber process typically operates at pressures between 70 and 200 times atmospheric pressure with temperatures in the range 400 °C to 600 °C.

Fritz Haber (1868–1934) applied theory to develop the process and made it work on a small scale. The key to success was to find the right conditions and to identify a suitable catalyst. Haber used platinum in his small-scale demonstration process. The engineering problems of manufacturing ammonia on a large scale were solved by Carl Bosch who worked for the chemical company BASF. Bosch's team also carried out

Figure 4.8.13 ◀ *The main uses of ammonia are: making fertilisers (80%), making nylon (7%), making nitric acid (5%)*

Test yourself

7 Write an equation for the reaction used to remove carbon dioxide from the mixture of gases being prepared for ammonia synthesis. The product is potassium hydrogencarbonate.

8 Which of the conditions used to manufacture ammonia help to increase:

a) the proportion of a required product at equilibrium

b) the rate at which products are formed?

9 Why does ammonia condense to a liquid much more easily than any of the other gases in the Haber process?

thousands of experiments to find a cheaper catalyst. They eventually developed a catalyst based on iron oxide mixed with small amounts of the oxides of other metals such potassium, aluminium and magnesium. The first industrial plant went into production in 1913 near Mannheim in Germany.

Nitric acid manufacture

Nitric acid manufacture is a process for converting ammonia to nitric acid in two stages:

Stage 1 Oxidation of ammonia

Oxygen from the air oxidises ammonia on the surface of a catalyst gauze made of an alloy of platinum and rhodium. The exothermic reaction keeps the catalyst glowing red hot at about 900 °C. The pressure is typically about 10 times atmospheric pressure.

$$4NH_3(g) + 5O_2(g) \rightleftharpoons 4NO(g) + 6H_2O(g) \quad \Delta H = -909\ \text{kJ mol}^{-1}$$

Figure 4.8.14 ▶
This man is holding up a large circular gauze of rhodium–platinum wire for use as a catalyst in the commercial production of nitric acid by the oxidation of ammonia

The conditions chosen are a compromise between the kinetic factors tending to increase the rate of reaction, equilibrium factors which determine the yield and economic factors tending to lower costs.

Increasing the pressure compresses the gases so the reaction rate increases. The higher pressure also reduces the size of piping and equipment for a given scale of production. These kinetic and economic factors outweigh the fact that at equilibrium the yield of NO would be slightly less.

The hot gases from the reactor pass through a heat exchanger where they cool. The energy converts water to steam to generate electricity.

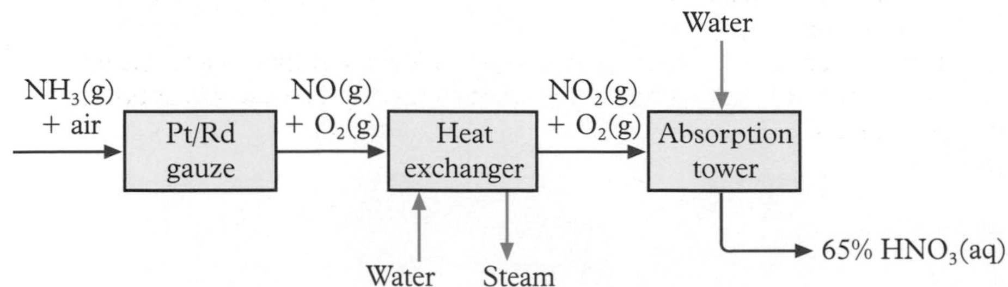

Figure 4.8.15 ▶
Flow diagram for the manufacture of nitric acid

Figure 4.8.16 ◀
About 80% of the 3–4 million tonnes of nitric acid produced each year in the UK is converted to ammonium nitrate. Ammonium nitrate is largely used as a nitrogen fertiliser. It is also widely used as part of most explosives for mining and quarrying. Nitric acid is used to make other explosives such as nitrocellulose and nitroglycerine. Nitric acid is also important reagent in the production of intermediates for making plastics, especially nylon (used in climbing ropes) and polyurethanes

Stage 2 Absorption in water in the presence of oxygen

Adding more air to the cool gas mixture from stage 1 converts NO to N_2O_4. Increasing the pressure favours this change.

$$2NO(g) + O_2(g) \rightleftharpoons 2NO_2(g) \quad \Delta H = -115 \text{ kJ mol}^{-1}$$

$$2NO_2(g) \rightarrow N_2O_4(g) \quad \Delta H = -58 \text{ kJ mol}^{-1}$$

The gases now pass up a tower where they meet a stream of water flowing the opposite way.

$$3N_2O_4(g) + 2H_2O(l) \rightarrow 4HNO_3(aq) + 2NO(g) \quad \Delta H = -103 \text{ kJ mol}^{-1}$$

Extra oxygen from the air converts the NO to N_2O_4 so by the time the gases reach the top of the tower, conversion to nitric acid is almost complete.

Fertilisers

Fertilisers supply plants with the mineral salts they need for growth. Plants need three major nutrients – nitrogen, N, phosphorus, P, and potassium, K – which must be available in the soil in a soluble form so that they can be taken up by roots.

Intensive agriculture relies on fertilisers made from minerals and the air. 'Straight fertilisers' contain one of the three elements nitrogen, phosphorus or potassium. 'Compound fertilisers' contain two or more of these elements.

Test yourself

10 What conditions of pressure and temperature tend to increase the proportion of products at equilibrium during the oxidation of ammonia? How does your answer compare with the conditions chosen to operate the process in industry?

11 Explain why lowering the temperature and raising the pressure favours the conversion of NO to N_2O_4.

12 Show that the reaction of $N_2O_4(g)$ with water is a disproportionation reaction (see page 153).

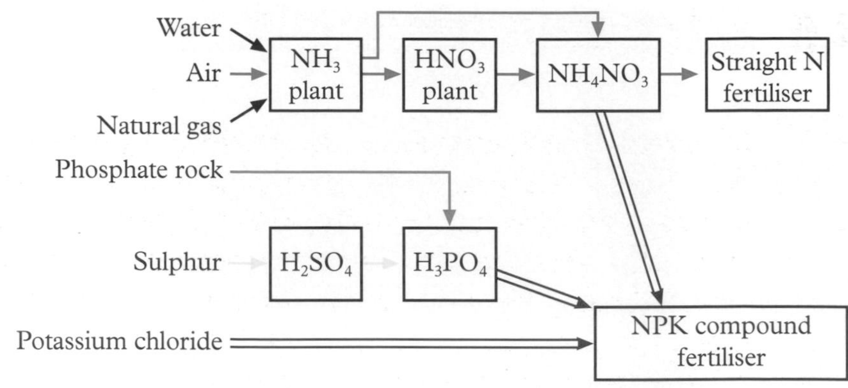

Figure 4.8.17 ▶ *Flow diagram for manufacture of nitrogen and NPK compound fertilisers*

Test yourself

13 Write an equation for the reaction of nitric acid with ammonia to make ammonium nitrate.

14 Compare the percentage by mass of nitrogen in each of these fertilisers: ammonium nitrate, NH_4NO_3, ammonium sulfate $(NH_4)_2SO_4$, and urea, NH_2CONH_2.

15 Trace the changes in oxidation number for nitrogen in all the reactions which convert nitrogen from the air into an ammonium nitrate fertiliser.

The use of manufactured fertilisers greatly increases crop yields. However the use of large amounts of fertiliser increases the risk of nutrients leaching from the soil in the rivers, lakes and ground water. This can lead to eutrophication.

Eutrophication means 'good for growth'. Nitrates from farmland or phosphates from sewage add nutrients to the water in rivers and lakes. Algae grow rapidly in eutrophic water. Thick layers of algae block out sunlight so that plants growing below the surface cannot produce oxygen as fast as usual. Then bacteria start to break down the mass of algae using up the remaining oxygen in the water. Under extreme conditions, fish and other animals die because they are starved of oxygen.

4.9 Extraction of metals

Most metals occur in combination with other elements in the minerals of the Earth's crust. These minerals contain variable proportions of oxides, sulfides, halides, carbonates and other compounds. The task of the mineral industry would be impossible if these compounds were evenly spread throughout the rocks of the crust. Fortunately natural processes have created local concentrations of valuable minerals which can be mined as metal ores. An ore is a mineral mass which can be profitably mined and processed to produce a metal.

Extraction methods

Extraction involves reducing metals from positive oxidation states in their compounds to the zero oxidation state of the free element.

The examples of metal extraction described in this section all involve reactions at high temperature often above 1000 °C. The high-temperature methods (pyrometallurgy) include:

- electrolysis of a molten compound (for example aluminium manufacture)
- chemical reduction by coke in a blast furnace (for example iron manufacture)
- chemical reduction by a more reactive metal (for example chromium and titanium manufacture).

There are other methods for metal extraction used for metals such as copper, zinc, tungsten and gold. These methods use aqueous solutions at quite low temperatures.

The methods based on aqueous reagents (hydrometallurgy) include:

- electrolysis of an aqueous solution (for example zinc manufacture)
- displacement of a less reactive metal by a more reactive metal (for example using iron to displace copper from a solution of copper(II) sulfate).

The choice of extraction method depends partly on the reactivity of the metal and partly on the cost of the reducing agent, the energy requirements and the required purity of the metal.

Figure 4.9.1 ▲
Aluminium is a silvery metal. It is very useful because it has a relatively low density while being resistant to corrosion and a good conductor of thermal energy and electricity. The aluminium ingots in this photo are made from recycled metal

Aluminium extraction

Aluminium extraction is based on the electrolysis of a solution of aluminium oxide in molten cryolite, Na_3AlF_6. Pure aluminium oxide for the process is obtained by purifying bauxite.

Bauxite consists of impure aluminium oxide. Impurities are removed by heating powdered bauxite with sodium hydroxide solution. Aluminium oxide, which is amphoteric (see page 143), dissolves but other oxides such as iron(III) oxide and titanium(IV) oxide do not. After filtering, seed crystals are added and hydrated aluminium oxide crystallises as the solution cools. Heating the hydrated crystals at 1000 °C produces anhydrous aluminium oxide, Al_2O_3, ready for aluminium extraction.

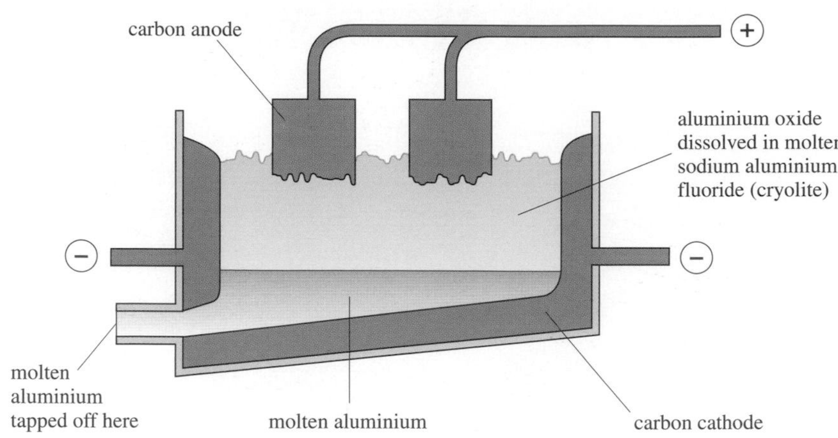

Figure 4.9.2 ► *Cross-sectional diagram of an electrolysis cell for extracting aluminium*

The discovery that aluminium oxide dissolves in molten cryolite was essential to the development of the process, because the pure oxide melts at 2015 °C – much too high for an industrial process.

Electrolysis takes place in carbon-lined steel tanks called 'pots'. The carbon lining is the cathode of the cell. The anodes are blocks of carbon. The currents used are high, of the order of 200 000 A, so the process is generally carried out where electricity is relatively cheap, often close to a source of hydroelectric power.

Reduction at the cathode:

$$Al^{3+} + 3e^- \rightarrow Al$$

The aluminium is liquid at the temperature of the molten electrolyte (970 °C) and it collects at the bottom of the pot. The molten metal is tapped off from time to time.

Oxidation at the anode:

$$2O^{2-} \rightarrow O_2 + 4e^-$$

Much of the oxygen reacts with the carbon of the anodes forming carbon dioxide.

The anodes burn away and have to be replaced regularly.

The waste gases from the cell contain fluorides and have to be thoroughly cleaned to avoid pollution of the region surrounding the plant.

Test yourself

1. Use oxidation numbers to show that aluminium ions are reduced to aluminium at the cathode during the extraction of aluminium.
2. Why are aluminium smelters usually built close to sources of hydroelectric power?
3. Why is graphite a suitable material for lining the electrolysis cells and for making the anodes?

Iron making

Iron extraction produces the metal from its oxide ores in large blast furnaces. The iron is fed into the top of the furnace with coke and limestone. A blast of hot air is pumped into the bottom of the furnace. This is a continuous process. A blast furnace operates day and night for years before being shut down for repairs to the furnace lining.

Coke burning in air heating the lower part of the furnace to about 2000 °C.

$$C(s) + O_2(g) \rightarrow CO_2(g) \quad \Delta H = -394 \text{ kJ mol}^{-1}$$

Coke also produces the main reducing agent by reacting with carbon dioxide further up the furnace to make carbon monoxide.

$$C(s) + CO_2(g) \rightarrow 2CO(g) \quad \Delta H = +173 \text{ kJ mol}^{-1}$$

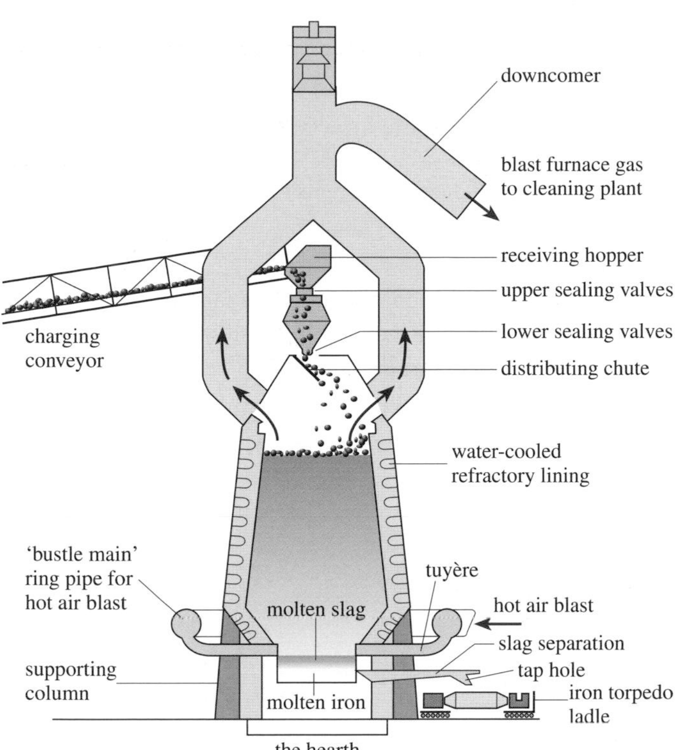

Figure 4.9.3 ◀
Diagram of a blast furnace for extracting iron

The carbon monoxide reduces the ore to iron.

$$Fe_2O_3(s) + 3CO(g) \rightarrow 2Fe(l) + 3CO_2(g) \quad \Delta H = -27\ kJ\ mol^{-1}$$

Where the furnace is hot enough, carbon too can act as the reducing agent.

Limestone, $CaCO_3$, decomposes to calcium oxide, CaO, which combines with silicon dioxide and other impurities to make a liquid slag. For example:

$$CaO(s) + SiO_2(s) \rightarrow CaSiO_3(l)$$

The molten metal and slag run to the bottom of the furnace where the slag floats on top of the metal so that it can be tapped off separately.

The iron from a blast furnace is not pure. It contains about 4% carbon, 0.2–0.3% silicon, 0.03–0.08% sulfur and about 0.1% phosphorus. Blast furnace iron solidifies to cast iron which is very brittle and unsuitable for most modern applications. Instead the molten iron is taken direct from the blast furnace to a steel making plant to remove most of these impurities.

The slag solidifies when it cools. It can then be crushed and used for road building and to make cement. By blowing air into molten slag it is possible to make a fibrous material called rock wool which is manufactured for insulating the walls and lofts of homes.

Steel making

Steels are alloys of iron with carbon or with other metals. Mild steel contains about 0.2% carbon as iron carbide. Mild steel is used for car bodies. Crystals of iron carbide in the metal structure make the steel strong and yet it is still malleable. As the carbon content increases, the steel becomes stronger and harder so that it is suitable for rail and tram lines.

Test yourself

4 In a modern blast furnace the iron ore, coke and limestone are ground up and mixed before they are fed into the furnace. What is the advantage of processing the raw materials in this way?

5 Write an equation for the reduction of iron(III) oxide by coke.

6 Explain why a blast furnace is lined with refractory ceramic bricks (see page 147).

7 Write an equation for the decomposition of limestone in a blast furnace.

8 Identify examples of exothermic and endothermic reactions in a blast furnace.

9 What steps can be taken to save energy resources when running a blast furnace (see Figure 4.9.3)?

Definition

Alloys are usually made by melting a metal with other metals or carbon and then allowing the mixture to cool. Alloys are often more useful than pure metals. Varying the composition of an alloy makes it possible to vary its properties.

There is a limit, however, and at around 4% carbon the metal becomes very brittle.

Alloy steels consist of iron with small amounts of carbon together with up to 50% of one or more of these metals: aluminium, chromium, cobalt, manganese, molybdenum, nickel, titanium, tungsten and vanadium. The presence of other metals distinguishes alloy steels from carbon steels. Examples of alloys steels are:

- stainless steel which include chromium and nickel
- tool steels with tungsten or manganese which make the alloy harder, tougher and able to keep its properties at higher temperatures so that it is suitable for drill bits and cutting tools.

Steel is made by the basic oxygen steelmaking (BOS) process which removes impurities from the molten iron formed in a blast furnace. This is a batch process which treats a large ladle full of hot iron each time. The first step is to remove sulfur by injecting powdered magnesium into the molten metal through a vertical tube. A highly exothermic reaction converts the sulfur to magnesium sulfide which floats to the surface and can be scraped off the molten metal.

Figure 4.9.4 ▶ *Diagram of a basic oxygen converter for making steel*

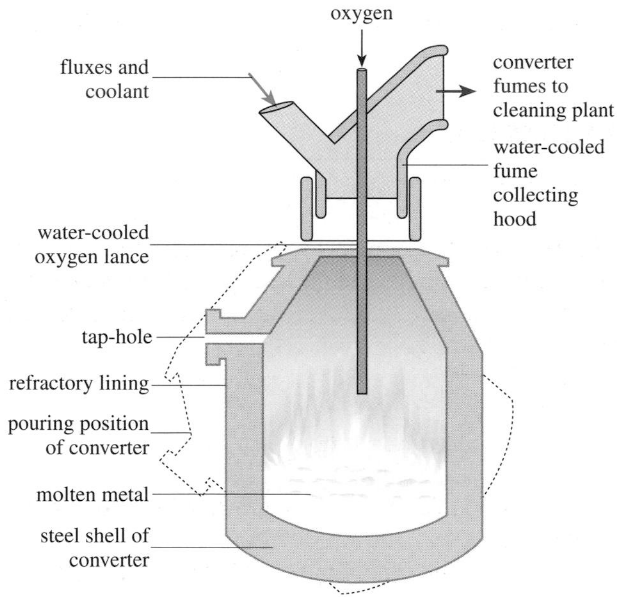

Test yourself

10 Suggest advantages of adding cold scrap steel to the converter before starting the oxygen blow.

11 Write equations for the reactions used to remove sulfur and silicon from impure iron.

12 Why is aluminium a suitable metal for removing excess dissolved oxygen at the end of the process?

Next the molten metal is poured into a converter and cold scrap steel is added. A supersonic blast of oxygen then converts impurities in the liquid metal, such as carbon, silicon and phosphorus, into their oxides. The changes are very violent and over a period of 20 minutes or so, the molten metal froths and foams inside the converter. Carbon monoxide escapes as a gas. The acidic oxides of the non-metals silicon and phosphorus (SiO_2 and P_4O_{10}, respectively) are converted to a molten slag by adding the basic oxides of calcium and magnesium. The slag floats on the surface of the liquid steel and can be poured off separately.

$$SiO_2 + CaO \rightarrow CaSiO_3(l)$$

Figure 4.9.5 ▶
This man is taking a sample from a blast furnace in a steel foundry

Excess oxygen dissolves in the metal during the process. This can be removed by throwing ingots of aluminium into the molten metal as it is poured from the converter into a ladle. Also at this point other metals are added such as chromium, manganese and tungsten. In this way the steelmaker adjusts the composition of the steel to match the demands of the customer who ordered the batch of metal.

Titanium extraction

The main titanium ores are rutile, TiO_2, and ilmenite, $FeTiO_3$. Titanium is the fourth most abundant metal in the Earth's crust and would be more widely used if the methods of extraction were less difficult and expensive. In theory it should be possible to extract titanium from its oxide with carbon, but in practice some of the titanium reacts with carbon forming carbides which make the metal brittle.

After purifying the ore it is heated with carbon in a stream of chlorine gas at about 1100 K.

$$2TiO_2 + 2Cl_2 + 2C \rightarrow TiCl_4 + 2CO_2$$

The titanium(IV) chloride condenses as a liquid which can be purified by fractional distillation.

In the UK, sodium is the reducing agent for producing titanium from its chloride. In most other countries the preferred reducing agent is magnesium. Either way the production is a batch process. The process is expensive whichever reducing agent is chosen because of the high cost of extracting sodium or magnesium by electrolysis.

In the UK method, titanium chloride passes into a reactor containing molten sodium at 500 °C in an inert argon atmosphere. Exactly the right amount of the chloride is added to react with all the sodium. The reaction is exothermic and so the temperature rises.

$$TiCl_4 + 4Na \rightarrow Ti + 4NaCl$$

The reactor is kept hot for about two days then it is removed from the furnace and allowed to cool. The solid product is crushed and leached with dilute hydrochloric acid which dissolves the sodium chloride leaving the titanium metal which is then washed and dried.

Test yourself

13 Use a table to compare continuous and batch methods of metal extraction.

14 Write equations for the electrode processes during the extraction of sodium by electrolysis of molten sodium chloride.

15 Suggest reasons why the extraction of sodium is costly.

16 Why is an argon atmosphere necessary during the extraction of titanium from its chloride?

17 Why does the temperature of the reaction mixture rise to as much as 1000 °C during the extraction of titanium from its chloride?

Figure 4.9.6 ▲
A titanium impeller for a jet engine. The green mould pattern is made of plastic. Titanium is a very strong metal which is much less dense than steel. It melts at the very high temperature of 1675 °C. It does not corrode because, like aluminium, it is protected by a thin layer of oxide on the surface of the metal. Titanium is mainly used to make aircraft engines and airframes. Other major uses are the production components of chemical plants such as heat exchangers

4.10 Environmental issues

Mining, mineral processing and metal extraction create wealth and produce the materials needed by industrialised societies. These activities also have an impact on some of the most beautiful parts of the world and they create wastes. In the UK for example, there are, conflicts in the Peak District between the companies quarrying limestone for the chemical and metallurgical industries and those who want to protect the environment.

Waste from mining

Mining nearly always produces a large quantity of waste rock and can leave very large holes in the ground. Miners use explosives to blast the rock and break it into manageable pieces. This is noisy and produces dust and can result in flying rock fragments.

Figure 4.10.1 ► *A disused open cast copper mine in Anglesey, Wales*

Waste from processing ores

Many metal ores are high value but low grade. The ore in an open-pit copper mine may contain as little as 0.4% of the metal and still be profitable. This means that 99.6% of the rock dug from the ground becomes waste. Near any mine there are waste tips for these wastes and often large ponds where the very fine material settles out from process water. These wastes can be a serious hazard if they contain traces of toxic heavy metals such as lead or mercury.

Waste from metal extraction

Some important metals, such as iron, copper and lead occur as sulfide ores, and the first step in extracting the metal is to convert the sulfide to an oxide by roasting the ore in air. This produces huge tonnages of sulfur dioxide which is toxic and causes acid rain. Releasing the gas into the air can devastate the surrounding countryside.

Figure 4.10.2 ◀
Lodgepole pine planted on tin mine spoil at Wheal Jane in Cornwall

Nowadays governments regulate the industry and insist that only very low levels of toxic gases, smoke and dust escape from furnaces. Sulfur dioxide can either be converted to sulfuric acid (see pages 160–161) or it can be converted to calcium sulfate by reaction with calcium oxide. Meanwhile increasingly effective methods have been developed for removing dust particles before waste gases are released into the air.

The pressure to cut pollution means than for some metals, the industry is moving away from the high temperature methods of pyrometallurgy and adopting the aqueous methods of hydrometallurgy which do not release sulfur compounds as gases.

Recycling metals

One method of reducing waste is recycling. In the steel and aluminium industries, recycling is well established.

The energy needed to make steel is around 32 megajoules per kilogram (MJ kg^{-1}) compared to 146 MJ kg^{-1} for aluminium, 90 MJ kg^{-1} for plastics such as polythene and about 56 MJ kg^{-1} for cardboard.

Recycling steel

Mining iron ores has a growing environmental impact because many of the high grade ores have now been worked out, leaving the lower grade deposits with less usable iron ore for every tonne of rock mined. All the stages of iron and steel making give off air pollutants, as do many of the processes used to fabricate steel products.

Recycling steel can be done repeatedly because it is as good as new after reprocessing. For every tonne of steel recycled there is a saving of 1.5 tonnes of iron ore and half a tonne of coal. There is also a big reduction in the total amount of water needed, since large quantities of water are involved in mineral processing.

Worldwide the steel industry recycles over 430 million tonnes of the metal each year which is a recycling rate of over 50%. The basic oxygen process (see page 170) uses a minimum of 25% scrap steel. Electric arc furnaces make steel plates, beams and bars by re-melting nearly 100% scrap steel.

Much of the recycled steel is waste from various stages of manufacturing steel objects. Another large source of scrap is from old motor vehicles. Shredders can process old cars at the rate of one a minute and turn the metal into dense blocks for feeding into steel-making furnaces.

Some steel is recovered from the tin cans in household waste. In the UK the recovery rate is still relatively low but in Germany it is over 80%. The advantage of steel is that it is magnetic so that magnets can easily separate cans from other waste materials.

The existence of networks of scrap metal dealers helps to recycle steel. The pressure for recycling is increasing as more people are becoming aware of environmental issues. A growing number of governments is banning the dumping of old steel goods in landfill sites. Other countries have laws requiring the recycling of a minimum percentage of packaging materials, including steel cans.

Figure 4.10.3 ▶
Aluminium cans collected for recycling

Recycling aluminium

Open cast mining extracts bauxite in tropical countries such as Jamaica, Brazil and Surinam. Separating the pure aluminium oxide leaves behind large volumes of a red mud consisting largely of iron oxides. It takes four tonnes of bauxite to make a tonne of aluminium.

Recycling reduces the impact on the environment by cutting the use of raw materials and hence the extent of mining and mineral processing. Recycling is also cost effective because the energy needed to manufacture the metal from bauxite is so high. Using recycled cans instead of bauxite, for example, allows the industry to make 20 times as many cans for the same amount of energy.

Aluminium scrap from manufacturing processes is always recycled. Scrap metal merchants handle the waste from these and other sources such as car parts, old saucepans and from buildings. Less than 1% of the mass of household waste is aluminium, almost all of it in the form of drinks cans. Worldwide about 55% of drinks cans are recycled. In Europe the recycling rate is moving up steadily from the 40% level in 1997 and only 30% in 1994. The contribution to this total in the UK means that over 1.5 thousand million cans are recycled every year.

Aluminium is not magnetic but can be separated from a waste stream by a rapidly varying magnetic field which induces eddy currents in the metal of the can. The interaction of the external field and the magnetic effect of the eddy currents leads to a force which can push cans out of the waste stream.

Very little aluminium is reused without re-smelting. The one significant example is the beer keg.

Test yourself

1. Write an equation to show how calcium oxide removes sulfur dioxide from waste gases by reacting with them to make calcium sulfate.
2. What are the possible uses of calcium sulfate (see pages 145–146)?
3. Identify social and economic benefits of recycling steel and aluminium.

Review

This guidance will help you to organise your notes and revision. Check the terms and topics against the specification you are studying. You will find that some topics are not required for your course.

Key terms

Show that you know the meaning of these terms by giving examples. Consider writing the key term on one side of an index card and the meaning of the term with an example on the other side. Then you can easily test yourself when revising. Alternatively use a computer database with fields for the key term, the definition and the example. Test yourself with the help of reports which just show one field at a time.

- Atomic number
- Group
- Period
- Periodicity
- Physical properties
- Atomic radius
- Ionic radius
- Ionisation energy
- Electron configuration
- Shielding
- Trend
- Electronegativity
- Electron transfer
- Half-equation
- Oxidising agents
- Oxidation
- Reducing agents
- Reduction
- Oxidation numbers
- Oxidation states
- Redox reactions
- s-block element
- Basic oxide
- Solubility
- Thermal decomposition
- Halogen
- Halide ion
- Displacement reactions
- Disproportionation reaction
- Bleaching

Symbols and conventions

Make sure that you understand the symbols and conventions which chemists use when writing equations. Illustrate your notes with examples.

- Rules for naming common inorganic compounds
- Use of oxidation numbers in the names of inorganic compounds
- Use of oxidation numbers to balance redox equations

Facts, patterns and principles

Use tables, charts, concept maps or mind maps to summarise key ideas. Brighten your notes with colour to make them memorable.

- Periodic table, the periodicity of physical properties: ionisation energies and oxidation states
- Similarities and trends in the chemistry of the group 2 elements
- Similarities and trends in the chemistry of the group 2 compounds: oxides chlorides, hydroxides, carbonates and nitrates
- Similarities and trends in the chemistry of the group 7 elements
- Similarities and trends in the chemistry of the group 7 compounds: halide ions, hydrogen halides, silver halides

Section four study guide

Laboratory techniques

Draw diagrams to show these key steps in practical procedures.

- How to carry out a flame test and interpret the colours of flames.
- How to carry out and interpret tests for common gases.
- How to carry out test-tube tests for common cations and anions.

Calculations

Give your own worked examples, with the help of the Test Yourself questions to show that you can carry out calculations to work out the following from given data.

- Results of titrations to estimate the concentration of chlorine bleach with potassium iodide and sodium thiosulfate.

Chemical applications

Use notes with charts or flow diagrams to summarise the key stages and reactions for the industrial processes mentioned in the course specification you are studying.

- The use of chlorine compounds in water treatment.
- Bromine manufacture.
- Manufacture of chlorine, sodium hydroxide and hydrogen from salt solution (brine).
- Manufacture of sulfuric acid.
- Ammonia manufacture.
- Nitric acid manufacture and the production of fertilisers.
- Extraction and processing of metals: aluminium, iron and steel, titanium.
- Environmental protection and conservation of resources through waste management and the recycling of materials.

Key skills

Communication

Finding out about the sources and uses of the halogens or of metals requires you to select and read information from a range of sources. You can bring the information together as a coherent report or as an illustrated presentation to others. If writing a report you can meet key skill requirements by synthesising relevant information and organising it clearly and coherently using specialist vocabulary.

Information technology

CD-ROMs and web sites can be a rich source of numerical data and descriptive information about elements and their compounds. You can plot data from databases to discover periodic patterns. CD-ROM

Section five

Organic Chemistry

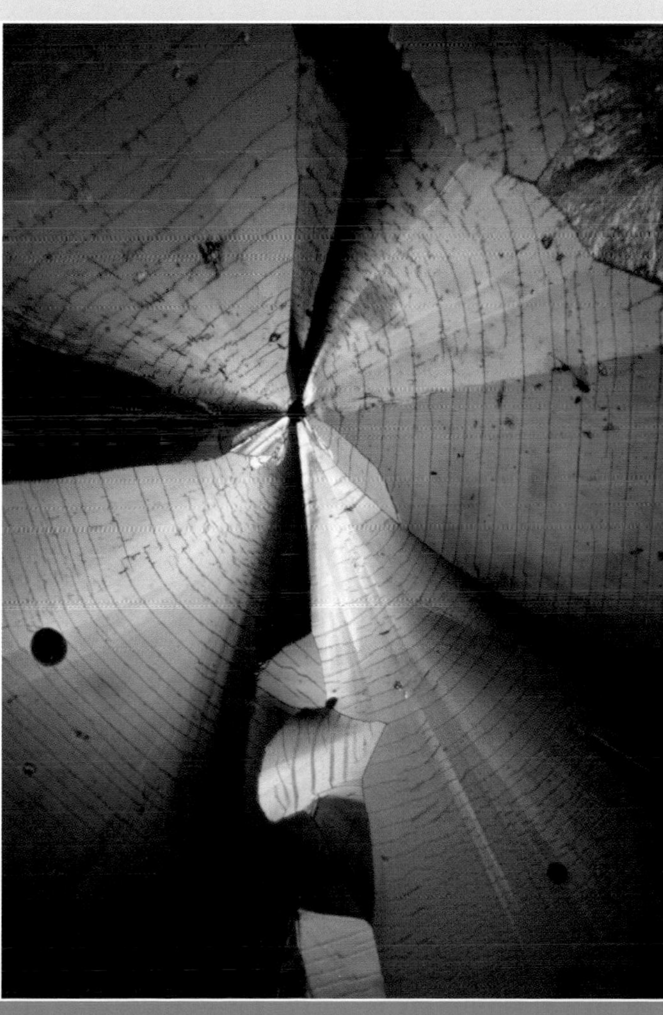

Contents

5.1 What is organic chemistry?

5.2 Organic molecules

5.3 Names of organic compounds

5.4 Types of organic reaction

5.5 Mechanisms of organic reactions

5.6 Hydrocarbons

5.7 Alkanes

5.8 Alkenes

5.9 Halogenoalkanes

5.10 Alcohols

5.11 Theoretical and percentage yields

5.12 Fuels and chemicals from crude oil

5.13 Chemicals from plants

5.14 Organic chemicals and the environment

5.1 What is organic chemistry?

Carbon is an amazing element. There are more compounds of carbon than there are of all the other elements put together. The chemistry of carbon compounds is so important that it forms a separate branch of chemistry. Organic chemistry includes the study of all carbon compounds except the very simplest ones, such as carbon dioxide and carbonates, which are usually treated as part of the study of the element in inorganic chemistry. This section gives you an overview of the main themes of organic chemistry which you are going to study.

Definitions

Catenation is the ability of the atoms of an element to join together in long chains or rings. Carbon atoms have an exceptional ability to catenate – hence the wide range of organic compounds.

A special element

Carbon can form so many compounds because carbon atoms can join together in different ways forming chains, branched chains and rings. This makes possible a great variety of carbon compounds. No other element can form such long chains of its atoms.

The linked carbon atoms in an organic compound are the skeleton to which the atoms of other elements can join. In organic chemistry, carbon often forms bonds with hydrogen, oxygen, nitrogen and halogen atoms. (See pages 182–187 for more detail about the bonding and structure of organic molecules.)

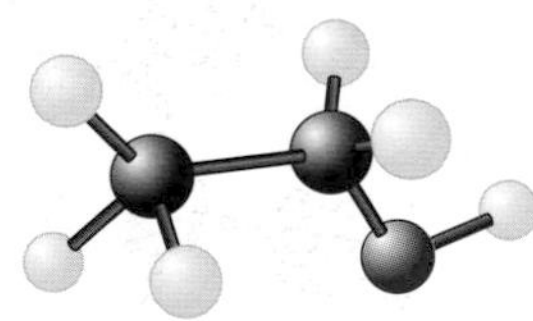

Figure 5.1.1 ▲
The structure of ethanol showing the carbon skeleton with carbon atoms linked to each other and to other atoms by covalent bonds. Ethanol is an alcohol molecule with the —OH functional group

Series of carbon compounds

Organic chemists have several ways to organise the mass of information about the millions of organic compounds.

One approach is to classify compounds into series, each of which has a distinctive group of atoms. Examples are the alkanes (see page 197), alkenes (see page 201), alcohols (see page 212) and halogenoalkanes (see page 207). Each of these families of compounds is an homologous series. The compounds in each homologous series have a distinctive functional group as explained on page 183.

Types of organic reaction

Another important way of making sense of carbon chemistry is to classify reactions. Organic compounds, like inorganic compounds, can be oxidised, reduced, hydrolysed or neutralised by acids or bases.

Another classification describes what happens to the molecules (see pages 191–193):

- addition reactions add atoms to a molecule
- substitution reactions replace one or more atoms with different atoms
- elimination reactions split off atoms from within a molecule.

An alternative classification of organic reactions is based on the mechanisms of the reaction, showing the types of reagents, the method of bond breaking and the nature of any intermediates. (See pages 194–195 for examples.)

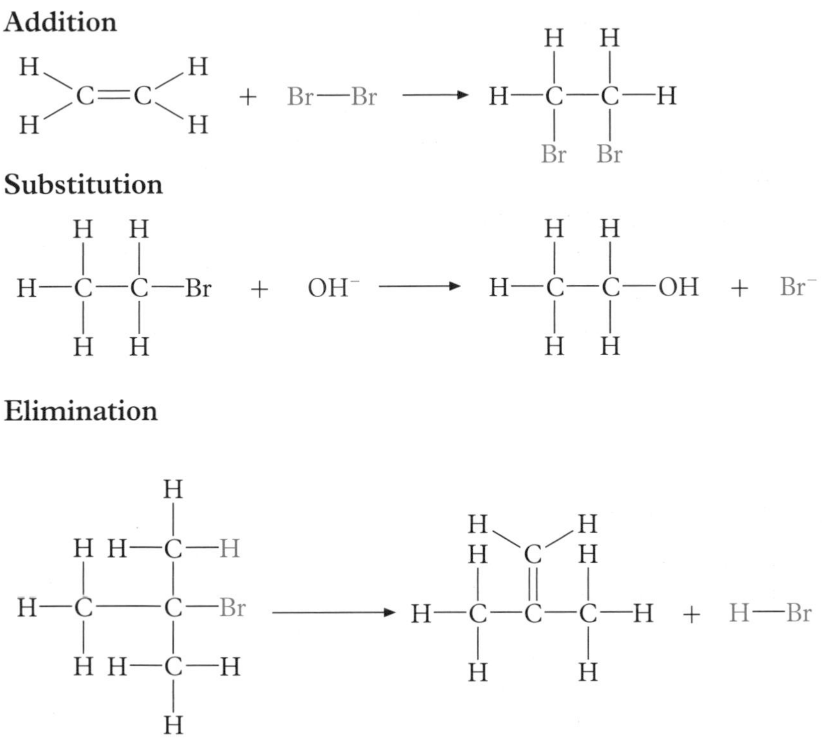

Addition

$$CH_2{=}CH_2 + Br{-}Br \longrightarrow CH_2Br{-}CH_2Br$$

Substitution

$$CH_3{-}CH_2{-}Br + OH^- \longrightarrow CH_3{-}CH_2{-}OH + Br^-$$

Elimination

$$(CH_3)_3C{-}Br \longrightarrow CH_2{=}C(CH_3)_2 + H{-}Br$$

Figure 5.1.2 ◄
Three ways of changing molecules

Big molecules

Most molecules in living things are carbon compounds, so biochemistry and molecular biology are important applications of organic chemistry. The compounds found in living cells include carbohydrates, fats, proteins and nucleic acids. The molecules of these compounds are large. Some of them are very large. The carbohydrate cellulose in cotton, for example, is a natural polymer made of very long chains of glucose units linked together.

Organic chemists make other long-chain molecules by linking together millions of small molecules to make polymers. Familiar synthetic polymers include polythene, pvc, polystyrene and nylon.

Figure 5.1.3 ▼
A short section of a polythene molecule

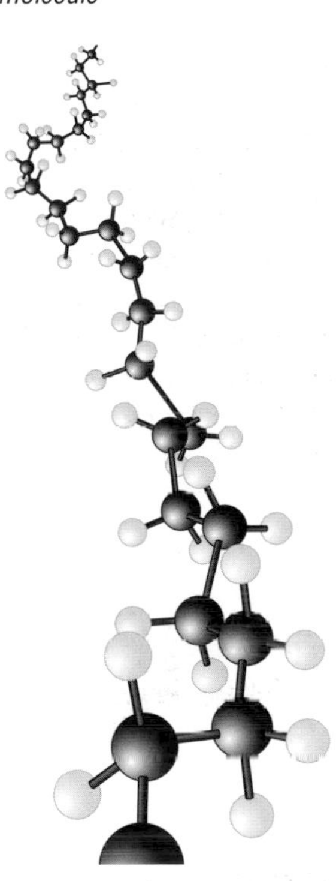

Practical investigations

Analysis

Working out the structures of carbon compounds has been a challenge for chemists. Traditionally they used chemical tests to recognise the groups of atoms in a structure. Now chemists rely more and more on instrumental methods, especially the various types of spectroscopy (see pages 53–55 and 90–91).

Synthesis

Organic chemists have proved their understanding of organic structures and reactions by devising ways to synthesise increasingly complex molecules. The job of the organic chemist is to start with a proposed structure and devise a practicable method of creating it from smaller molecules.

Definitions

Analysis means breaking down chemicals to find out what they are made of.
Synthesis means joining chemicals together to make new products.

5.2 Organic molecules

This section introduces you to organic molecules, their formulae, structures and bonding. The properties of an organic compound are partly determined by the strong covalent bonding between the atoms in its molecules. Also very important are the weak intermolecular forces between the molecules. CD-ROM

Note

Combustion means burning in oxygen or air. Complete combustion in plenty of oxygen ensures that the elements combine with as much oxygen as possible as they burn.

Empirical formulae

The empirical formula is the formula of a compound found by experiment. Experimental results give the combining masses of elements in a compound and chemists calculate the empirical formula from these results.

The empirical formula shows the simplest ratio of the amounts of elements in the compound; it therefore gives the simplest ratio of the numbers of atoms.

Combustion analysis is one method for determining the empirical formulae of organic compounds. The analyst burns a weighed sample of the compound in excess oxygen mixed with helium. This converts the carbon to carbon dioxide and the hydrogen to water. A catalyst ensures that combustion is complete.

The inert helium carries the products of combustion and the excess oxygen through a tube which contains chemicals to remove any volatile halogen, sulfur or phosphorus compounds. Oxides of nitrogen are converted to nitrogen gas and the excess oxygen combines with copper.

The water vapour is absorbed in magnesium chlorate(VII). Carbon dioxide is absorbed in soda lime. In a modern instrument, measurements of the thermal conductivity of the helium before and after absorption make it possible to determine the masses of water, carbon dioxide and nitrogen formed by burning the sample.

From the results it is possible to calculate the percentage composition of the compound. Any mass of the sample not accounted for is assumed to be due to oxygen.

Figure 5.2.1 ▶ *Researchers involved in drug development studying a computer model of a protein. The purple ribbon is the general structure of the molecule. The green pattern shows the atoms in the active site of the molecule which helps cells respond to hormones*

Worked example

Complete combustion of 0.15 g of a liquid compound produced 0.22 g of carbon dioxide and 0.09 g water. What is the empirical formula of the compound?

Notes on the method

The molar mass of carbon dioxide, CO_2 = 44 g mol^{-1} of which carbon is 12 g mol^{-1}.

The molar mass of water, H_2O = 18 g mol^{-1} of which hydrogen is 2 g mol^{-1}.

Answer

The mass of carbon in the sample = $\frac{12}{44} \times 0.22$ g = 0.06 g.

The mass of hydrogen in the sample = $\frac{2}{18} \times 0.09$ g = 0.01 g.

The total mass of carbon and hydrogen = 0.07 g in a sample with a mass of 0.15 g.

The difference gives the mass of oxygen in the sample which is 0.08 g

These are the amounts of the elements in the sample.

carbon:	0.06 g ÷ 12 g mol^{-1}	= 0.005 mol
hydrogen:	0.01 g ÷ 12 1 g mol^{-1}	= 0.01 mol
oxygen:	0.08 g ÷ 12 16 g mol^{-1}	= 0.005 mol

The ratios C:H:O are 1:2:1

The empirical formula of the compound is CH_2O.

Test yourself D

1 Calculate the empirical formulae of these compounds:

a) complete combustion of a sample of a hydrocarbon produced 1.69 g of carbon dioxide and 0.346 g of water.

b) complete combustion of 0.292 g of a compound formed 0.748 g carbon dioxide and 0.308 g of water.

Molecular formulae

The molecular formula shows the number of atoms of each element in a molecule. The molecular formula of chlorine is Cl_2, of ammonia is NH_3 and of ethanol is C_2H_5OH. The term 'molecular formula' only applies to substances which consist of molecules.

For molecular compounds, the relative molecular mass shows whether or not the empirical formula is the same as the molecular formula.

A molecular formula is always a simple multiple of the empirical formula. Analysis shows, for example, that the empirical formula of hexane is C_3H_7. The mass spectrum shows that the relative molecular mass of hexane is 86. The relative mass of the empirical formula, $M_r(C_3H_7) = (3 \times 12) + (7 \times 1) = 43$. So the molecular formula is twice the empirical formula. The molecular formula of hexane is C_6H_{14}.

Test yourself D

2 Determine the empirical and molecular formulae of:

a) a compound consisting of 38.7% carbon, 9.68% hydrogen with the remainder being oxygen; the relative molecular mass of the compound is 62.

b) a hydrocarbon which consists of 82.8% carbon, with a relative molecular mass estimated to be between 50 and 60.

Structural formulae

Structural formulae show the arrangements of atoms and functional groups in molecules.

Sometimes it is enough to show structures in a condensed form such as CH_3CH_2OH for ethanol.

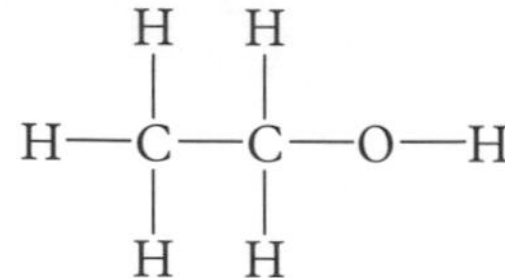

Figure 5.2.2 ▲
The displayed or graphical formula of ethanol showing all the atoms and all the bonds

Often it is clearer to write the full structural formula showing all the atoms and all the bonds. This type of formula is also called a displayed formula or a graphical formula.

Strong bonds within molecules

The atoms in molecules are linked by strong covalent bonds. A few simple rules about bonding help to work out ways in which atoms link together. Figure 5.2.3 lists the number of bonds each type of atom forms with other atoms.

Element	No. of bonds	Colour code
carbon, C	4	black
hydrogen, H	1	white
oxygen, O	2	red
nitrogen, N	3	blue
chlorine, Cl	1	green
bromine, Br	1	green
iodine, I	1	green

Figure 5.2.3 ▶

Figure 5.2.4 shows how these patterns of bonding apply in some examples of hydrocarbons. Note that the structures written flat on paper do not give the right idea about the shapes of molecules – molecular models are better.

Name and molecular formula	Displayed (graphical) formula	Ball-and-stick model	Space-filling model
propane C_3H_8	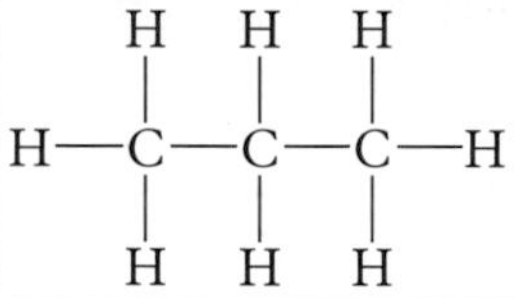 H H H H—C—C—C—H H H H	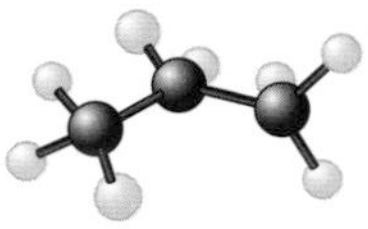	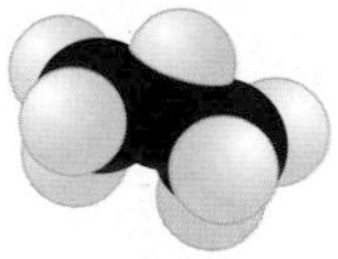
butane C_4H_{10}	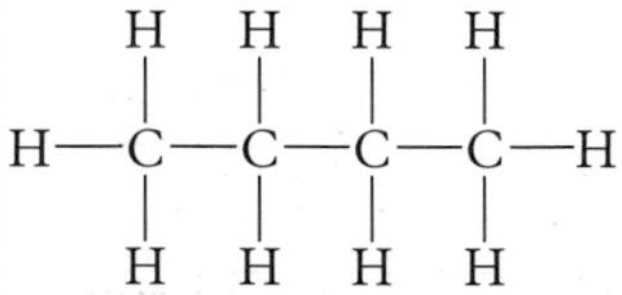 H H H H H—C—C—C—C—H H H H H		
2-methylpropane C_4H_{10}	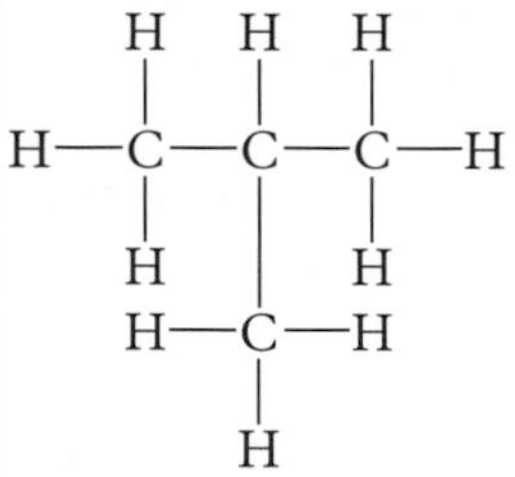 H H H H—C—C—C—H H H H—C—H H	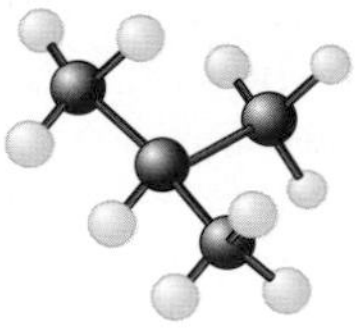	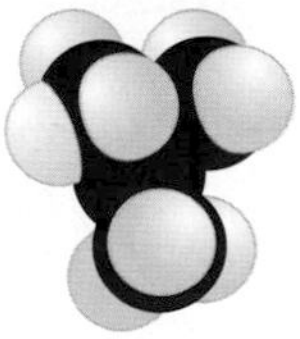

Figure 5.2.4 ▲
Ways of representing the bonding and shape of carbon compounds. Count the numbers of bonds formed by each atom and check that they match the numbers for the elements listed in Figure 5.2.3

CD-ROM

Ball-and-stick models show the numbers of bonds and the bond angles clearly. Space-filling models give a more realistic picture of the molecular shapes.

Functional groups

A functional group is the group of atoms and bonds which give a series of organic compounds its characteristic properties. The functional group in a molecule is responsible for most of its reactions. The hydrocarbon chain which makes up the rest of any organic molecule is generally inert to most common reagents such as acids and alkalis.

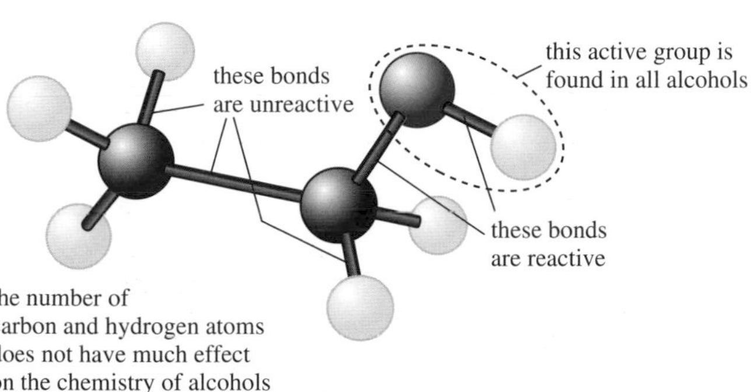

Figure 5.2.5 ◀ *The structure of ethanol labelled to show the reactive functional group and the hydrocarbon skeleton which does not change in many of the alcohol's reactions*

Examples of functional groups include:

>C=C< in alkenes, —Cl in chloroalkanes,

—OH in alcohols, >C—O in aldehydes and ketones,

—C(=O)—O—H in carboxylic acids

Some organic molecules have two or more functional groups. Lactic acid in sour milk, for example, has both an —OH group and a $-CO_2H$ group. In its reactions, lactic acid sometimes acts like an alcohol, sometimes like an acid and sometimes it shows the properties of both types of compound.

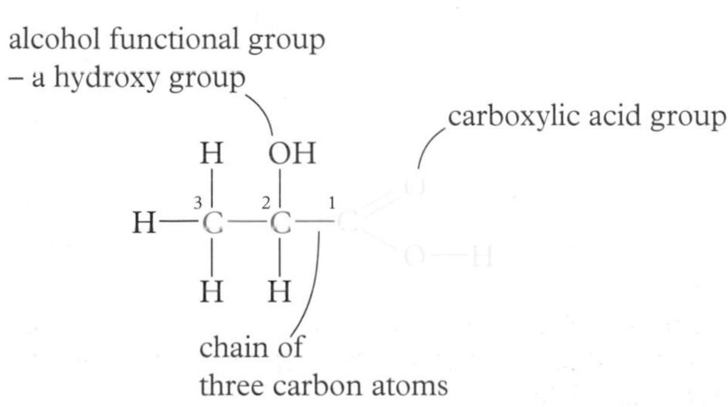

Figure 5.2.6 ◀ *The structure of lactic acid (2-hydroxypropanoic acid)*

Homologous series

Note

Homologous means 'related'. Homologous compounds have related structures and chemical properties.

An homologous series is a series of closely-related organic compounds. The compounds in an homologous series have the same functional group and can be described by a general formula. The formula of one member of the series differs from the next member by CH_2. The general formula of the alkane series is C_nH_{2n+2}

Figure 5.2.7 ▶ *The first six members of the homologous series of alkanes showing the change in structure from one to the next*

CD-ROM

Name	Formula	Displayed (graphical) formula
methane	CH_4	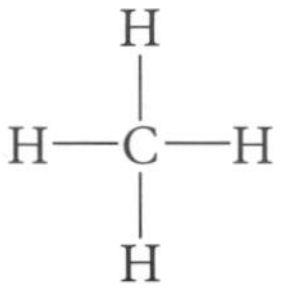 H–C–H with one H above and one H below the C
ethane	C_2H_6	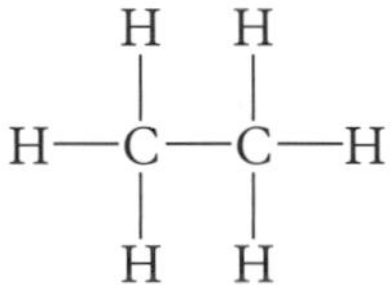 H–C–C–H with H above and below each C
propane	C_3H_8	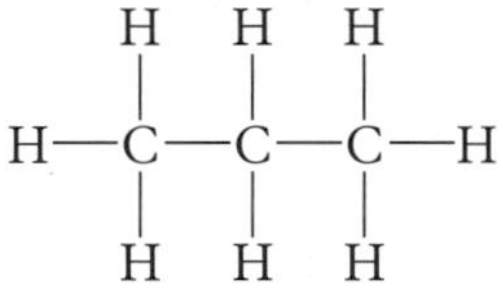 H–C–C–C–H with H above and below each C
butane	C_4H_{10}	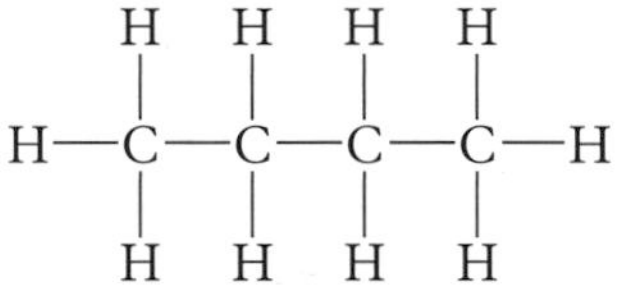 H–C–C–C–C–H with H above and below each C
pentane	C_5H_{12}	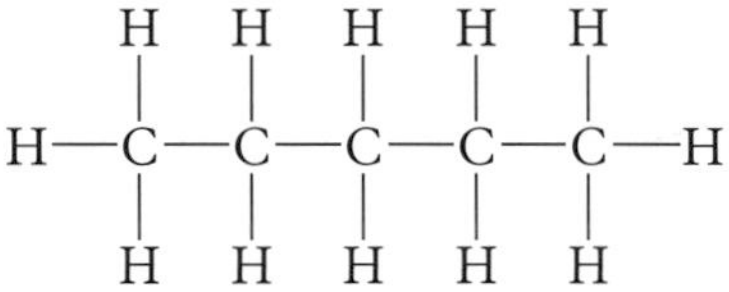 H–C–C–C–C–C–H with H above and below each C
hexane	C_6H_{14}	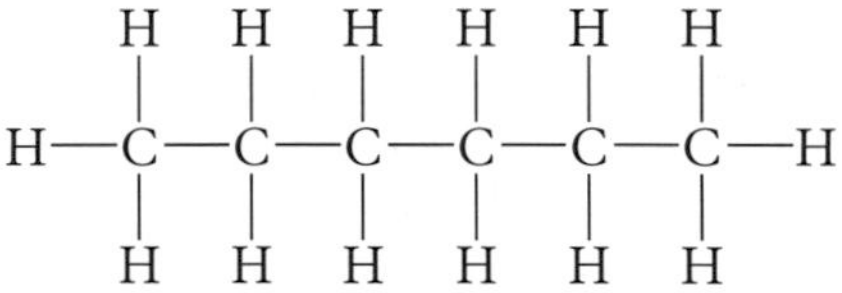 H–C–C–C–C–C–C–H with H above and below each C

Physical properties, such as the boiling point, show a steady trend in values along an homologous series.

The members of an homologous series have very similar chemical properties because they all have the same functional group.

Test yourself D

3 Use a set of molecular models, or computer software, to make three dimensional models of propene, ethanol, ethanoic acid, 1-chlorobutane, propanone and ethylamine.

4 Which of these molecular formulae are alkanes: C_2H_2, C_3H_8, C_4H_8, C_6H_{12}, C_7H_{16}?

5 What is the general formula of the series of alkenes?

6 Use a data book or database to find the boiling points of the series of alkanes. Plot a graph of boiling point against number of carbon atoms (electronically or on graph paper) for at least the first 10 members of the series. What does your graph show?

Isomerism

Isomers are compounds with the same molecular formula but different structures. They arise because of changes to the arrangement of atoms in molecules and changes to the shapes of molecules.

Structural isomers

Structural isomerism may exist because:

- the chain of carbon atoms is branched in different ways,

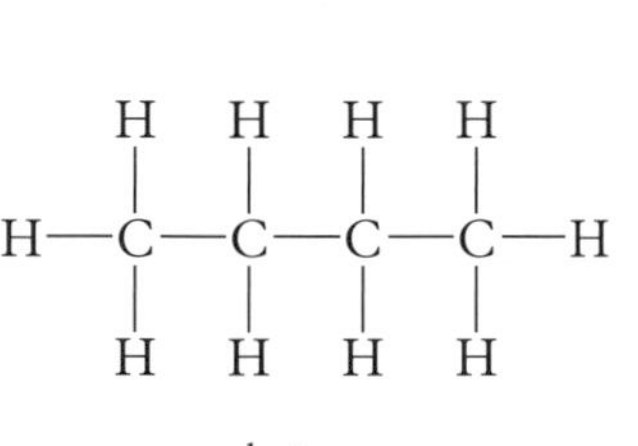

butane

2-methylpropane

Figure 5.2.8 ◀ *Isomers of C_4H_{10}*

- the functional group is in a different position,

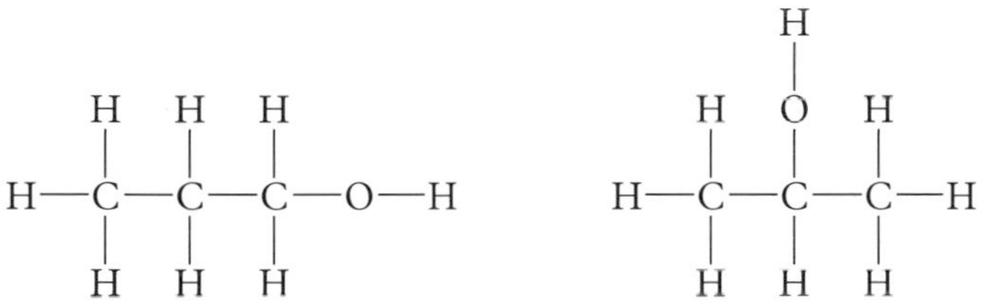

propan-1-ol

propan-2-ol

Figure 5.2.9 ◀ *Isomers of C_3H_8O*

- the functional groups are different,

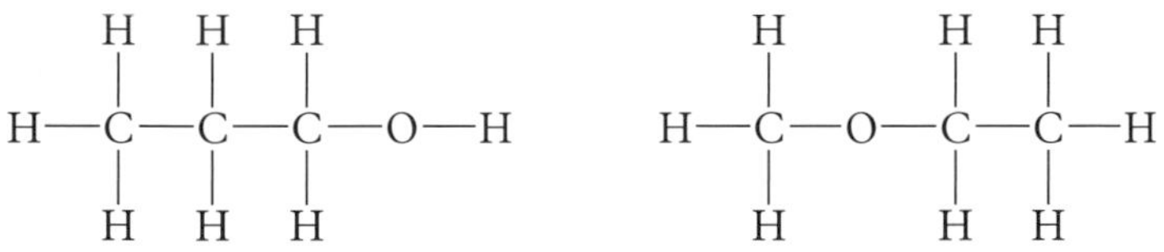

propan-1-ol
(an alcohol)

methyoxyethane
(an ether)

Figure 5.2.10 ◀ *Isomers of C_3H_8O*

Geometrical isomerism

Where there is a single bond, one part of a molecule can rotate relative to the rest of the molecule.

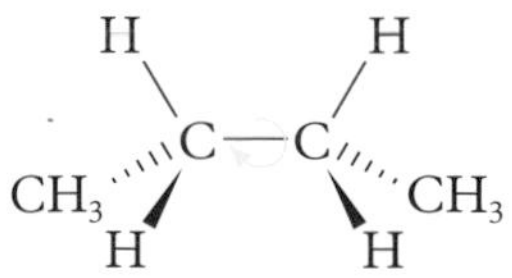

Just one compound because the ends of the molecule can rotate freely around the single bond

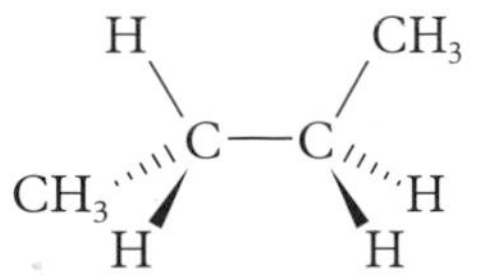

Figure 5.2.11 ◀ *Diagram to show rotation about a single bond*

Double bonding stops free rotation and this can account for the existence of geometric isomers.

Geometric isomers are molecules with the same molecular formulae and structural formulae but different shapes (geometries). Alkenes and other compounds with C=C double bonds may have geometrical isomers because there is no rotation about the double bond (Figure 5.2.12).

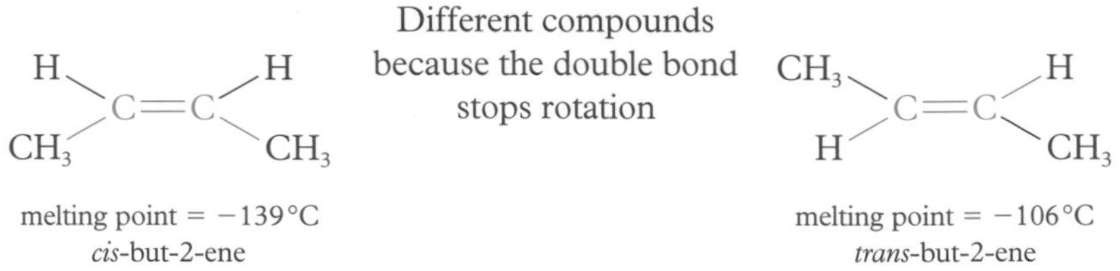

Figure 5.2.12 ► *Geometric isomers of but-2-ene. They are distinct compounds with different melting points, boiling points and densities* CD-ROM

H, CH_3 / C=C / H, CH_3	Different compounds because the double bond stops rotation	CH_3, H / C=C / H, CH_3
melting point = −139 °C *cis*-but-2-ene		melting point = −106 °C *trans*-but-2-ene

Note

'Trans' is a prefix in many words meaning 'across', such as transport, transplant and transmute.

Test yourself

7 Make models and draw the structures of the five isomers of C_6H_{14}.

8 Draw the structures of the two isomers with the molecular formula C_2H_6O. Which of the isomers is an alcohol? CD-ROM

9 Draw the structures of the alcohols with the molecular formula $C_4H_{10}O$.

10 Draw the structures of the three isomers of $C_2H_2Br_2$. Identify and make models of the two geometric isomers.

The existence of a ring of atoms in a structure can also stop free rotation and give rise to geometric isomerism.

The isomers are labelled *cis* and *trans*. In the *cis* isomer, similar functional groups are on the same side of the double bond. In the *trans* isomer, similar functional groups are on opposite sides of the double bond (Figure 5.2.12).

Weak bonds between molecules

Intermolecular forces are weak attractive forces between molecules (see page 83). Without intermolecular forces there could be no organic liquids or solids.

Generally intermolecular forces get bigger as molecules get longer with more electrons and a larger area of contact over which intermolecular forces can operate. This accounts for the rise in melting and boiling points as the number of carbon atoms increases in the series of straight chain alkanes.

Non-polar molecules with weak intermolecular forces tend to mix and dissolve in non-polar solvents (see page 89). The liquid alkanes, for example, mix freely with each other in petrol. Hydrocarbons do not, however, mix or dissolve in water; they float on it, forming an oily upper layer. The hydrogen bonding between polar water molecules is 10 times stronger than other intermolecular forces. Molecules which cannot form hydrogen bonds are unable to get in between the molecules of water.

Hydrogen bonding is an attraction between molecules which is much

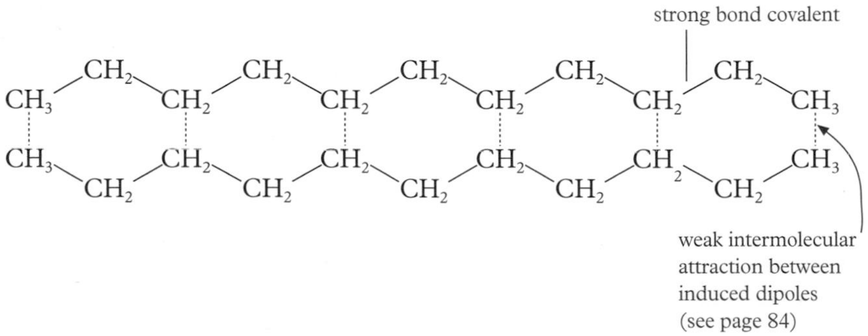

strong bond covalent

CH_3–CH_2–CH_2–CH_2–CH_2–CH_2–CH_2–CH_2–CH_2–CH_2–CH_3

CH_3–CH_2–CH_2–CH_2–CH_2–CH_2–CH_2–CH_2–CH_2–CH_2–CH_3

weak intermolecular attraction between induced dipoles (see page 84)

Figure 5.2.13 ► *The longer the molecules, the stronger the intermolecular forces*

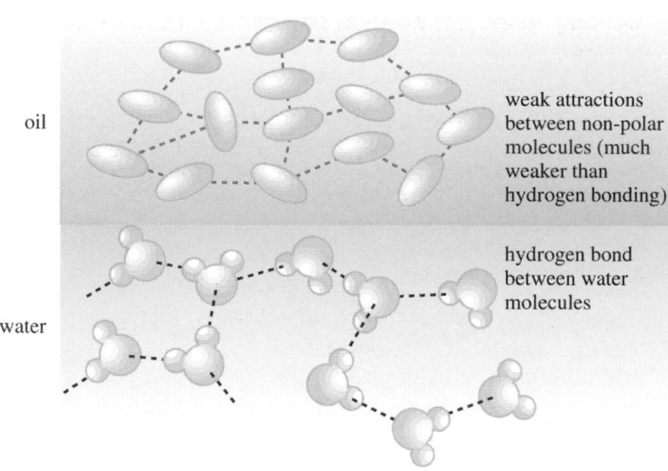

Figure 5.2.14 ◀
Oil and water do not mix. Strong bonding between water molecules keeps out the non-polar hydrocarbon molecules

stronger than other types of intermolecular force, but much weaker than covalent bonding (see page 85).

Hydrogen bonding affects molecules in which hydrogen is covalently bonded to one of the three highly electronegative elements nitrogen, oxygen and fluorine. In organic chemistry, this means that the properties of alcohols, carboxylic acids and amines are affected by hydrogen bonding.

Hydrogen bonding accounts for the solubility of alcohols and carboxylic acids in water. Solubility is associated with the polar functional groups in these compounds. As the length of the non-polar hydrocarbon chain gets longer and makes up a higher proportion of the molecules, the solubility decreases.

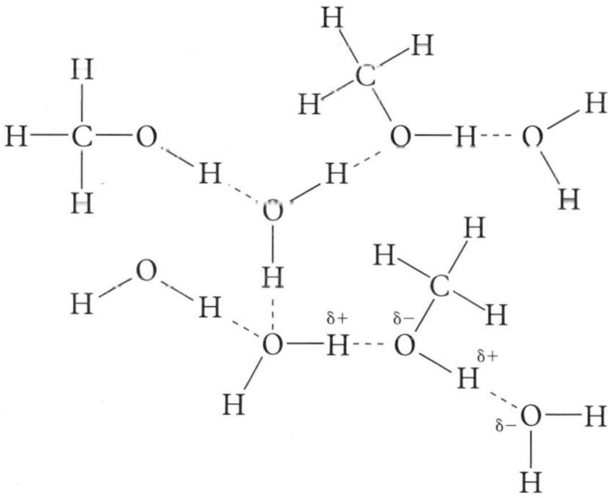

Figure 5.2.15 ◀
Like dissolves like. Methanol molecules hydrogen bonding with water molecules

Test yourself

11 Draw one possible structure for each of these molecules and decide which of them are polar and which are non-polar (see page 83): C_5H_{12}, CH_2O, C_3H_7OH, CCl_4.

12 Which of the molecules you have drawn in **11** would you expect to mix with water molecules?

5.3 Names of carbon compounds

This section gives you an outline of the systematic way that chemists name organic compounds. Given the systematic name you can work out the structure and vice versa.

IUPAC

The International Union of Pure and Applied Chemistry (IUPAC) is the recognised authority for the names of chemical compounds. IUPAC names are systematic names based on a set of rules which make it possible to work out the chemical structure of a compound from its name. Chemists increasingly use approved IUPAC names for simpler compounds but stick to traditional names when the systematic name is complex. The systematic name 2-hydroxypropane-1,2,3-tricarboxylic acid, for example, describes the structure but it is cumbersome compared to the traditional name citric acid. Chemists chose this name originally because the acid was found in citrus fruits.

Systematic names

Systematic names make it possible to work out the name from the formula and the formula from the name.

The name of an organic compound is based on the longest straight chain or main ring of carbon atoms in the carbon skeleton. If the main part is a straight chain, the name is based on the corresponding alkane.

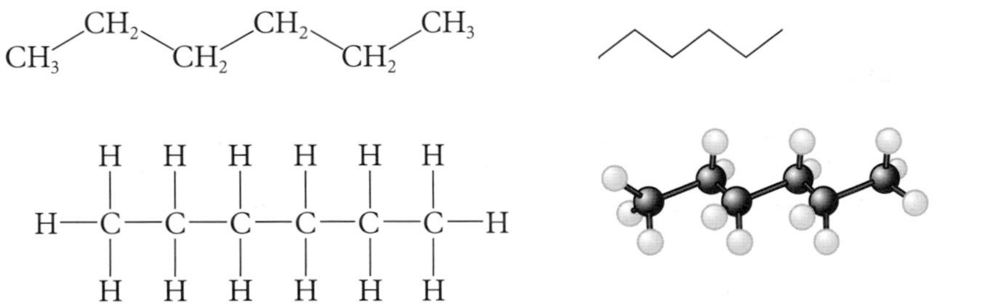

CH_3 CH_2 CH_2 CH_2 CH_2 CH_3

$$\begin{array}{ccccccccccccc} & & H & & H & & H & & H & & H & & H \\ & & | & & | & & | & & | & & | & & | \\ H & - & C & - & C & - & C & - & C & - & C & - & C & - & H \\ & & | & & | & & | & & | & & | & & | \\ & & H & & H & & H & & H & & H & & H \end{array}$$

Figure 5.3.1 ▶ *Four ways of representing the structure of hexane. A 'straight' chain alkane molecule with no branches* CD-ROM

Numbering the carbon atoms identifies the positions of side chains and functional groups, with the number repeated if there are two side groups on the same carbon atom. Chemists number the chains from the end that gives the lowest possible numbers in the name. They leave out the numbers when there is no doubt about the position of the side chains or functional groups as in ethanol, propanoic acid and butanone.

Alkyl groups

Alkyl groups are groups of carbon and hydrogen atoms which form part of the structure of molecules. The simplest example is the methyl group CH_3— which is methane with one hydrogen atom removed. In general, alkyl groups are alkane molecules minus one hydrogen atom.

Definitions

Skeletal formulae are outline formulae for carbon compounds which are a useful shorthand for complex molecules. The formulae need careful study because they only represent the hydrocarbon part of the molecule with lines for the bonds between carbon atoms while leaving out the symbols for the carbon and hydrogen atoms. Functional groups are included.

Alkyl group	Formula
methyl	$CH_3—$
ethyl	$CH_3CH_2—$
propyl	$CH_3CH_2CH_2—$
butyl	$CH_3CH_2CH_2CH_2—$

Figure 5.3.2 ◄ *Structures of alkyl groups*

A useful shorthand for any alkyl group is the capital letter R. Other alkyl groups are then represented by R′ or R″.

Prefixes (in front) and suffixes (following) the hydrocarbon name identify the side chains and functional groups.

Where there are two or more of the same side chain or functional group the number is stated as di-, tri-, tetra- and so on.

Prefixes are used for:

- alkyl groups, such as 2,3-dimethylbutane

```
     H   CH3 CH3  H
     |    |   |   |
    1|   2|  3|  4|
 H — C — C — C — C — H   chain of four
     |    |   |   |      carbon atoms
     H    H   H   H
```

- halogenoalkanes, such as 1-bromobutane, 2-iodopropane or trichloromethane.

```
    H   H   H   H                  H   H   H                H
    |   |   |   |                  |   |   |                |
   4|  3|  2|  1|                 1|  2|  3|                |
H — C — C — C — C — Br        H — C — C — C — H      Cl — C — Cl
    |   |   |   |                  |   |   |                |
    H   H   H   H                  H   I   H                Cl
```

Suffixes are used for:

- double bonds in alkenes, such as but-1-ene (with a double bond between the first and second carbon atoms) and but-2-ene (with a double bond between the second and third carbon atoms)

```
    H   H   H                          H   H   H   H
    |   |   |        H                 |   |   |   |
   4|  3|  2|   1   /                 4|  3|  2|  1|
H — C — C — C = C                  H — C — C = C — C — H
    |   |           \                  |           |
    H   H            H                 H           H
```

- alcohols, such as propan-1-ol and propan-2-ol

```
    H   H   H                          H   H   H
    |   |   |                          |   |   |
   3|  2|  1|                         3|  2|  1|
H — C — C — C — O — H              H — C — C — C — H
    |   |   |                          |   |   |
    H   H   H                          H   O   H
                                           |
                                           H
```

Alkane	No. of C atoms
methane	1
ethane	2
propane	3
butane	4
pentane	5
hexane	6

- aldehydes, such as propanal

$$\mathrm{H{-}\overset{\displaystyle H}{\underset{\displaystyle H}{C}}{-}\overset{\displaystyle H}{\underset{\displaystyle H}{C}}{-}C{\overset{\displaystyle \diagup\!\!\!\!{=}O}{\underset{\displaystyle \diagdown H}{}}}}$$

the aldehyde group has to be at the end of the chain

- ketones, such as pentan-2-one

$$\mathrm{H{-}\overset{\displaystyle H}{\underset{\displaystyle H}{\overset{1}{C}}}{-}\overset{\displaystyle O}{\overset{\|}{\overset{2}{C}}}{-}\overset{\displaystyle H}{\underset{\displaystyle H}{\overset{3}{C}}}{-}\overset{\displaystyle H}{\underset{\displaystyle H}{\overset{4}{C}}}{-}\overset{\displaystyle H}{\underset{\displaystyle H}{\overset{5}{C}}}{-}H}$$

- carboxylic acids, such as hexanoic acid (with six carbon atoms), or ethanedioic acid (with two carboxylic acid groups).

$$\mathrm{H{-}\overset{\displaystyle H}{\underset{\displaystyle H}{C}}{-}\overset{\displaystyle H}{\underset{\displaystyle H}{C}}{-}\overset{\displaystyle H}{\underset{\displaystyle H}{C}}{-}\overset{\displaystyle H}{\underset{\displaystyle H}{C}}{-}\overset{\displaystyle H}{\underset{\displaystyle H}{C}}{-}C(=O){-}O{-}H} \qquad \mathrm{H{-}O{-}C(=O){-}C(=O){-}O{-}H}$$

the carboxylic acid group has to be at the end of the chain

You will find more examples to illustrate the rules for naming compounds on pages 197, 201, 207 and 212.

Test yourself

1 Name the compounds with these structures:

$$CH_3{-}\underset{\displaystyle CH_3}{\underset{|}{CH}}{-}CH_2{-}CH_3$$

$$CH_3{-}CH_2{-}CH_2{-}Br$$

$$CH_3{-}\underset{\displaystyle CH_3}{\underset{|}{C}}{=}CH_2$$

CD-ROM

$$CH_3{-}CH_2{-}CH_2{-}CH_2{-}CH_2{-}OH$$

$$CH_3{-}CH_2{-}C(=O){-}OH$$

2 Write out the structures of these compounds: 2,2-dimethylbutane, 2-bromobutane, 2-methylbut-2-ene, 2-methylpropan-2-ol, ethanal, butanone, pentanoic acid.

3 Show why there is no need to include numbers in the names of ethanol, propanoic acid and butanone.

5.4 Types of organic reaction

Organic compounds can be involved in all the same types of reactions as inorganic chemicals, including acid–base, redox and hydrolysis reactions (see pages 28–34). Chemists also find it helpful to classify organic reactions by looking at what happens to the carbon molecules. Does the reaction replace some atoms by others, add bits to the molecules or split bits off?

Acid–base

Organic acids show the same characteristic properties as other acids. A familiar example of an organic acid is ethanoic acid which gives vinegar its flavour and smell.

Organic acids form solutions in water with a pH below 7. They also change the colours of acid–base indicators and react with bases such as ammonia to produce salts.

$$\underset{\text{ammonia}}{NH_3(aq)} + \underset{\text{ethanoic acid}}{CH_3CO_2H(aq)} \rightarrow \underset{\text{ammonium ethanoate}}{CH_3CO_2^-NH_4^+(aq)}$$

Organic acids are weak acids. What this means is that they only partially ionise when they dissolve in water. Only about one in a hundred molecules split up into ions in a dilute (0.1 mol dm^{-3}) solution of ethanoic acid. So the pH of the solution is about 3 – less acidic than a similar solution of a strong acid such as hydrochloric acid (pH = 1).

$$CH_3CO_2H(aq) \rightleftharpoons CH_3CO_2^-(aq) + H^+(aq)$$

Redox

Oxidation originally meant gaining oxygen or losing hydrogen but the term now covers all reactions in which atoms, molecules or ions lose electrons. The definition is further extended to cover molecules as well as ions by defining oxidation as a change which makes the oxidation number of an element more positive, or less negative (see page 133).

Oxidation number rules apply in principle in organic chemistry but it is often easier to use the older definitions.

Oxidation and reduction always go together in redox reactions. The orange dichromate(VI) ions turn to green chromium(III) ions when an acid

Test yourself

1 With the help of the examples on page 30, write word and symbol equations for the reactions of ethanoic acid with:

a) sodium carbonate, Na_2CO_3
b) the metal, magnesium.

2 Decide whether the reagent for each of these changes needs to be an acid, a base, an oxidising agent, a reducing agent or water for hydrolysis:

a) $CH_3CH_2CH_2OH \rightarrow CH_3CH_2CHO$
b) $CH_3CH_2CHO \rightarrow CH_3CH_2CO_2H$
c) $CH_3CO_2H \rightarrow CH_3CO_2^-Na^+$
d) $CH_3CH_2CHO \rightarrow CH_3CH_2CH_2OH$

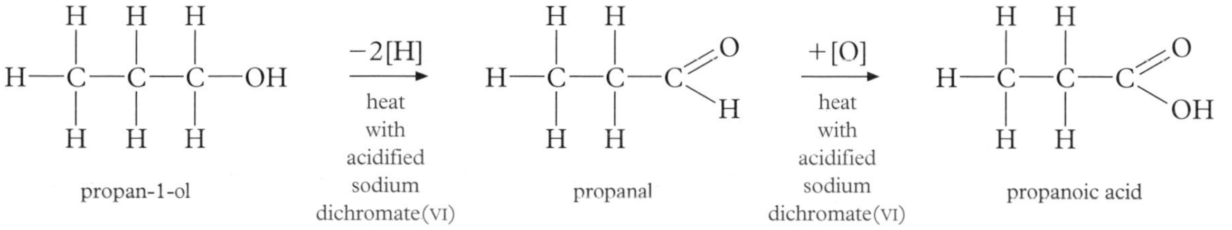

$$\underset{\text{propan-1-ol}}{CH_3CH_2CH_2OH} \xrightarrow[\text{heat with acidified sodium dichromate(VI)}]{-2[H]} \underset{\text{propanal}}{CH_3CH_2CHO} \xrightarrow[\text{heat with acidified sodium dichromate(VI)}]{+[O]} \underset{\text{propanoic acid}}{CH_3CH_2COOH}$$

Figure 5.4.1 ▲
Two stage oxidation of a primary alcohol: first to an aldehyde (losing hydrogen) and then to a carboxylic acid (gaining oxygen) (see also page 215)

Figure 5.4.2 ▶ Reduction of a ketone to an alcohol by the addition of hydrogen

$$CH_3COCH_3 \text{ (propanone)} + 2[H] \text{ (from reducing agent)} \longrightarrow CH_3CH(OH)CH_3 \text{ (propan-2-ol)}$$

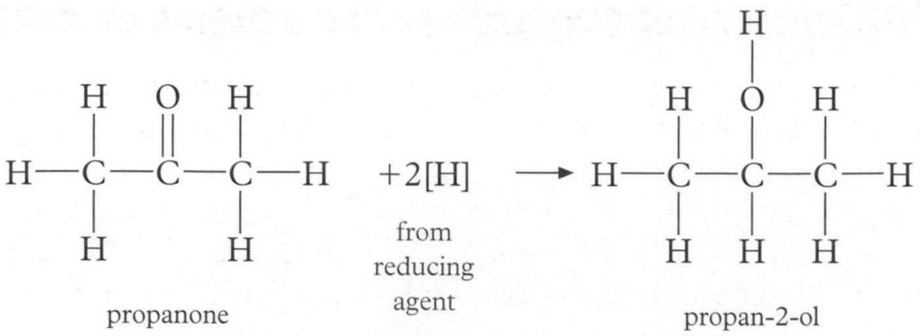

solution of sodium dichromate(VI) oxidises an alcohol. The reaction reduces chromium from the +6 to the +3 state.

Reduction is the opposite of oxidation. According to the older definitions, reduction meant removal of oxygen or the addition of hydrogen.

Hydrolysis – splitting apart with water

Hydrolysis is any reaction which splits apart a compound with water. Hydrolysis reactions are often catalysed by acids or alkalis. One example is the hydrolysis of an ester (see page 215), splitting it up into an organic acid and an alcohol.

Figure 5.4.3 ▼ Hydrolysis of the ester ethyl ethanoate catalysed by acid

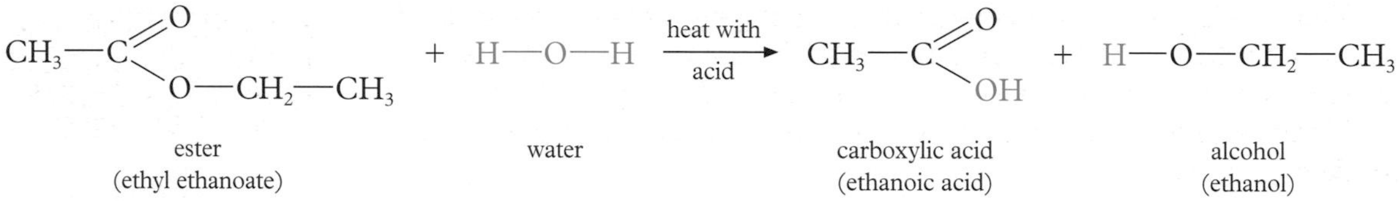

$$CH_3-C(=O)-O-CH_2-CH_3 + H-O-H \xrightarrow[\text{acid}]{\text{heat with}} CH_3-C(=O)-OH + H-O-CH_2-CH_3$$

ester (ethyl ethanoate) + water → carboxylic acid (ethanoic acid) + alcohol (ethanol)

Note

'Hydro-lysis' literally means 'water-splitting'. The term is based on two Greek words.

Hydrolysis is used to make soap from fats and oils. This is a further example of ester hydrolysis. The catalyst used in soap making is an alkali.

Another example of hydrolysis in organic chemistry converts halogenoalkanes to alcohols (see page 208).

Addition – adding bits to molecules

Note

Unsaturated compounds have double or triple bonds and can use their 'spare bonding' when they react to add on extra atoms.

During an addition reaction, two molecules add together to form a single product. Bromine, for example, adds to ethene to form the addition product 1,2-dibromoethane.

Addition reactions are characteristic of unsaturated compounds with double bonds, especially alkenes and carbonyl compounds.

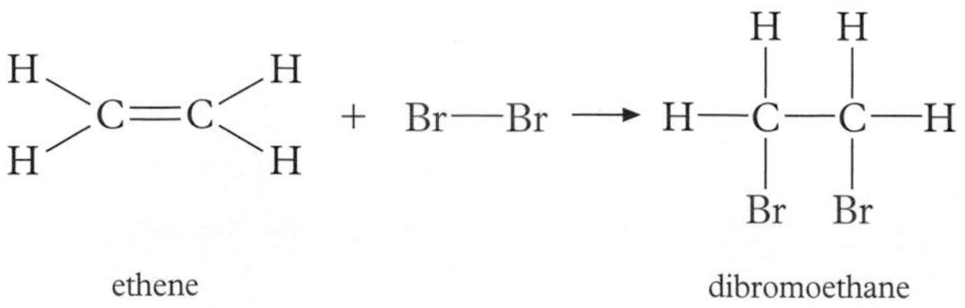

Figure 5.4.4 ▶ Addition of bromine to ethene (see page 203)

$$H_2C=CH_2 \text{ (ethene)} + Br-Br \longrightarrow CH_2Br-CH_2Br \text{ (dibromoethane)}$$

Substitution – replacing one or more atoms by others

Substitution reactions replace an atom or group of atoms by another atom or group of atoms. An example is the reaction of butan-1-ol with hydrogen bromide.

The hydrogen bromide is often made in the reaction mixture by warming sodium bromide with concentrated sulfuric acid. The mixture turns yellow as some bromine forms but bromine itself does not react with the alcohol.

Substitution reactions are characteristic of halogenoalkanes (see page 208). Other examples of substitution reactions include the replacement of hydrogen atoms by chlorine or bromine atoms (see page 199).

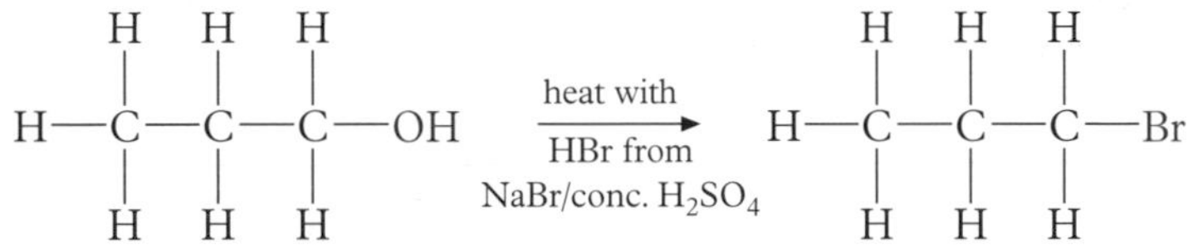

```
      H   H   H                                   H   H   H
      |   |   |              heat with            |   |   |
  H — C — C — C — OH      ─────────────▶      H — C — C — C — Br
      |   |   |              HBr from             |   |   |
      H   H   H          NaBr/conc. H2SO4         H   H   H
```

Figure 5.4.5 ◀
Substitution of a bromine atom for an —OH group

Elimination – splitting bits off from molecules

An elimination reaction splits off a simple molecule from within a larger molecule to form a double bond.

An example of an elimination reaction is the removal of water from an alcohol to form an alkene. This is a useful reaction in synthesis for introducing double bonds into molecules.

The conditions for reaction are either to pass the vapour of the alcohol over a hot solid catalyst such as aluminium oxide, or to dehydrate the alcohol by heating it with concentrated sulfuric acid (see page 214).

Elimination of a hydrogen halide from a halogenoalkane also produces in alkene (see page 210).

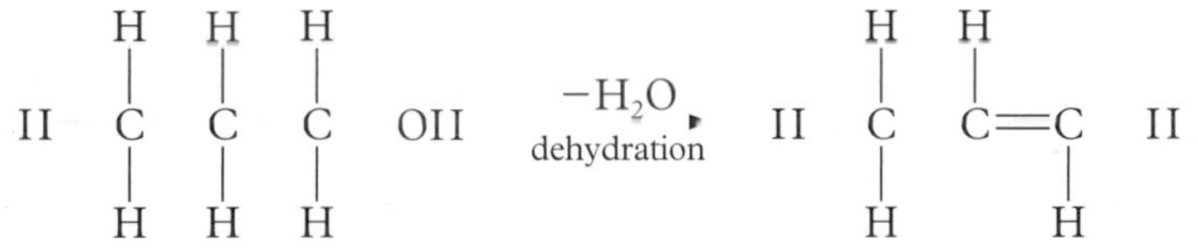

```
      H   H   H                            H   H
      |   |   |                            |   |
  II  C   C   C   OII      −H2O      II    C   C ═ C   II
      |   |   |        dehydration         |       |
      H   H   H                            H       H
```

Figure 5.4.6 ▲
Formation of an alkene from an alcohol

Test yourself

3 Decide which type of change is involved in each of these conversions (addition, substitution or elimination):

a) $CH_3CH_2OH \rightarrow CH_2{=}CH_2$
b) $CH_3CH_2Br \rightarrow CH_3CH_2OH$
c) $CH_2{=}CH_2 \rightarrow CH_3CH_2Br$
d) $CH_3{-}CH_2I \rightarrow CH_3CH_2CN$

5.5 Mechanisms of organic reactions

The mechanism of a reaction shows, step by step, the bonds which break and the new bonds which form as reactants turn into products. Chemists have used much ingenuity to work out mechanisms. They began with simpler examples but are now using their knowledge to explain what happens during industrial processes and in living cells where enzymes control the biochemical changes.

Homolytic bond breaking

A covalent bond involves a shared pair of electrons. There are two ways in which a bond can break. In one way, each atom keeps one electron as the bond breaks. This is 'equal splitting' (homolytic fission).

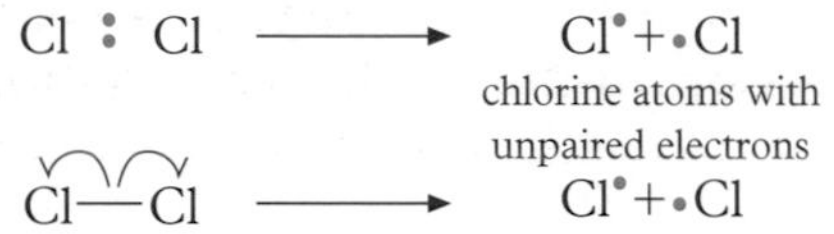

$$\text{Cl} : \text{Cl} \longrightarrow \text{Cl}^{\bullet} + {}^{\bullet}\text{Cl}$$

chlorine atoms with unpaired electrons

$$\text{Cl—Cl} \longrightarrow \text{Cl}^{\bullet} + {}^{\bullet}\text{Cl}$$

Figure 5.5.1 ▶ *Homolytic bond breaking. Note the use of single-headed curly arrows to show what happens to the electrons as the bond breaks. The covalent bond breaks so that the atoms joined by the bond separate, each taking one of the shared pair of electrons*

Note

The prefix 'homo-' always means 'the same'. Chemical terms which include this prefix include: homolytic fission, homogeneous catalyst and homologous series.

The prefix 'hetero-' means 'different'. Chemical terms which include this prefix include: heterolytic fission and heterogeneous catalyst.

This type of splitting produces fragments with unpaired electrons. Chemists call these fragments free radicals. They exist for a short time during the reaction but quickly react to form new products. So free radicals are usually very short-lived. They are intermediates which come into being during a reaction but then disappear as the reaction carries on.

The symbol for a free radical generally shows the unpaired electron as a dot. Other paired electrons in the outer shells are often not shown.

Free radicals are intermediates in reactions taking place either in the gas phase or in a non-polar solvent. Shining ultra-violet light on a reaction mixture can speed up free-radical reactions.

Examples of free-radical processes include the thermal cracking of hydrocarbons (see page 222), the burning of petrol in an engine cylinder (see page 219) and the substitution reactions of alkanes with halogens (see page 199). Free-radical reactions are important high in the atmosphere where gases are exposed to intense ultra-violet radiation from the Sun. The reactions which form and destroy the ozone layer are free-radical reactions (see pages 228–229).

Heterolytic bond breaking

There is another type of bond breaking in which one atom takes both of the electrons from a covalent bond, leaving the other atom with none.

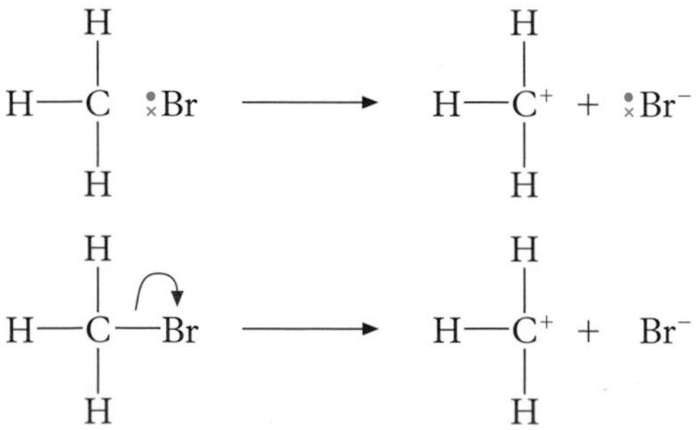

$$\text{H}_3\text{C} \; {}^{\bullet}_{\times}\text{Br} \longrightarrow \text{H}_3\text{C}^{+} + {}^{\bullet}_{\times}\text{Br}^{-}$$

$$\text{H}_3\text{C—Br} \longrightarrow \text{H}_3\text{C}^{+} + \text{Br}^{-}$$

Figure 5.5.2 ▶ *Heterolytic bond breaking. Note the use of double-headed curly arrows to show what happens to the electrons as the bond breaks. The covalent bond breaks so that the atoms joined by the bond separate with one atom taking both electrons of the shared pair*

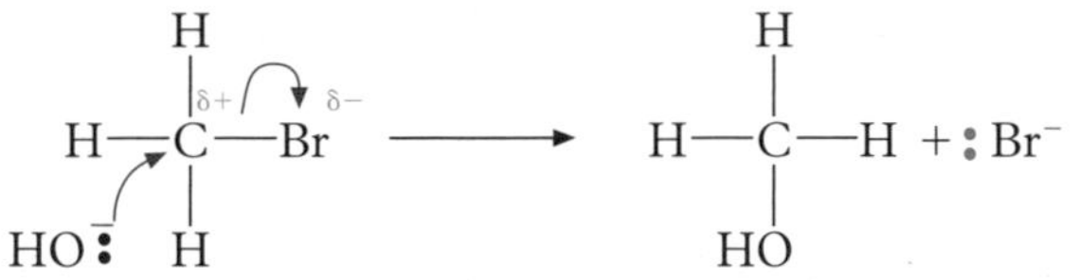

$$\mathrm{H{-}C(H)_2{-}Br}\ (\delta+\ \mathrm{C},\ \delta-\ \mathrm{Br}) + \mathrm{HO}^{-} \longrightarrow \mathrm{H{-}C(H)_2{-}OH} + \mathrm{:Br^-}$$

Figure 5.5.3 ◀
An ionic reagent attacking a polar bond leading to heterolytic bond breaking

Heterolytic bond breaking produces ionic intermediates in reactions. This type of bond breaking is favoured when reactions take place in polar solvents such as water. Often the bond which breaks is already polar (see page 81) with a δ+ end and a δ− end.

Some of the reagents which start reactions seek out the δ+ end of polar bonds. These are nucleophiles.

Other reagents seeks out the δ− end of polar bonds. These are electrophiles.

Nucleophiles

Nucleophiles are molecules or ions with a lone pair of electrons which can form a new covalent bond. They are 'electron-pair donors'. Nucleophiles are reagents which attack molecules where there is a partial positive charge, δ+, so they seek out positive charges – they are 'nucleus-loving'.

The substitution reactions of halogenoalkanes involve nucleophiles.

Electrophiles

Electrophiles are reactive ions and molecules which attack parts of molecules which are rich in electrons. They are 'electron-loving' reagents. Electrophiles form a new bond by accepting a pair of electrons from the molecule attacked during a reaction.

An example of an electrophile is the H atom at the δ+ end of the H—Br bond in hydrogen bromide. See, for example, the electrophilic addition reactions of alkenes (page 205).

Figure 5.5.4 ▼
Examples of nucleophiles

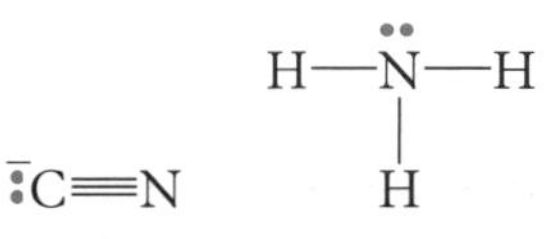

$:\mathrm{C{\equiv}N}^-$ — cyanide ion

$\mathrm{H{-}\ddot{N}(H){-}H}$ — ammonia molecule

Note

Curly arrows describe the movement of electrons as bonds break and form in the steps which describe the mechanism of a reaction. A curly arrow with both halves of the arrow head shows the movement of a pair of electrons. Note that the tail of the arrow starts where the electron pair begins. The head of the arrow points to where the electron pair will be after the change.

A curly arrow with only half an arrow head indicates the movement of a single electron.

Test yourself

1 In each of these examples decide whether the reagent attacking the carbon compound is a free radical, a nucleophile or an electrophile:

a) $CH_3CH_2I + H_2O \rightarrow CH_3CH_2CH + HI$

b) $CH_2{=}CH_2 + Br_2 \rightarrow CH_2BrCH_2Br$

c) $Cl\bullet + CH_4 \rightarrow CH_3Cl + H\bullet$

d) $CH_3CH_2Br + CN^- \rightarrow CH_3CH_2CN + Br^-$

5.6 Hydrocarbons

Hydrocarbons are important because they make up the majority of crude oil – the source of most fuels and the main raw material for the chemical industry. Hydrocarbons are compounds which consist of just carbon and hydrogen.

Types of hydrocarbon

Aliphatic hydrocarbons

Aliphatic hydrocarbons are those with no rings of carbon atoms. The chains of carbon atoms may be branched or unbranched. Alkanes, alkenes and alkynes are all aliphatic compounds.

Alicyclic hydrocarbons

Alicyclic hydrocarbons are those with rings of carbon atoms. Examples are cycloalkanes and cycloalkenes.

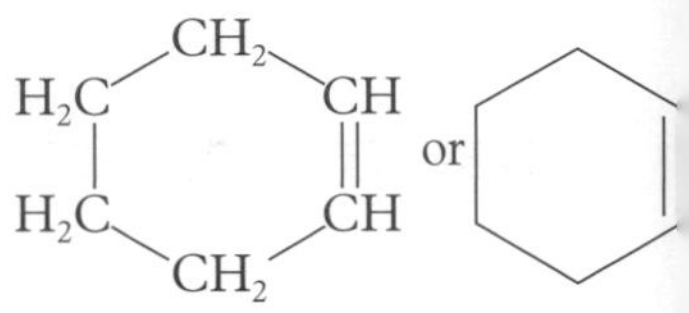

cyclohexene or C_6H_{10}

Figure 5.6.1 ▲ *Structures of a cyclic hydrocarbon: cyclohexene*

Saturated compounds

Saturated compounds contain only single bonds between the atoms in their molecules. Examples of saturated hydrocarbons are the alkanes.

The term 'saturated' is also used for compounds with saturated hydrocarbon chains such as the saturated fats and fatty acids. Saturated compounds do not undergo addition reactions.

Unsaturated compounds

Unsaturated compounds contain one or more double or triple bonds between atoms in their molecules. This term is often applied to the hydrocarbons alkenes and alkynes which typically undergo addition reactions. The term is also commonly used to describe unsaturated fats and fatty acids which have double C=C bonds in their hydrocarbon side chains.

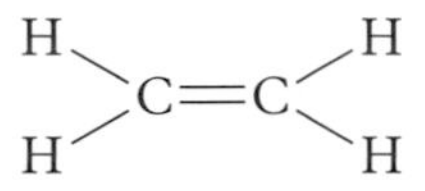

ethene
an alkene

H—C≡C—H

ethyne
an alkyne

Figure 5.6.2 ▲ *Examples of unsaturated hydrocarbons*

Arenes

Arenes are hydrocarbons such as benzene, methylbenzene and naphthalene. They are ring compounds in which there are delocalised electrons. Traditionally chemists called the arenes 'aromatic hydrocarbons' ever since the German organic chemist, Friedrich Kekulé (1829–1896) was struck by the fragrant smell of oils such as benzene. In the modern name the 'ar-' comes from aromatic and the ending '-ene' means that these hydrocarbons are unsaturated compounds. They are, however, much less reactive than alkenes.

Benzene is more stable and less reactive than expected for a compound with three double bonds.

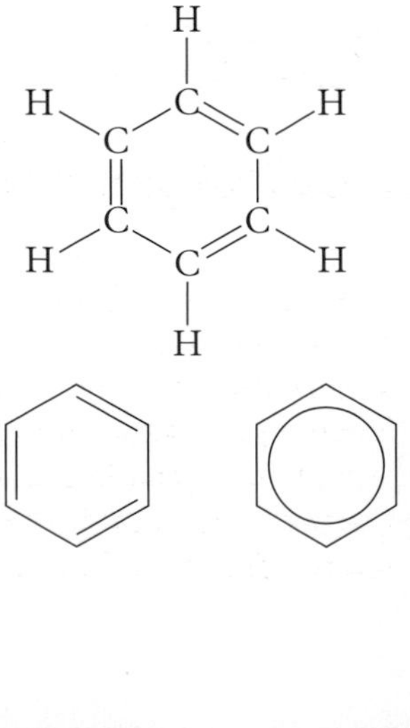

Figure 5.6.3 ► *Representations of a benzene molecule. Benzene does not behave like a cyclic alkene with three double bonds. The ring in the third structure signifies that each carbon contributes one electron to a cloud of delocalised electrons*

5.7 Alkanes

Alkanes are the hydrocarbons which make up most of crude oil and natural gas. They are saturated hydrocarbons with the general formula C_nH_{2n+2}. The carbon atoms in alkane molecules may be in straight chains or branched chains but all the bonds are single bonds.

CD-ROM

Structures and names

The names of branched alkanes are based on the longest straight chain in the molecule with the positions of the side-chain alkyl groups identified by numbering the carbon atoms (see page 188).

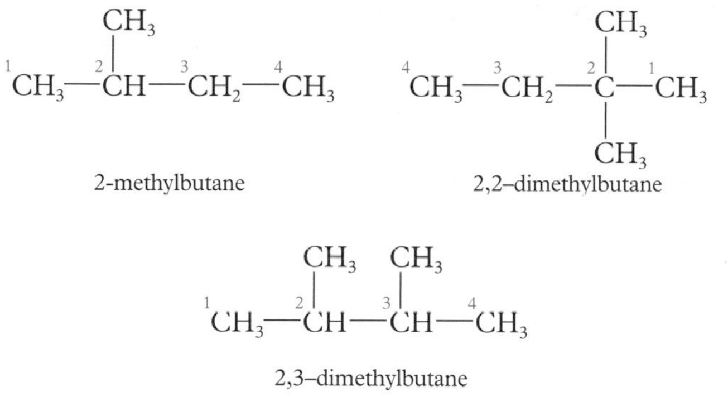

$$\overset{1}{CH_3}-\overset{2}{CH}(CH_3)-\overset{3}{CH_2}-\overset{4}{CH_3}$$

2-methylbutane

$$\overset{4}{CH_3}-\overset{3}{CH_2}-\overset{2}{C}(CH_3)_2-\overset{1}{CH_3}$$

2,2–dimethylbutane

$$\overset{1}{CH_3}-\overset{2}{CH}(CH_3)-\overset{3}{CH}(CH_3)-\overset{4}{CH_3}$$

2,3–dimethylbutane

Figure 5.7.1 ◀ *Names and structures of three branched alkanes*

Physical properties

Alkane molecules are non-polar so they do not mix with or dissolve in polar solvents such as water. The molecules are only held together by weak intermolecular forces (see pages 83–85 and 186). The longer the molecules, the greater the attraction between them. The boiling points rise as the number of carbon atoms per molecule increases. Alkanes in the range C_1 to C_4 are gases at room temperature and pressure. Under the same conditions, alkanes in the range C_5–C_{17} are liquids while those with more than 17 carbon atoms per molecule are solids. Liquid alkanes with longer chain lengths are viscous liquids used as lubricants.

Chemical properties

The bond enthalpies for C—C and C—H bonds are high so the bonds are relatively hard to break. In addition the bonds are non-polar. This means that alkanes are very unreactive with ionic reagents in water such as acids and alkalis as well as oxidising and reducing reagents. Three important reactions of alkanes are: combustion, halogenation and cracking. Cracking and the halogenation of alkanes are examples of free-radical chain reactions. In all these reactions, the bond breaking is homolytic and the intermediates are free radicals (see page 194).

Test yourself D

1. Look up the boiling points of the three isomers of C_5H_{12}. How does chain branching affect the boiling points of these isomers?
2. Compare the boiling points of a number of 2-methyl alkanes with the unbranched isomer. Do your findings confirm your findings in **1**?
3. Use what you know of intermolecular forces to suggest an explanation for the effect of chain branching on the boiling points of alkanes.

Burning

Many common fuels consist mainly of alkanes. The hydrocarbons burn in air forming carbon dioxide and water. If the air is in short supply, the products may include particles of carbon (soot) and the toxic gas carbon monoxide. Burning is highly exothermic.

Figure 5.7.2 ► *Cooking using a propane burner* CD-ROM

Test yourself

4 Write an equation for the complete combustion of butane.

5 Why is the incomplete combustion of alkane fuels hazardous to health?

6 Write an equation to represent the cracking of ethane to make ethene and hydrogen.

7 What is the evidence in Figure 5.7.3 that the product molecules are smaller and more reactive than the starting materials?

Cracking

Oil refineries use cracking to get the best value from the fractions distilled from crude oil. Cracking is a process which uses heat and catalysts to break up large molecules into smaller and more useful molecules. Cracking also converts alkanes to alkenes which are feedstocks for making petrochemicals.

Cracking takes place at high temperatures. Cracking with steam produces high yields of alkenes which are useful to the chemical industry. Cracking with a catalyst is the process which oil refineries adopt to make as much petrol as possible from oil fractions.

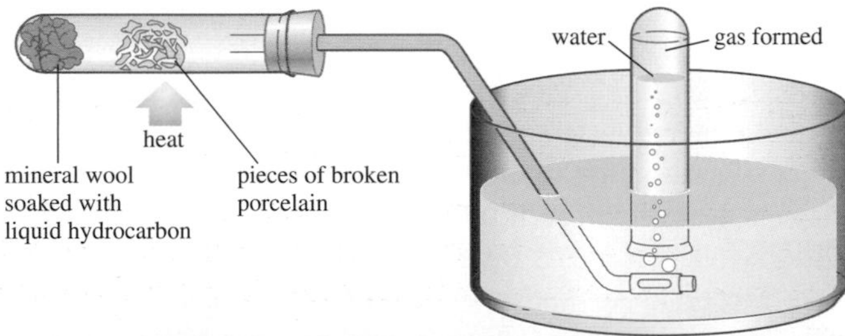

Figure 5.7.3 ► *Small scale cracking of a mixture of hydrocarbons*

Property	Liquid hydrocarbon	Gas formed by cracking
colour	colourless	colourless
smell	no smell	sweetish smell
does it burn?	burns on strong heating	burns easily with a yellow flame
reaction with aqueous bromine, which is orange	no reaction	the bromine solution becomes colourless on shaking

Product	Starting material ethane	Starting material propane
hydrogen	5	2
methane	9	27
ethene	78	42
propene	3	19
buta-1,3-diene	2	3
petrol	3	7

Figure 5.7.4 ◀
Products of steam cracking giving the percentages of the products by mass

Reactions with chlorine and bromine

The reactions of alkanes with chlorine and bromine are important because they can be the first step in making other valuable chemicals. The products, halogenoalkanes, also have a variety of important uses (see page 207). Alkanes react with chlorine or bromine either on heating or when exposed to ultra-violet light. These are substitution reactions in which hydrogen atoms are replaced by halogen atoms (see page 193). Any of the hydrogen atoms in an alkane may be replaced. The reaction can continue until all the hydrogen atoms have been substituted for all the halogen atoms. Thus the product is a mixture of compounds.

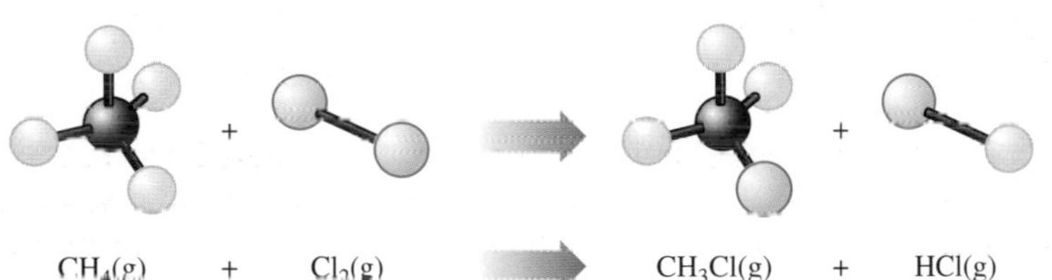

Figure 5.7.5 ◀
Models representing one of the possible substitution reactions of methane with chlorine

High temperatures or ultra-violet light can break covalent bonds producing free radicals. Free-radical chain reactions involve three stages:

- initiation – the step which produces free radicals
- propagation – steps which produce products and more free radicals
- termination – steps which remove free radicals by turning them into molecules.

Initiation: $Cl{-}Cl \rightarrow Cl\bullet + Cl\bullet$

Propagation: $CH_4 + Cl\bullet \rightarrow CH_3\bullet + HCl$
$CH_3\bullet + Cl_2 \rightarrow CH_3Cl + Cl\bullet$

Termination: $CH_3\bullet + CH_3\bullet \rightarrow CH_3CH_3$
$CH_3\bullet + Cl\bullet \rightarrow CH_3Cl$

The reaction of an alkane with bromine in ultra-violet light is a free-radical chain reaction. The main products are bromomethane and hydrogen bromide. The presence of some ethane in the mixture of products is evidence for the termination step.

Figure 5.7.6 ▶ *The effect of light on a solution of bromine in hexane*

Test yourself

8 Draw and name the structures of all the possible products when ethane reacts with chlorine.

9 Explain why a solution of bromine in hexane in a test tube remains orange in a dark place but soon fades and becomes colourless in sunlight. Why can acid fumes be detected above the solution once the colour has faded?

10 Write an equation for the initiation step of the catalytic cracking of ethane which produces methyl radicals.

5.8 Alkenes

Alkenes, such as ethene and propene, are invaluable to chemists because they react in many ways to make useful products. These alkenes come from cracking oil fractions. They are important starting points for synthesis because of the reactivity of the double bonds in their molecules. CD-ROM

Alkenes are unsaturated hydrocarbons with the general formula C_nH_{2n}. The characteristic functional group of the alkenes is a carbon–carbon double bond. The presence of the double bond makes alkenes more reactive than alkanes.

Structures and names

The name of an alkene is based on the name of the corresponding alkane with the ending change to -ene.

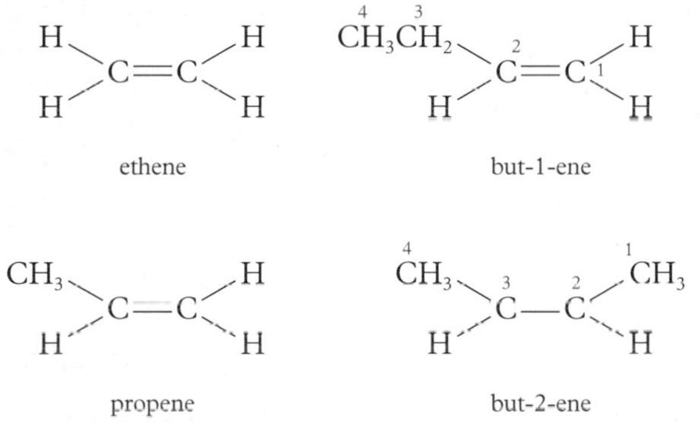

ethene: $H_2C{=}CH_2$

but-1-ene: $\overset{4}{C}H_3\overset{3}{C}H_2(H)\overset{2}{C}{=}\overset{1}{C}H_2$

propene: $CH_3(H)C{=}CH_2$

but-2-ene: $\overset{4}{C}H_3(H)\overset{3}{C}{=}\overset{2}{C}(H)\overset{1}{C}H_3$

Figure 5.8.1 ◄ *Names and structures of alkenes* CD-ROM

Where necessary, a number in the name shows the position of the double bond in the structure as in the two structural isomers but-1-ene and but-2-ene. Counting starts from the end of the chain that will give the lowest possible number in the name. The number in the name shows the first of the two atoms connected by the double bond. In but-1-ene, for example the double bond is between the first and the second atoms in the carbon chain.

Physical properties

The boiling points of alkenes increase as the number of carbon atoms in the molecules increase. Ethene, propene and the butenes are gases at room temperature. Alkenes with more than four carbon atoms are liquids or even solids. Alkenes, like other hydrocarbons, do not mix with or dissolve in water.

The double bond in alkenes

Chemists have extended the theory of atomic orbitals (see page 59) to describe the distribution of electrons in molecules. This molecular orbital theory is helpful when discussing the reactivity of alkenes.

Test yourself

1. Which of these unsaturated compounds have geometric isomers (page 186): but-1-ene, buta-1,3-diene, pent-2-ene? CD-ROM
2. Draw the displayed formulae for *cis* hex-2-ene and *trans* hex-2-ene.
3. Predict the bond angles in ethene (see page 80).
4. Make a summary of the reactions which produce alkenes (see pages 198, 210 and 211).

Molecular orbitals are the result of atomic orbitals overlapping and interacting as atoms bind together. The shapes of molecular orbitals show the regions in space where there is a high probability of finding electrons.

A sigma (σ) bond is a single covalent bond formed by a pair of electrons in an orbital in a molecule with the electron density concentrated between the two nuclei. Free rotation is possible about single bonds.

Sigma bonds can form by overlap of two s-orbitals, an s-orbital and a p-orbital, or two p-orbitals.

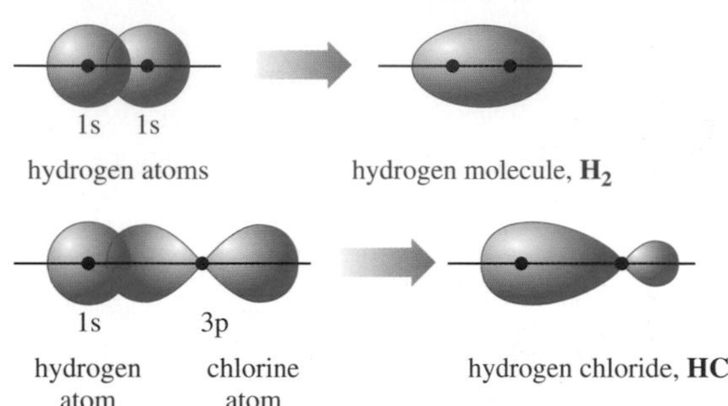

Figure 5.8.2 ▶ *Examples of sigma bonds in molecules*

A pi(π)-bond is the type of bond found in molecules with double and triple bonds. The bonding electrons are in a π-orbital formed by sideways overlap of two atomic p-orbitals. In a π-bond, the electron density is concentrated on either side of the line between the nuclei of the two atoms joined by the bond.

π-bonds prevent rotation about the double bond, giving rise to geometrical isomerism.

The reagents which attack alkenes are electrophiles attracted by the high electron density in the double bond.

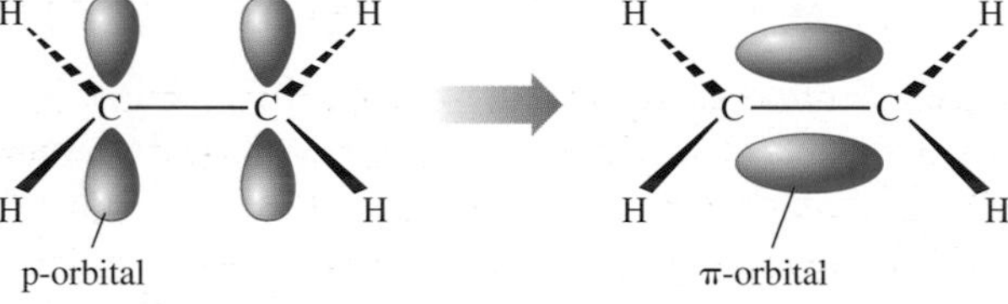

Figure 5.8.3 ▶ *π-bond in ethene*

Chemical reactions

The characteristic reactions of alkenes are addition reactions (see page 192).

> **Note**
>
> When describing an organic reaction, always write an equation, name the reagents and state the conditions (temperature, pressure, catalysts).

Addition of hydrogen

Hydrogen adds to C=C double bonds at room temperature in the presence of a platinum or palladium catalyst or on heating in the presence of a nickel catalyst. This is an example of hydrogenation.

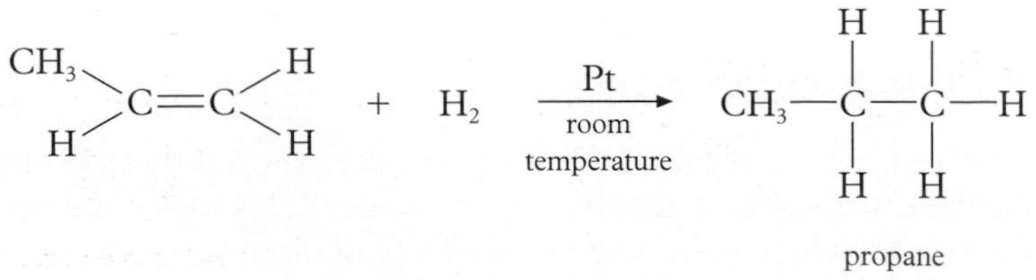

$$CH_3(H)C{=}CH_2 + H_2 \xrightarrow[\text{room temperature}]{Pt} CH_3{-}CH_2{-}CH_3$$

propane

Figure 5.8.4 ▶ *Addition of hydrogen to propene*

The advantage of using the heterogeneous catalyst, nickel, is that the metal can be held in a reaction vessel as the reactants flow in and the products flow out. There is no difficulty in separating the products from the catalyst.

Heterogeneous catalysts work by adsorbing reactants at active sites on the surface of the solid. Nickel acts as a catalyst for the addition of hydrogen to C=C in alkenes by adsorbing hydrogen molecules which probably split up into single atoms held on the surface of the metal crystals.

Definitions

Hydrogenation is a reaction which adds hydrogen to a compound.

A **heterogeneous catalyst** is a catalyst which is in a different phase from the reactants. Generally a heterogeneous catalyst is a solid while the reactants are gases or in solution.

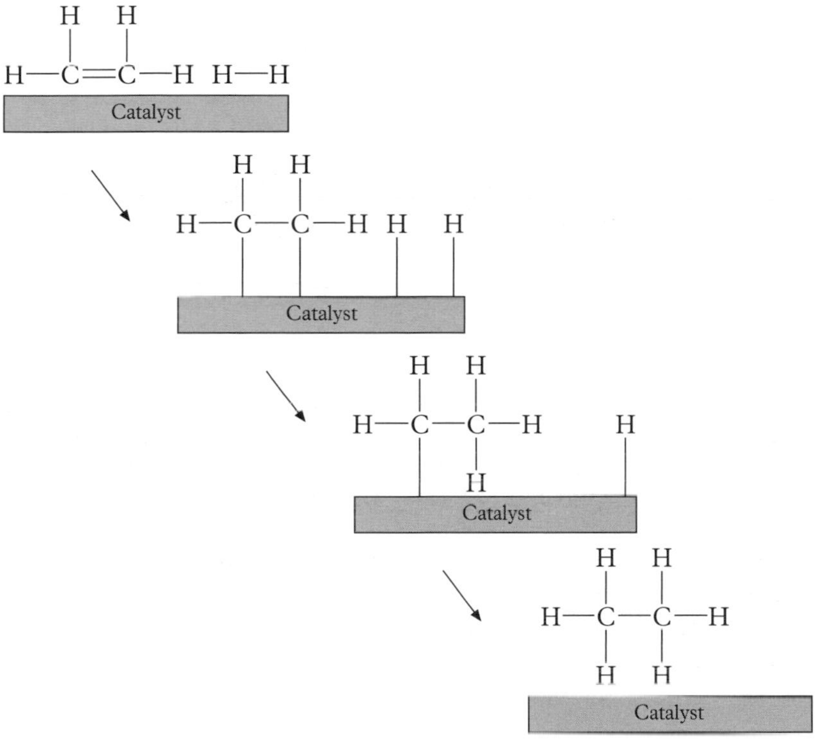

H—C(H)=C(H)—H H—H
Catalyst

H—C(H)—C(H)—H H H
Catalyst

H—C(H)—C(H)(H)—H H
Catalyst

H—C(H)(H)—C(H)(H)—H
Catalyst

Figure 5.8.5 ◀ *Possible mechanism for the hydrogenation of an alkene in the presence of a catalyst*

If a metal is to be a good catalyst for a hydrogenation reaction, it must not adsorb the hydrogen so strongly that the hydrogen atoms become unreactive. This happens with tungsten. Equally if adsorption is too weak there will not be enough adsorbed atoms for the reaction to go at a useful rate, as is the case with silver. The strength of adsorption must have an intermediate value. Suitable metals are nickel, platinum and palladium.

Hydrogenation converts liquid vegetable oils to the solid fats used as ingredients of margarine. In these reactions, hydrogen adds to double bonds in the hydrocarbon chains of unsaturated fatty acids.

Note

The reaction of aqueous bromine with a hydrocarbon is a useful test for double bonds. If a hydrocarbon is unsaturated, it will quickly decolourise aqueous bromine. Orange bromine in water adds to double bonds to form colourless products. CD-ROM

Addition of bromine or chlorine

Chlorine and bromine add rapidly to alkenes at room temperature.

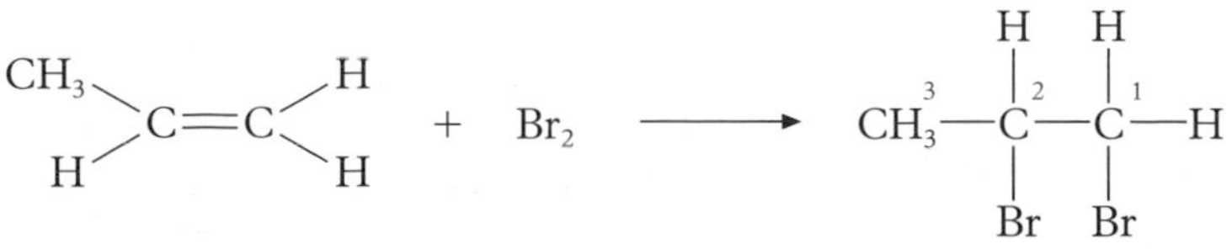

$$CH_3CH{=}CH_2 + Br_2 \longrightarrow \overset{3}{C}H_3—\overset{2}{C}H(Br)—\overset{1}{C}H_2(Br)$$

1,2-dibromopropane

Figure 5.8.6 ◀ *Addition of bromine to propene*

Addition of hydrogen halides

Hydrogen bromide reacts with ethene to form 1-bromoethane. The reaction happens at room temperature.

The other hydrogen halides, HCl and HBr react in a similar way.

Figure 5.8.7 ▶
Addition of hydrogen bromide to ethene

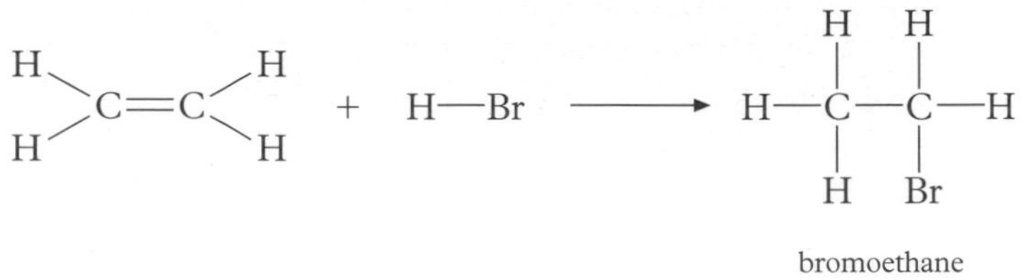

$$H_2C{=}CH_2 + H{-}Br \longrightarrow CH_3{-}CH_2Br$$

bromoethane

Addition of water

The addition of water to ethene is the method used to manufacture ethanol in the UK. A mixture of ethanol and steam passes over an acid catalyst at 300 °C under pressure. The catalyst is phosphoric acid supported on an inert solid.

Figure 5.8.8 ▶
Addition of water to ethene to make ethanol

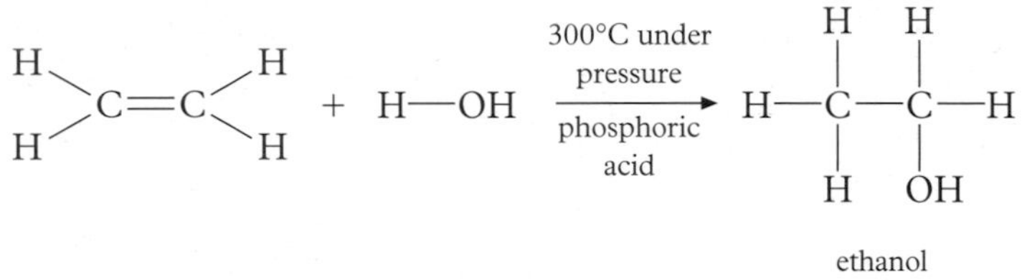

$$H_2C{=}CH_2 + H{-}OH \xrightarrow[\text{phosphoric acid}]{\text{300°C under pressure}} CH_3{-}CH_2OH$$

ethanol

Absorbing ethene in cold, concentrated sulfuric acid and then diluting the product with water can achieve the same effect on a laboratory scale. First the sulfuric acid adds to ethene. Then a hydrolysis reaction converts the addition product to ethanol and sulfuric acid.

Figure 5.8.9 ▼
Two step conversion of ethene to ethanol – addition followed by hydrolysis

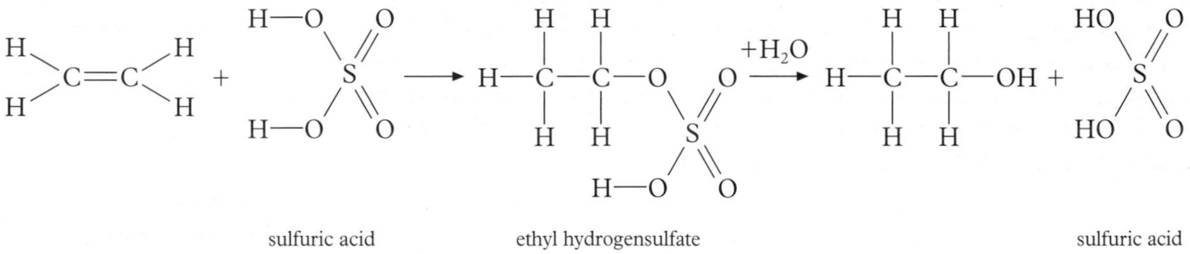

$$H_2C{=}CH_2 + (HO)_2SO_2 \longrightarrow CH_3{-}CH_2{-}O{-}SO_2{-}OH \xrightarrow{+H_2O} CH_3{-}CH_2{-}OH + (HO)_2SO_2$$

sulfuric acid ethyl hydrogensulfate sulfuric acid

Addition polymerisation

Addition polymerisation is a process for making polymers from compounds containing double bonds. Many molecules of the monomer add together to form a long chain polymer. Ethene, for example, polymerises to form poly(ethene).

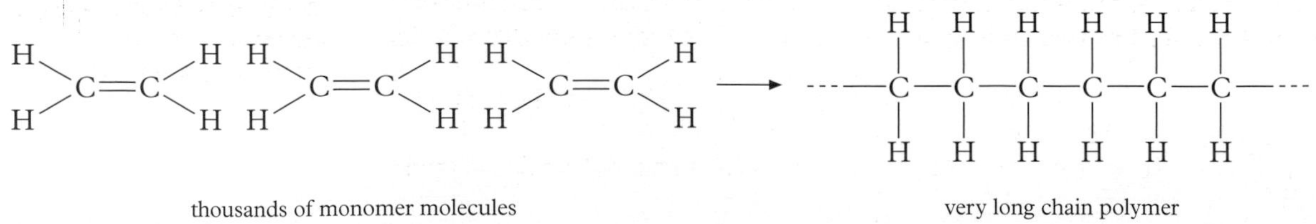

$$H_2C{=}CH_2 \;\; H_2C{=}CH_2 \;\; H_2C{=}CH_2 \longrightarrow \cdots{-}CH_2{-}CH_2{-}CH_2{-}CH_2{-}CH_2{-}CH_2{-}\cdots$$

thousands of monomer molecules very long chain polymer

Figure 5.8.10 ▲
Addition polymerisation of ethene

Oxidation

Potassium manganate(VII) oxidises alkenes. The products depend on the conditions. A dilute, acidified solution of potassium manganate(VII) converts an alkene to a diol at room temperature.

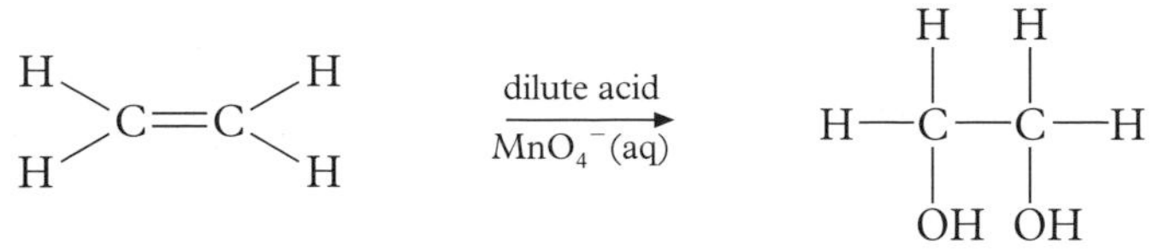

$$H_2C{=}CH_2 \xrightarrow[MnO_4^-(aq)]{\text{dilute acid}} HOCH_2{-}CH_2OH$$

ethane-1,2-diol

Figure 5.8.11 ▲
The reaction of ethene with dilute, acidified manganate(VII) ions producing ethane-1,2-diol

Definitions

A **monomer** (one-part) is a small molecule which can polymerise to make a long chain molecule or **polymer** (many-parts).
Addition polymerisation is the process for making polymers from unsaturated compounds with double bonds.

A solution of manganate(VII) ions is purple. The colour disappears as it reacts with an alkene. So, like the reaction with aqueous bromine, the reaction with cold MnO_4^-(aq) ions can be used to distinguish between unsaturated and saturated hydrocarbons.

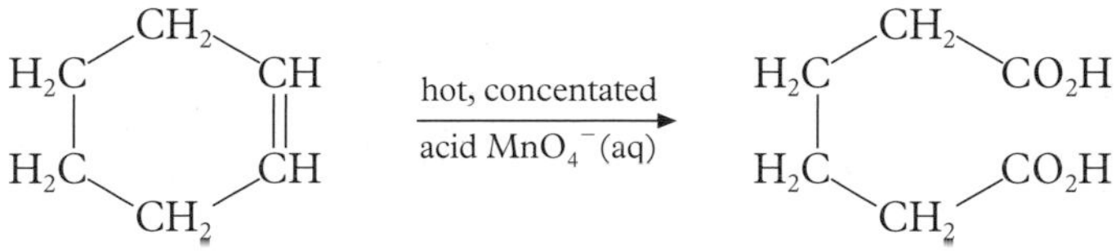

hot, concentated acid MnO_4^-(aq)

hexane-1,6-dioic acid

Figure 5.8.12 ◄
Using hot, acidified potassium manganate(VII) to break a carbon chain at a double bond. In this example, the reaction converts cyclohexene to hexane-1,6-dioic acid

A hot and concentrated solution of acidic potassium manganate(VII) breaks apart the double bond in an alkene.

Electrophilic addition

Most of the reactions of alkenes are electrophilic addition reactions. The electrophile (see page 195) attacks the electron-rich region of the double bond between two carbon atoms. Electrophiles which add to alkenes include hydrogen bromide, bromine and water in the presence of an acid catalyst.

Hydrogen bromide molecules are polar. The hydrogen atom, with its δ+ charge is the electrophilic end of the molecule.

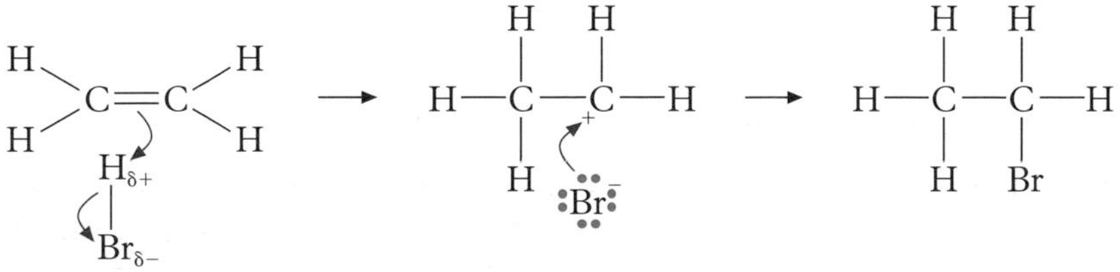

Figure 5.8.13 ◄
Electrophilic addition of hydrogen bromide to ethene. The reaction takes place in two steps. The intermediate has a positive charge on a carbon atom. It is a carbocation. Curly arrows show the movement of a pair of electrons. Bond breaking is heterolytic and the intermediates are ions

Bromine molecules are not polar but they become polarised as they approach the electron-rich double bond. Electrons in the double bond repel electrons in the bromine molecule. The δ+ end of the molecule is electrophilic.

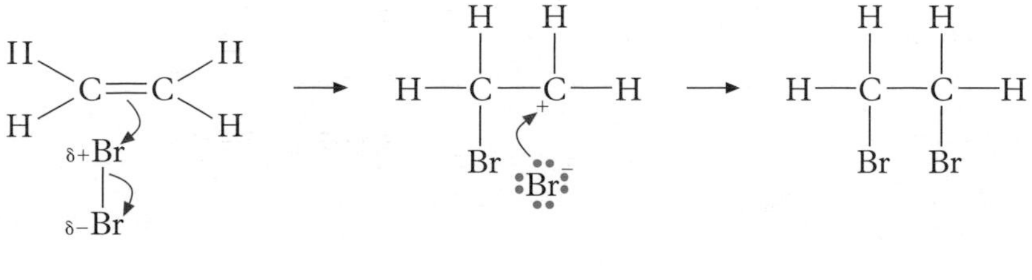

Figure 5.8.14 ◄
Electrophilic addition of bromine to ethene

Addition to unsymmetrical alkenes

When a compound HX (such as H—Br or H—OH) adds to an unsymmetrical alkene (such as propene), the hydrogen atom mainly adds to the carbon atom of the double bond which already has more hydrogen atoms attached to it (see Figure 5.8.15). This pattern was first reported by the Russian chemist Vladimir Markovnikov who studied a great many alkene addition reactions during the 1860s. Hence the name Markovnikov's rule.

The mechanism for electrophilic addition helps to account for this rule. On adding HBr to propene there are two possible intermediate carbocations (see Figure 5.8.16). The carbocation with the positive charge in the middle of the carbon chain is preferred because it is slightly more stable than the carbocation with the charge at the end of the chain.

Definitions

Carbocations are intermediates formed during organic reactions in which a carbon atom carries a positive charge.

Intermediates in reactions are atoms, molecules, ions or free radicals which do not appear in the balanced equation but which are formed during one step of a reaction but then used up in the next step.

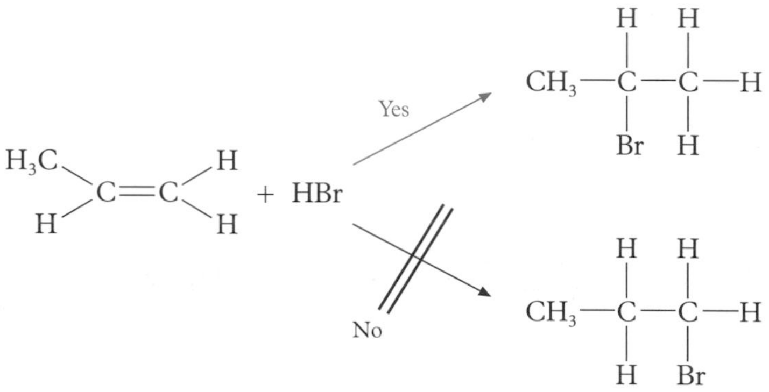

$H_3C(H)C{=}CH_2$ + HBr → (Yes) $CH_3{-}CHBr{-}CH_3$; (No) $CH_3{-}CH_2{-}CH_2Br$

Figure 5.8.15 ▶ *Markovnikov's rule in action*

The more stable ion has two alkyl groups pushing electrons towards the positively-charged carbon atom. This helps to stabilise the ion by 'spreading' the charge over the ion. This is an example of the inductive effect.

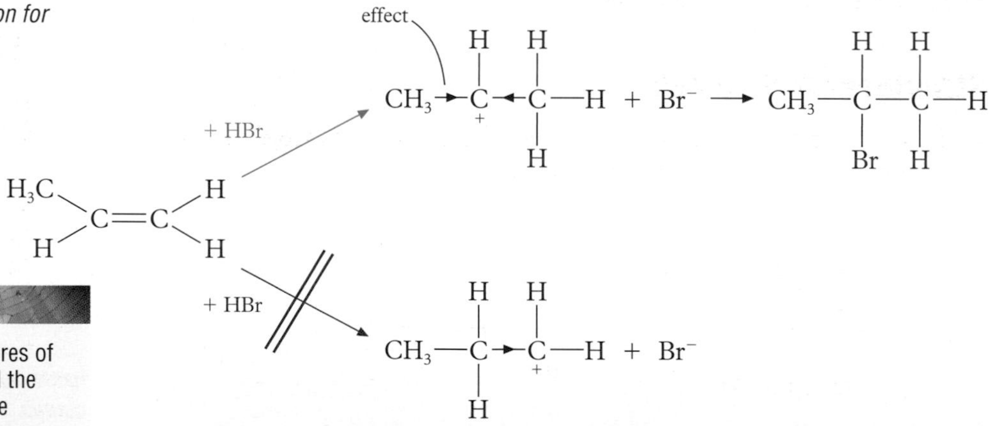

Figure 5.8.16 ▶ *A possible explanation for Markovnikov's rule*

$H_3C(H)C{=}CH_2$ + HBr → $CH_3{\rightarrow}\overset{+}{C}H{\leftarrow}CH_3$ (inductive effect) + Br^- → $CH_3{-}CHBr{-}CH_3$

$H_3C(H)C{=}CH_2$ + HBr ↛ $CH_3{-}CH_2{\rightarrow}\overset{+}{C}H_2$ + Br^-

Test yourself

5 Write the structures of the products and the conditions for the reaction when propene reacts with:

- **a)** hydrogen
- **b)** chlorine
- **c)** potassium manganate(VII).

6 Identify the products when hot, acidified potassium manganate(VII) oxidises pent-2-ene.

Definition

The **inductive effect** describes the extent to which electrons are pulled away from, or pushed towards, a carbon atom by the atoms or group to which it is bonded.

Alkyl groups have a slight tendency to push electrons towards the carbon atom to which they are bonded. One of the effects of this kind of inductive effect is that the more alkyl groups that are attached to the carbon atom with the positive charge, the more stable the carbocation.

5.9 Halogenoalkanes

Halogenoalkanes are important to organic chemists both in laboratories and in industry. The reason is that they are reactive compounds which can be converted to other more valuable products. This makes them useful as intermediates when converting one chemical to another.

There are growing restrictions on the uses of many halogenoalkanes because of concern about their hazards to health, their persistence in the environment and their effect on the ozone layer.

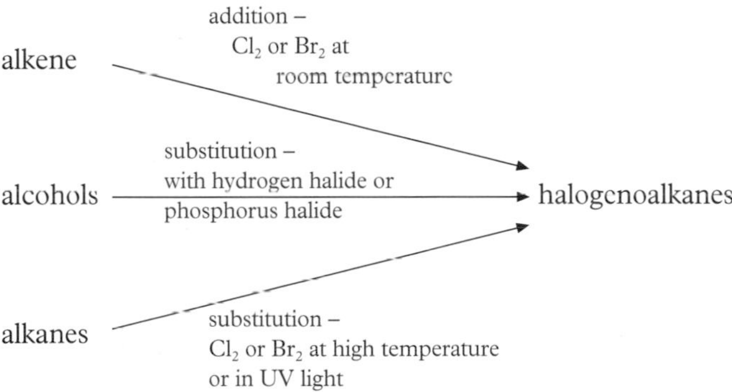

Figure 5.9.2 ▲
Reactions used to make halogenoalkanes

Structures and names

In the structure of a halogenoalkane, one or more of the hydrogen atoms in an alkane structure is replaced with halogen atoms.

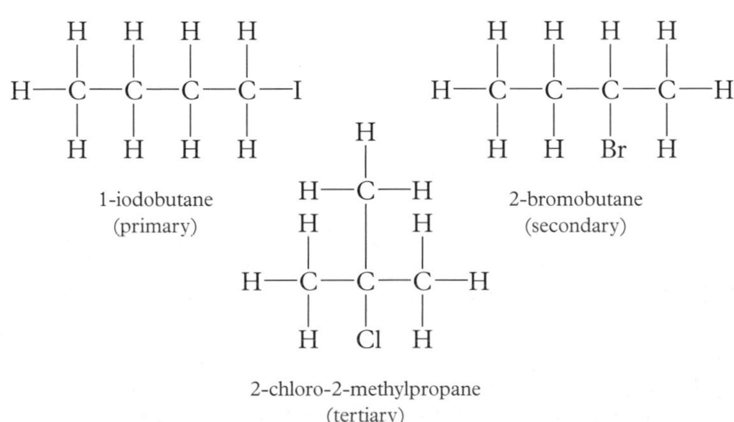

Figure 5.9.3 ▲
Names and structures of some halogenoalkanes 

Definition

The terms **primary**, **secondary** and **tertiary** are used to label organic compounds to show the position of functional groups. A primary halogenoalkane has the halogen atom at the end of the chain. A secondary compound has the halogen atom somewhere along the chain but not at the ends. A tertiary halogenoalkane has the halogen atom at a branch in the chain.

Figure 5.9.1 ▲
Some halogenoalkanes are valuable for their own sake. They are also used as:

- *solvents (for example dichloromethane),*
- *refrigerants (for example hydrochlorofluorocarbons, such as $CHClF_2$, which are replacing CFCs),*
- *pesticides (for example bromomethane) and*
- *fire extinguishers (e.g. CBr_2ClF)*

Test yourself

1 Draw the structures of these compounds and identify the primary, secondary and tertiary compounds.

a) 1-iodopropane
b) 2-chloro-2-methylbutane
c) 3-bromopentane

2 Give examples of the three methods for making halogenoalkanes. Name the reactants and products and state the conditions required for each reaction.

Physical properties

Chloromethane, bromomethane and chloroethane are gases at room temperature. Most other halogenoalkanes are colourless liquids which do not mix with water.

Chemical reactions

The two main types of reaction of halogenoalkanes are substitution reactions and elimination reactions (see page 193).

Substitution by hydroxide ions

Cold water slowly hydrolyses halogenoalkanes replacing the halogen atoms with an —OH group to form an alcohol (see page 212).

$$CH_3CH_2CH_2I(l) + H_2O(l) \rightarrow CH_3CH_2CH_2OH(l) + HI(aq)$$

This type of reaction is much quicker with an aqueous solution of an alkali such as sodium or potassium hydroxide. Heating increases the rate further.

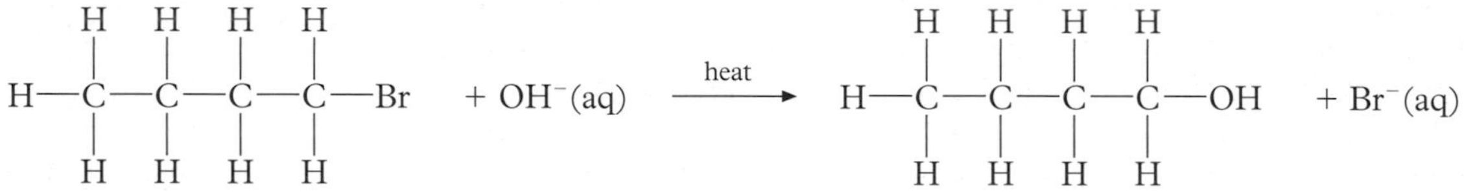

$$\begin{array}{c} \text{H}\;\;\text{H}\;\;\text{H}\;\;\text{H} \\ \text{H—C—C—C—C—Br} \\ \text{H}\;\;\text{H}\;\;\text{H}\;\;\text{H} \end{array} + OH^-(aq) \xrightarrow{\text{heat}} \begin{array}{c} \text{H}\;\;\text{H}\;\;\text{H}\;\;\text{H} \\ \text{H—C—C—C—C—OH} \\ \text{H}\;\;\text{H}\;\;\text{H}\;\;\text{H} \end{array} + Br^-(aq)$$

1-bromobutane — butan-1-ol

Figure 5.9.4 ▲
Reaction of a halogenoalkane with an alkali on heating

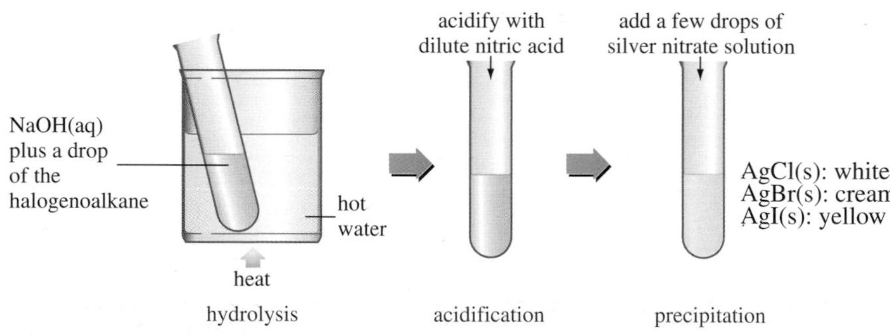

Figure 5.9.5 ▲
The hydrolysis of halogenoalkanes makes it possible to distinguish between chloro-, bromo- and iodo- compounds. Heating the compound with an alkali releases halide ions. Acidifying with nitric acid and then adding silver nitrate produces a precipitate of the silver halide

CD-ROM

Substitution by cyanide ions

The reactions between halogenoalkanes and cyanide ions are useful because they make it possible to add an extra carbon atom to the carbon chain. The products are nitriles which are not useful in themselves but can be converted to organic acids and amines which are valuable.

The reagent for converting a halogenoalkane to a nitrile is a solution of potassium cyanide in ethanol. Using ethanol instead of water as the solvent prevents hydrolysis.

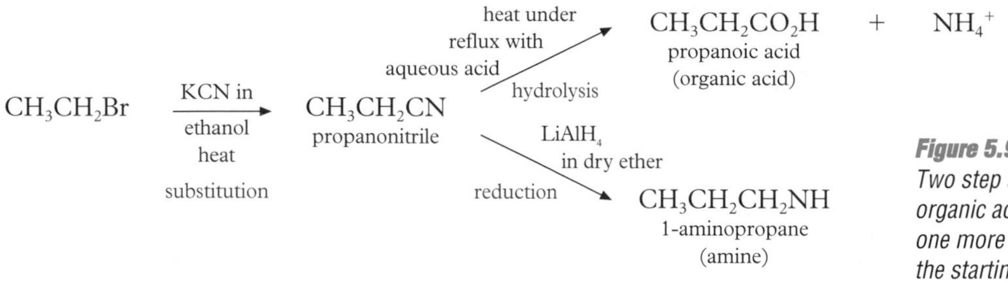

$$CH_3CH_2Br \xrightarrow[\text{ethanol, heat; substitution}]{\text{KCN in}} CH_3CH_2CN \text{ (propanonitrile)}$$

$$CH_3CH_2CN \xrightarrow[\text{hydrolysis}]{\text{heat under reflux with aqueous acid}} CH_3CH_2CO_2H \text{ (propanoic acid (organic acid))} + NH_4^+$$

$$CH_3CH_2CN \xrightarrow[\text{reduction}]{LiAlH_4 \text{ in dry ether}} CH_3CH_2CH_2NH \text{ (1-aminopropane (amine))}$$

Figure 5.9.6 ◀
Two step synthesis of an organic acid or an amine with one more carbon atom than the starting material

Substitution reaction with ammonia

Warming a halogenoalkane with a solution of ammonia in ethanol produces an amine. The other product is a hydrogen halide which reacts with excess ammonia to form an ammonium salt.

This happens because there is a lone pair of electrons on the ammonia molecule (see page 78). The problem is that there is also a lone pair on the nitrogen atom of the product which is even more reactive. So the product can react with the halogenoalkane. Fortunately it is possible to limit this reaction by using an excess of concentrated ammonia solution so that there is a much greater chance of halogenoalkane molecules reacting with ammonia molecules.

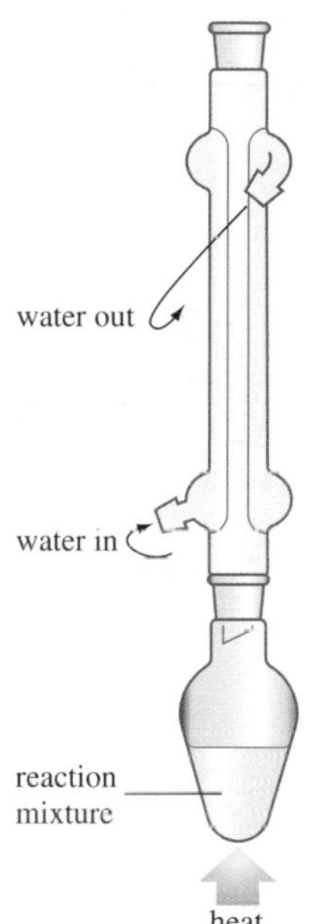

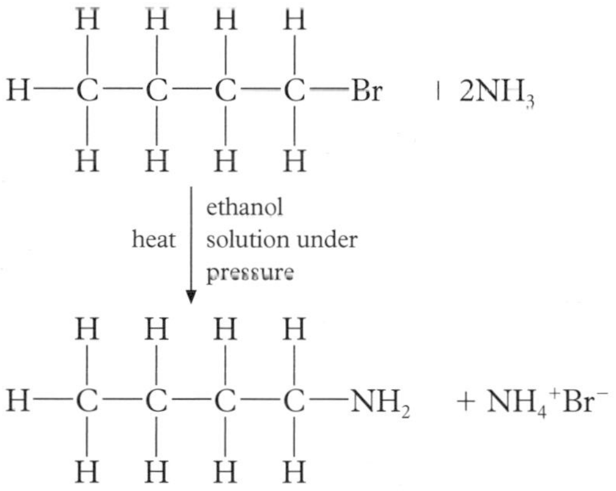

Figure 5.9.7 ▲
Heating a solution of potassium cyanide in ethanol with 1-bromobutane to make a nitrile. The condenser prevents the liquids escaping. They condense and flow back into the flask. Hence the term 'reflux' meaning 'flow back'

$$CH_3CH_2CH_2CH_2Br + 2NH_3 \xrightarrow[\text{heat}]{\text{ethanol solution under pressure}} CH_3CH_2CH_2CH_2NH_2 + NH_4^+Br^-$$

Figure 5.9.8 ▲
The reaction of 1-bromobutane with ammonia to make 1-aminobutane (butylamine)

Nucleophilic substitution

The reagents which react with halogenoalkanes are nucleophiles (see page 195). Carbon–halogen bonds are polar because halogen atoms are more electronegative than carbon atoms. So the characteristic reactions of halogenoalkanes are nucleophilic substitution reactions.

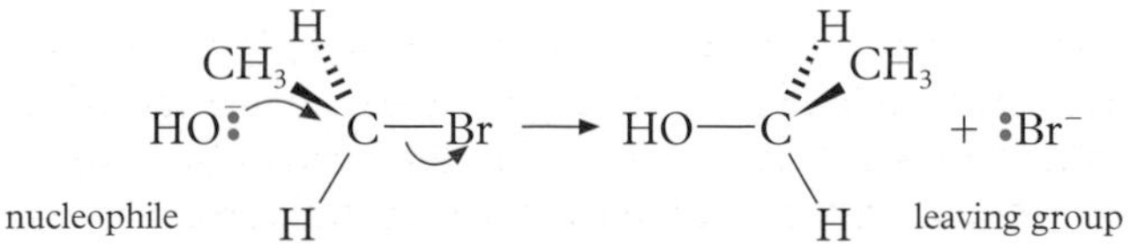

$$HO:^- \text{ (nucleophile)} + CH_3CH_2Br \rightarrow HO\text{—}CH(CH_3)H + :Br^- \text{ (leaving group)}$$

Figure 5.9.9 ◀
Attack by a nucleophile on a δ+ carbon atom leading to substitution. The nucleophile has a lone pair of electrons to form a new covalent bond with the carbon atom it attacks. The halogen atom leaves, taking the bonding electrons with it so that it leaves as a halide ion

The rates of reaction of halogenoalkanes are in the order: RI > RBr > RCl where R represents an alkyl group.

The reaction rates do not correlate with bond polarity. Chlorine is the most electronegative of the elements so the C—Cl bond is the most polar and the C—I bond the least polar. So bond polarity is not the factor which determines the rates of reaction.

The reaction rates do correlate with the strength of the bonds. The C—I bond is the longest and the weakest (as measured by the mean bond enthalpy). The C—Cl bond is the shortest and the strongest.

Elimination reactions

The same reagents which can act as nucleophiles can also form dative bonds with hydrogen ions, H^+. So they can also act as bases. This is particularly true of the hydroxide ion. Heating a halogenoalkane with a base can bring about elimination of a hydrogen halide instead of substitution.

Hydrolysis leading to substitution is much more likely using potassium or sodium hydroxide dissolved in water. Elimination is more likely if there is no water and the alkali is dissolved in ethanol.

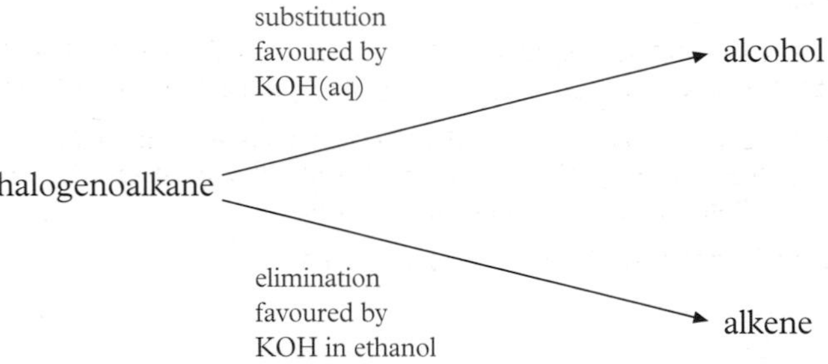

Figure 5.9.10 ▶
The alternative reactions with solutions of hydroxide ions

Often the result of these reactions is a mixture of products. Generally elimination happens more readily with secondary or tertiary halogenoalkanes.

```
HO:⁻
     H   Br   H                          Br⁻
     |   |    |          H2O              H
  H—C——C——C—H    →          H\           |
     |   |    |                  C=C——C—H
     H   H    H              H/    |    |
                                   H    H
```

Figure 5.9.11 ▲
Elimination of hydrogen bromide from 2-bromopropane

Test yourself D

3 Which of these molecules are polar and which are non-polar: $CHCl_3$, CH_2Cl_2, $CHCl_3$ and CCl_4?

4 Look up the boiling points of these compounds and suggest an explanation for the trend in values: 1-chlorobutane, 1-bromobutane, 1-iodobutane.

5 Compare the boiling points of the isomers 1-bromobutane, 2-bromobutane and 2-bromo-2-methylbutane. Suggest an explanation for the differences in boiling points of the primary, secondary and tertiary compounds.

6 Explain the use of the term hydrolysis to describe the reaction of a halogenoalkane with water or an alkali.

7 Refer to Figure 5.9.5.

a) Why is hydrolysis necessary before testing with silver nitrate?
b) Why must nitric acid be added before the silver nitrate solution?
c) Write the equations for the three reactions which take place when detecting bromide ions in 1-bromobutane by this method.

8 Write the equation for the reaction taking place in Figure 5.9.7. State the conditions required for the reaction.

9 Suggest the structure of the product when 1-aminobutane reacts with 1-bromobutane. Can this product also react with 1-bromobutane and if so, what will be formed?

10 Show the bond forming and bond breaking when a cyanide ion reacts with 1-iodopropane. Use 'curly arrows' to show the movement of pairs of electrons.

11 The products of the reaction of 2-bromobutane with potassium hydroxide solution depend on the conditions. Which conditions favour the formation of an alcohol? Which conditions favour the formation of an alkene?

12 Write equations to show the hydroxide ion acting:

a) as a base
b) as a nucleophile.

13 Draw the structure of the main product:

a) on heating 2-bromo-2-methylpropane under reflux with a solution of potassium hydroxide in ethanol
b) on heating 1-iodopropane with an aqueous solution of potassium hydroxide.

14 Complete this diagram to summarise the reactions of 1-bromopropane. Write the reagents and conditions for the reactions beside the arrows. Give the structures and names of the main products. Also show whether the reactions are substitution or elimination reactions.

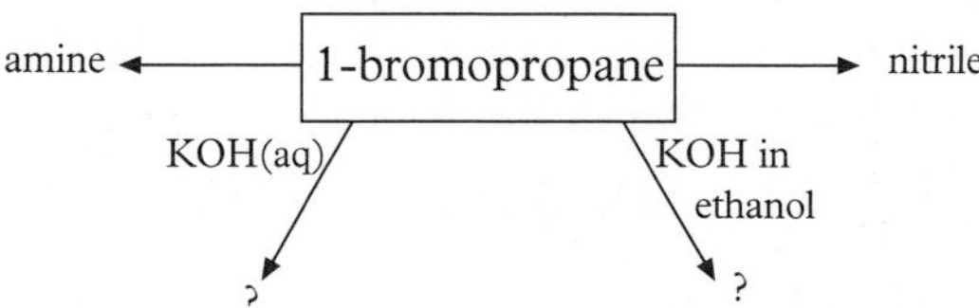

5.10 Alcohols

Ethanol is the best known member of the family of alcohols. It is the alcohol in beer, wine and spirits. Alcohols are useful solvents in the home, in laboratories and in industry. Understanding the properties of the —OH functional group in alcohols helps to make sense of the reactions of many important biological chemicals, especially carbohydrates such as sugars and starch. Alcohols are much more reactive than alkanes because the C—O and O—H bonds in the molecules are polar. CD-ROM

Structure and names

Alcohols are compounds with the formula R—OH where R represents an alkyl group. The hydroxy group —OH is the functional group which gives the compounds their characteristic reactions.

The IUPAC rules name alcohols by changing the ending of the corresponding alkane to -ol. So ethane becomes ethanol.

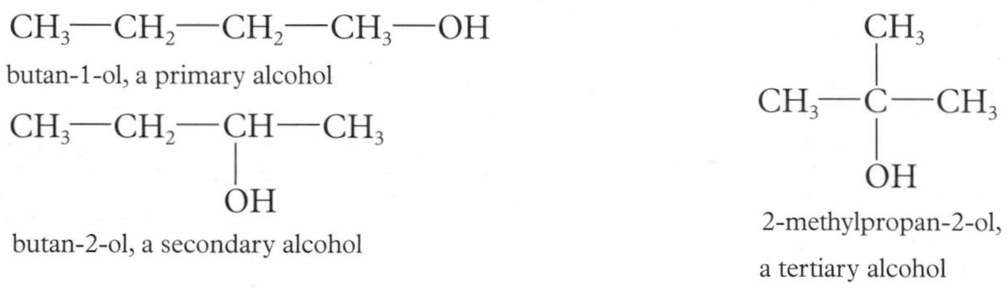

Figure 5.10.1 ▶ *Names and structures of alcohols* CD-ROM

$CH_3—CH_2—CH_2—CH_2—OH$

butan-1-ol, a primary alcohol

$CH_3—CH_2—CH(OH)—CH_3$

butan-2-ol, a secondary alcohol

$(CH_3)_3C—OH$

2-methylpropan-2-ol, a tertiary alcohol

Physical properties

Even the simplest alcohols, such as methanol and ethanol, are liquids at room temperature because of hydrogen bonding between the —OH groups. Alcohols are much less volatile than their equivalent hydrocarbons. For the same reason, alcohols with relatively short hydrocarbon chains mix freely with water.

Chemical properties

Combustion

Alcohols burn in air with a clean, colourless flame. Methanol and ethanol are both common fuels (see page 230) or fuel additives.

$$2CH_3OH(l) + 3O_2(g) \rightarrow 2CO_2(g) + 4H_2O(l)$$

Test yourself D

1 Look up and compare the boiling points of alkanes and alcohols of about the same relative molecular mass. Is it true that alcohols are less volatile than the equivalent alkanes?

Reaction with sodium

Alcohols react with sodium in a similar way to water. This is because both water molecules and alcohol molecules contain the —OH group. With water the products are sodium hydroxide and hydrogen. With ethanol the products are sodium ethoxide and hydrogen.

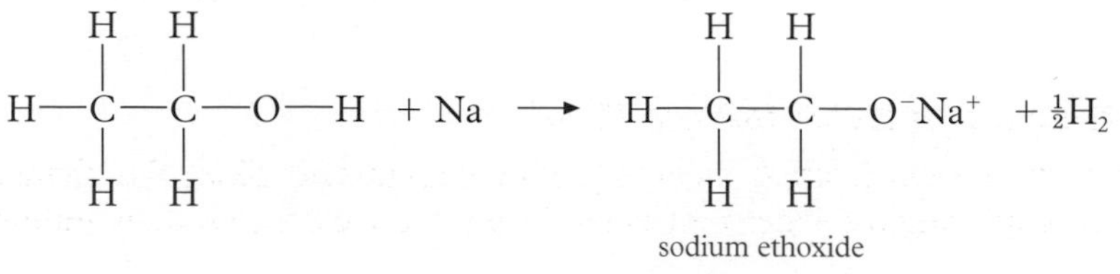

$$H_3C—CH_2—O—H + Na \longrightarrow H_3C—CH_2—O^-Na^+ + \tfrac{1}{2}H_2$$

sodium ethoxide

Figure 5.10.2 ▶ *The reaction of ethanol with sodium. Note the ionic bond in sodium ethoxide*

The reaction is dangerously rapid with water, but much slower with alcohols. A safe way of disposing of small amounts of waste sodium in a laboratory is to cut the metal into small pieces and then add it to propan-1-ol a little at a time.

Substitution of a halogen atom for an —OH group

Alcohols react rapidly at room temperature with phosphorus pentachloride to give chloroalkanes. The reaction also produces the gas hydrogen chloride and this makes it a useful test for the presence of —OH groups in molecules.

$$C_3H_7OH(l) + PCl_5(s) \rightarrow C_3H_7Cl(l) + POCl_3(l) + HCl(g)$$

A similar method converts alcohols to iodoalkanes. The reagent is a mixture of red phosphorus and iodine which combine to make phosphorus triodide.

The alternative method of converting an alcohol to a halogenoalkane is to use the hydrogen halide as the reagent. When making a bromoalkane it is convenient to make the hydrogen bromide in the reaction vessel from a mixture of sodium bromide and concentrated sulfuric acid.

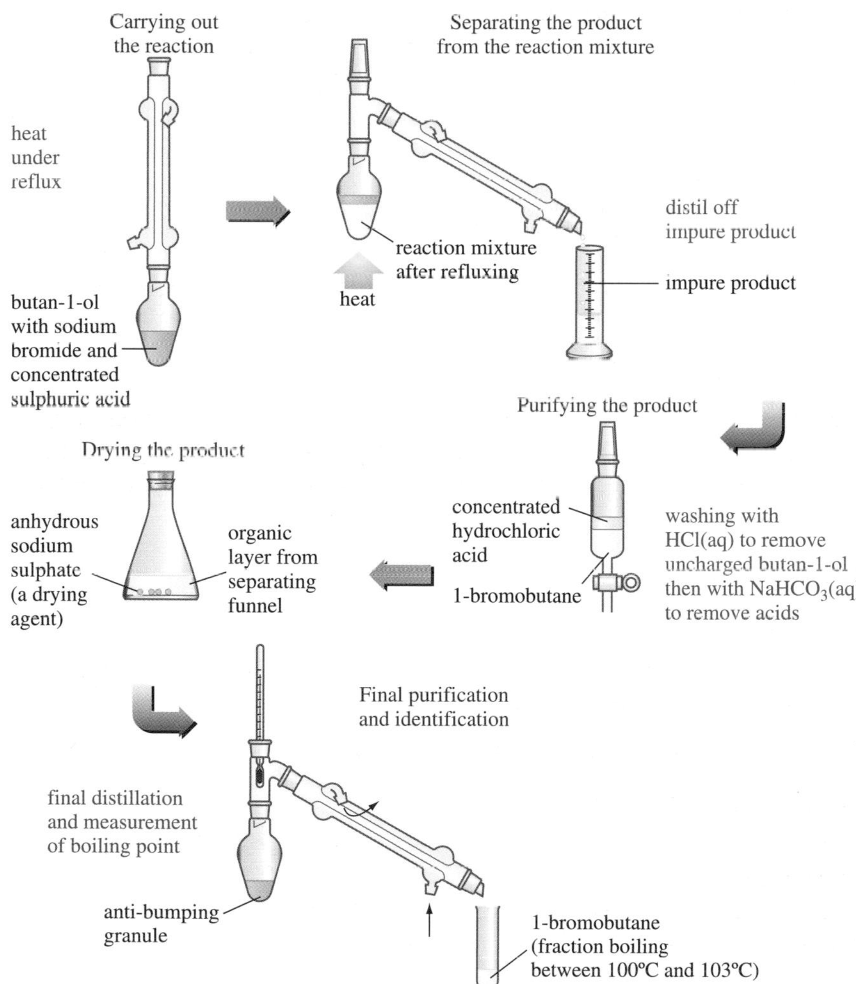

Figure 5.10.3 ◀ *Steps in the synthesis of 1-bromoethane from butan-1-ol* CD-ROM

Definition

Esters are the fruity compounds which contribute to the flavour of bananas, pineapples and many other fruits. The smell of esters also contributes to the scent of perfumes. Fats and vegetable oils are esters. The general formula for an ester is

$$R-C(=O)-O-R'$$

where R and R′ are alkyl groups.

Elimination of water (dehydration)

Eliminating water from an alcohol produces an alkene. Since this change involves splitting off water molecules, its alternative name is dehydration.

Definition

A **leaving group** is an atom or molecule which breaks away from a molecule during a reaction.

One method is to pass the vapour of the alcohol over hot powdered aluminium oxide.

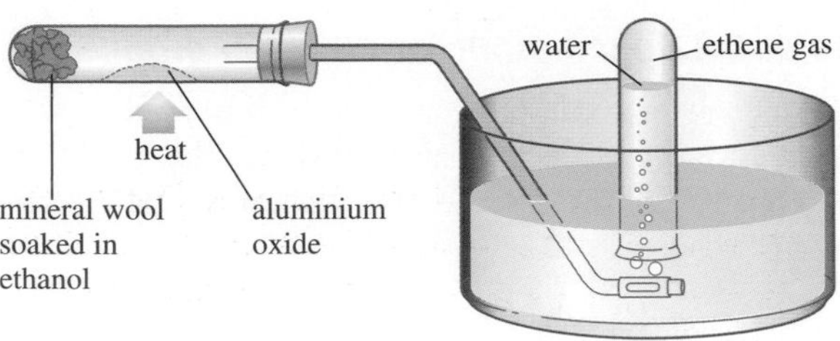

Figure 5.10.4 ▶ *Small-scale dehydration of ethanol to ethene* CD-ROM

The alternative method for dehydration is heating the alcohol with concentrated sulfuric or phosphoric acid. The advantage of phosphoric acid is that it is a non-oxidising acid and leads to fewer side reactions.

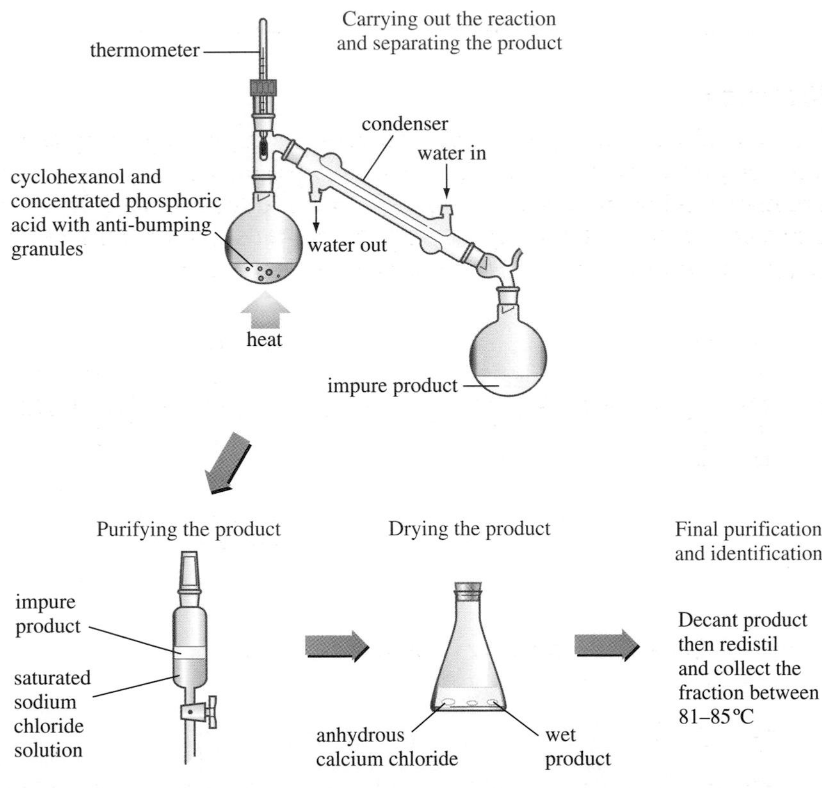

Figure 5.10.5 ▲ *Converting cyclohexanol to cyclohexene* CD-ROM

The acid acts as a catalyst for the elimination reaction. The first step of the mechanism is that the —OH group of the alcohol acts as a base accepting a proton, H^+, from the acid. This means that when the C—O bond breaks, the leaving group is a water molecule rather than a hydroxide ion. The proton is reformed in the process so the catalyst is involved in the reaction but is not used up.

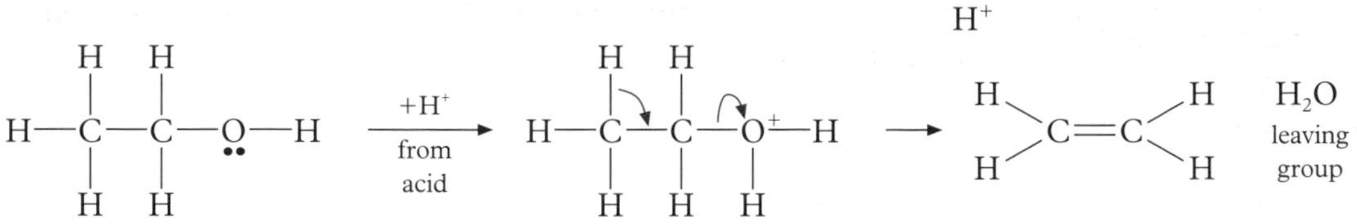

$$\mathrm{CH_3CH_2OH} \xrightarrow[\text{from acid}]{+\mathrm{H^+}} \mathrm{CH_3CH_2OH_2^+} \longrightarrow \mathrm{H_2C{=}CH_2} + \mathrm{H_2O}\ (\text{leaving group}) + \mathrm{H^+}$$

Figure 5.10.6 ▲ *Mechanism for the elimination of water from an alcohol*

Ester formation

Alcohols react with carboxylic acids to form esters. The reaction happens on gentle heating in the presence of a little concentrated sulfuric acid to act as a catalyst. This is an example of homogeneous catalysis (see page 116). The reaction is reversible.

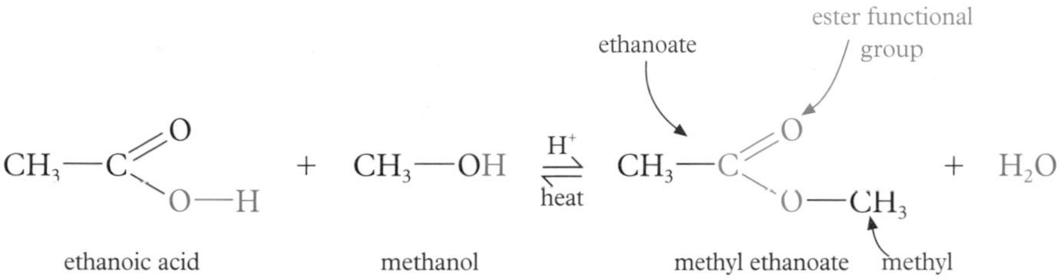

$$\underset{\text{ethanoic acid}}{\mathrm{CH_3{-}C({=}O){-}O{-}H}} + \underset{\text{methanol}}{\mathrm{CH_3{-}OH}} \underset{\text{heat}}{\overset{\mathrm{H^+}}{\rightleftharpoons}} \underset{\text{methyl ethanoate}}{\mathrm{CH_3{-}C({=}O){-}O{-}CH_3}} + \mathrm{H_2O}$$

Labels: ethanoate; ester functional group; methyl

Figure 5.10.7 ◄ *Formation of an ester from an acid and an alcohol*

Oxidation

An acidified solution of potassium dichromate(VI) oxidises primary and secondary alcohols. There is no reaction with tertiary alcohols. Oxidation of a primary alcohol takes place in two steps, first producing an aldehyde and then a carboxylic acid. During the reaction the reagent turns from the orange colour of $Cr_2O_7^{2-}$(aq) to the green colour of Cr^{3+}(aq).

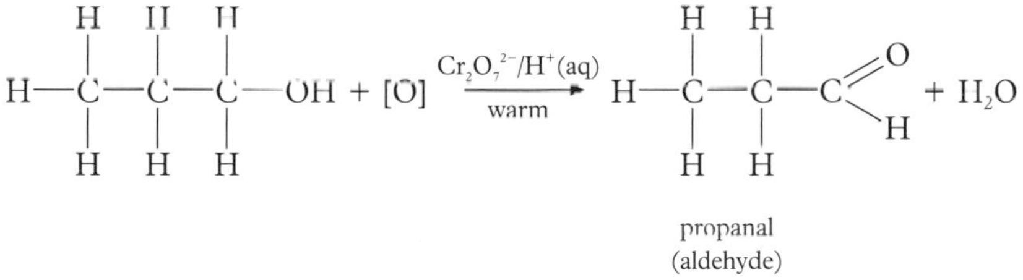

$$\mathrm{CH_3CH_2CH_2OH} + [\mathrm{O}] \xrightarrow[\text{warm}]{\mathrm{Cr_2O_7^{2-}/H^+(aq)}} \underset{\substack{\text{propanal}\\ \text{(aldehyde)}}}{\mathrm{CH_3CH_2CHO}} + \mathrm{H_2O}$$

Figure 5.10.8 ◄ *Oxidation of propan-1-ol by acidified $Cr_2O_7^{2-}$(aq). The full equations are complex. The symbol [O] is a shorthand way of balancing the equation. [O] represents the oxygen atoms from the oxidising agent*

Oxidation of a secondary alcohol produces a ketone.

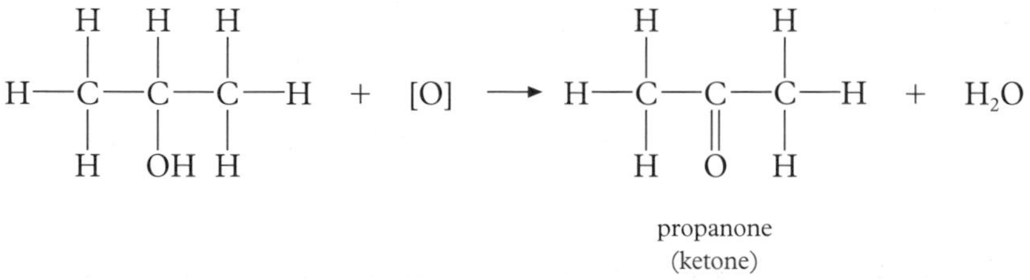

$$\mathrm{CH_3CH(OH)CH_3} + [\mathrm{O}] \longrightarrow \underset{\substack{\text{propanone}\\ \text{(ketone)}}}{\mathrm{CH_3COCH_3}} + \mathrm{H_2O}$$

Figure 5.10.9 ▲ *Oxidation of propan-2-ol to propanone*

These oxidation reactions help to distinguish between primary, secondary and tertiary alcohols.

Note

Reducing agents will reverse the oxidation of aldehydes and ketones. A suitable reducing agent is sodium tetrahydridoborate(III), $NaBH_4$, in aqueous solution.

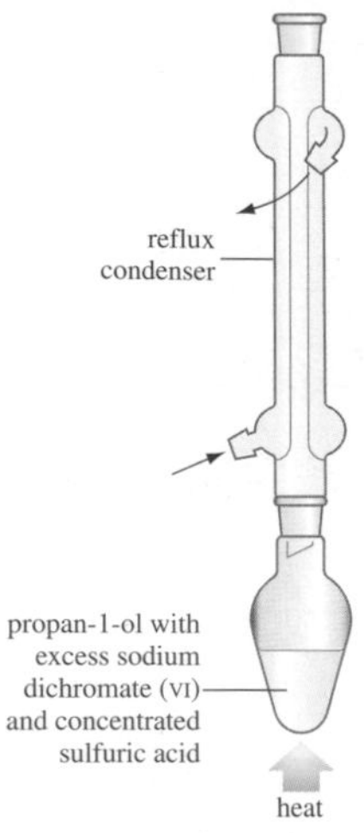

Figure 5.10.11 ▲
Apparatus for the oxidation of a primary alcohol to a carboxylic acid. The reflux condenser ensures that any volatile aldehyde condenses and flows back into the flask where excess oxidising agent ensures complete conversion to a carboxylic acid

CD-ROM

Two further test reagents, Fehling's solution and Tollen's reagent, help by distinguishing aldehydes (from a primary alcohol) and ketones (from a secondary alcohol). Both test reagents are oxidising agents which change colour as they oxidise an aldehyde to a carboxylic acid. Neither reagent will oxidise ketones.

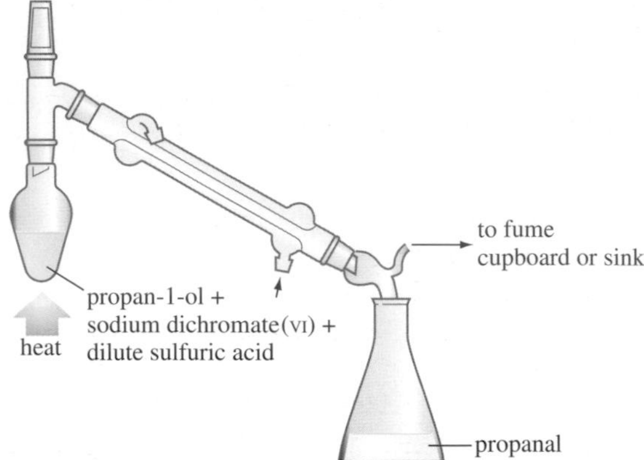

Figure 5.10.10 ▶
Apparatus for oxidising a primary alcohol to an aldehyde. The aldehyde distils off as it forms. This prevents further oxidation of the aldehyde

CD-ROM

Fehling's solution contains an alkaline solution of complex copper(II) ions. On warming with an aldehyde, the deep blue solution turns greenish and then loses its blue colour as an orange-red precipitate of copper(I) oxide appears.

Tollen's reagent also gives a positive result with aldehydes but not with ketones. A shiny silver mirror forms on the glass of a clean test tube on warming the reagent with an aldehyde.

An alternative to these chemical tests is to use infra-red spectra to distinguish between the functional groups in the products of the oxidation reactions (see pages 90–91).

Note

Fehling's solution does not keep, so it is prepared as needed by mixing two solutions. One solution is copper(II) sulfate in water. The other solution contains 2,3-dihydroxybutanedioate ions in strong alkali. The 2,3-dihydroxybutanedioate ions form a complex ion with copper(II) ions so that they do not precipitate as copper(II) hydroxide in alkali.

Tollen's reagent is unstable and can produce explosive products if stored. It too is made as required by dissolving a precipitate of silver oxide in ammonia solution. The reagent contains a complex formed between silver(I) ions and ammonia molecules.

The tri-iodomethane reaction

This reaction distinguishes alcohols with the $CH_3CHOH—$ group. The reaction conditions are to warm a drop of the alcohol with a solution of iodine in sodium hydroxide. If the alcohol contains a methyl group next to the —OH group, the result is a yellow precipitate of tri-iodomethane, CHI_3.

Test yourself

2 Write an equation for the reaction of water with sodium and compare it with the reaction of propan-1-ol with sodium.

3 How can you identify HCl(g) given off when PCl_5 reacts with an alcohol?

4 Write an equation for the reaction of PI_3 with propan-1-ol given that the inorganic product is the acid H_3PO_3.

5 What tests can be used to show that the gas formed in Figure 5.10.4 is an alkene?

6 Suggest conditions which will increase the yield of ester formed when a carboxylic acid reacts with an alcohol.

7 Is it the C—O or the O—H bond which breaks when an alcohol reacts with:

a) sodium?
b) PCl_5?
c) concentrated sulfuric acid?

8 Write the equation for the reduction of propanal to propan-1-ol by $NaBH_4$. Represent the hydrogen from the reducing agent as [H].

9 Which of these alcohols give a precipitate of CHI_3 on warming with iodine in alkali: ethanol, propan-1-ol, propan-2-ol?

5.11 Theoretical and percentage yields

A perfectly efficient reaction would convert all of the starting material to the required product. This would give a 100% yield.

Few reactions are completely efficient and most reactions, especially organic reactions, give low yields. There are several reasons why the overall yield may be low:

- the reaction may be incomplete (perhaps because it is slow or because it reaches an equilibrium state) so that a proportion of the starting chemicals fail to react
- there may be side reactions producing by-products
- recovery of all the product from the reaction is impossible
- some of the product is usually lost during transfer of the chemicals from one container to another as the product is separated and purified.

Definitions

The **theoretical yield** is the mass of product assuming that the reaction goes according to the chemical equation and the synthesis is 100% efficient.

The **actual yield** is the mass of product obtained.

The **percentage yield** is given by this relationship:

$$\text{percentage yield} = \frac{\text{actual mass of product}}{\text{theoretical yield}} \times 100\%$$

Definition

A **limiting reactant** is the chemical in a reaction mixture which is present in an amount which limits the theoretical yield. Often in a chemical synthesis some of the reactants are added in excess to make sure that the most valuable chemical is converted as far as possible to the required product. The limiting reactant is the one which is not in excess and so would be used up if the reaction went to completion.

Worked example

What is the theoretical yield of cyclohexene when dehydrating 10 g cyclohexanol by heating it with phosphoric acid? What is the percentage yield if the actual yield of cyclohexene is 7.1 g?

Notes on the method

Start by writing the equation for the reaction. This need not be the full equation so long as the equation includes the limiting reactant and the product. Here the theoretical yield depends on the amount of cyclohexanol.

Answer

The equation: $C_6H_{11}OH \xrightarrow{-H_2O} C_6H_{10}$

The molar mass of cyclohexanol = 100 g mol^{-1}

The amount of cyclohexanol at the start of the synthesis =

$$\frac{10\text{ g}}{100\text{ g mol}^{-1}} = 0.1\text{ mol}$$

1 mol of the alcohol produces 1 mol of the cycloalkene.

The molar mass of cyclohexene = 82 g mol^{-1}

The theoretical yield of cyclohexene = 0.1 mol × 82 g mol^{-1} = 8.2 g

$$\text{Percentage yield} = \frac{7.1\text{ g}}{8.2\text{ g}} \times 100\% = 87\%$$

Test yourself

1 Draw the structures of cyclohexanol and cyclohexene.

2 A synthesis of 1-bromobutane produced (Figure 5.10.3) 6.5 g product from 6.0 g of butan-1-ol using excess sodium bromide and concentrated sulfuric acid. Calculate the theoretical and percentage yield.

5.12 Fuels and chemicals from crude oil

The petrochemical industry converts crude oil and natural gas into useful products such as pharmaceuticals, fertilisers, detergents, paints and dyes. Crude oil is not, however, the only commercial source of organic raw materials. Increasingly industries are turning to plants and to micro-organisms as sources of fuels and chemicals because of the impact on the environment caused by the use of fossil fuels.

Definition

Fossil fuels are non-renewable energy resources which we burn to benefit from the energy from the Sun which was stored up millions of years ago by photosynthesis. The fossil fuels are coal, oil and natural gas.

Refining crude oil

Crude oil is a complex mixture of hydrocarbon molecules formed over millions of years. Oil contains no fossils but the evidence suggests that it formed from the remains of tiny sea creatures and plants accumulating deep under the sea where the water was free of oxygen. Bacteria feeding on these remains removed oxygen. Heat and pressure in the sediments slowly turned the chemicals into oil.

Crude oil is now the main source of fuels and organic compounds. The composition of crude oil varies from one oilfield to another. Some crudes contain significant quantities of sulfur and nitrogen compounds as well as traces of various metals. The challenge for refineries is to produce the various oil products in the quantities needed by industrial and domestic users. Generally crude oil contains too much of the high boiling point fractions with bigger molecules and not enough of the low boiling point fractions with smaller molecules needed for fuels such as petrol.

Fractional distillation of oil is the first step in refining to produce fuels and lubricants as well as feedstocks for the petrochemical industry.

Fractional distillation

Fractional distillation of oil is a large-scale, continuous process for separating crude oil into fractions. A furnace heats the oil which then flows

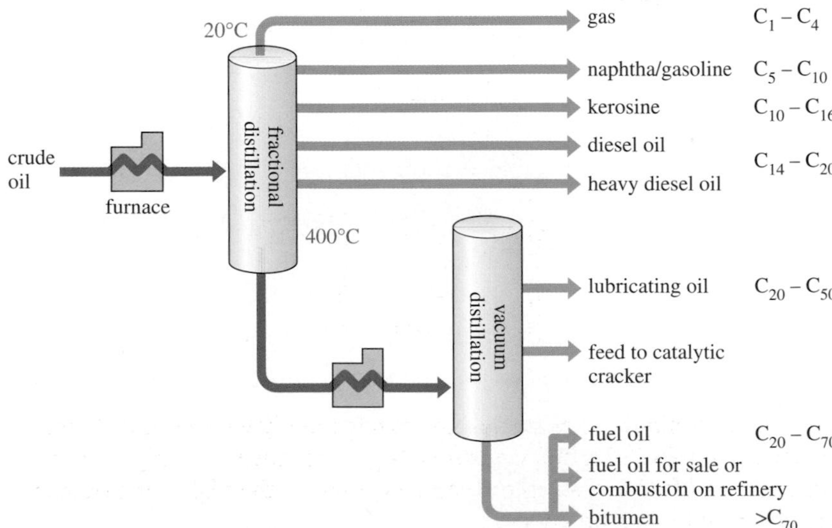

Figure 5.12.1 ▶ *Fractional distillation of crude oil*

into a fractionating tower containing about 40 'trays' pierced with small holes. Condensed vapour flows over the trays and runs down into the tray below. Rising vapour mixes with liquid on a tray as it bubbles up through the holes.

The column is hotter at the bottom and cooler at the top. Rising vapour condenses when it reaches the tray with liquid at a temperature below its boiling point. Condensing vapour releases energy which heats the liquid in the tray which in turn evaporates the more volatile compounds in the mixture.

With a series of trays the outcome is that the hydrocarbons with small molecules rise to the top of the column while larger molecules stay at the bottom. Fractions are drawn off from the column at various levels.

Some components of crude oil have boiling points too high for them to vaporise at atmospheric pressure. Lowering the pressure in a separate vacuum distillation column reduces the boiling points of the hydrocarbons and makes it possible to separate them.

Fuels from oil

Petrol is a blend of hydrocarbons based on the gasoline fraction (hydrocarbons with 5–10 carbon atoms). Jet fuel is produced from the kerosene, or paraffin, fraction (hydrocarbons with 10–16 carbon atoms). Fuel for diesel engines is made from diesel oil (hydrocarbons with 14–20 carbon atoms). Fuels have to be refined to remove components, such as sulfur compounds, which would harm engines or cause air pollution when they burn.

Petrol has to be carefully blended if modern engines are to start reliably and run smoothly. The proportion of volatile hydrocarbons added to petrol is higher in winter to help cold-starting but lower in summer to prevent vapour forming before the fuel gets to the carburettor.

Definitions

Knocking (or pinking) is a noise heard from a petrol engine when the mixture of fuel and air ignites too early while still being compressed by the piston. This is pre-ignition.

Compression heats up the mixture of fuel and air. **Pre-ignition** means that the fuel starts burning before being ignited by a spark from the spark plug. Internal combustion engines run powerfully if the fuel starts to burn when the piston is at the right point in the cylinder so that the expanding gases force the piston down smoothly. Knocking is a sign that a fuel with a higher octane number is needed. Prolonged knocking damages an engine.

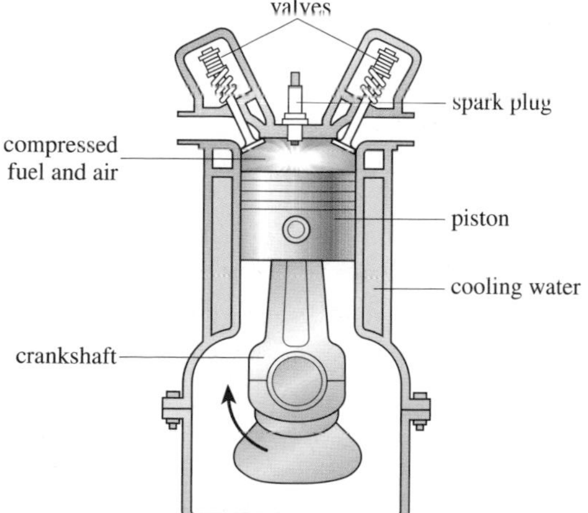

Figure 5.12.2 ◀ *The working parts of a cylinder of an internal combustion engine which runs on petrol*

For smooth running, the petrol has to burn smoothly without knocking. The octane number of a fuel measures its performance. The higher the compression of fuel and air in the engine cylinders, the higher the octane number has to be to stop knocking.

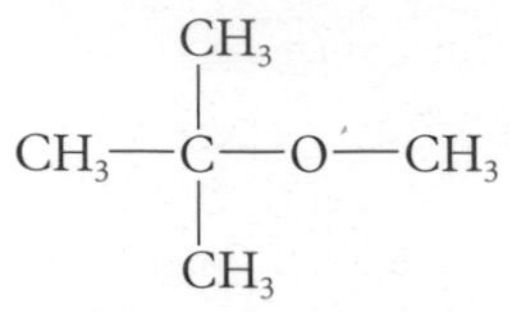

$$CH_3-\overset{\displaystyle CH_3}{\underset{\displaystyle CH_3}{\overset{|}{\underset{|}{C}}}}-O-CH_3$$

Figure 5.12.3 ▲ *Structure of the ether MTBE which has an octane number of 120. Adding MTBE and other methods raise the octane number of gasoline from about 70 to 95 as required for unleaded premium petrol*

The octane number scale was devised by the American inventor Thomas Midgley (1889–1944). He discovered anti-knock additives based on lead which were used for many years. Leaded fuel is currently being phased out and now the oil companies produce high-octane fuel by increasing the proportions of both branched alkanes and arenes. They may also blend in some oxygen compounds. The four main approaches are:

- cracking, which not only makes more small molecules but also forms hydrocarbons with branched chains
- isomerisation, which turns straight-chain alkanes into branched-chain compounds by passing them over a platinum catalyst
- reforming, which turns cyclic alkanes into arenes such as benzene and methyl benzene
- adding alcohols and ethers such as MTBE (initials based on its older name methyl tertiary butyl ether, now called 2-methoxy-2-methylpropane).

Catalytic cracking

Catalytic cracking converts heavier fractions, such as diesel oil or fuel oil, from the fractional distillation of crude oil into more useful hydrocarbons for fuels by breaking up larger molecules into smaller ones.

Cracking converts the longer chain alkanes with a dozen or more carbon atoms into smaller molecules which are a mixture of branched alkanes, cycloalkanes, alkenes and branched alkenes.

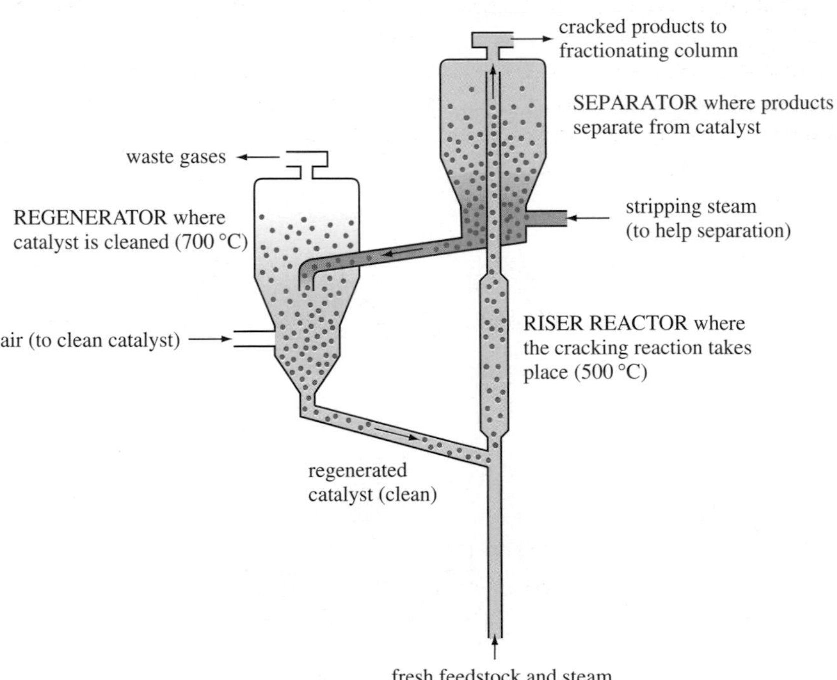

Figure 5.12.4 ► *Catalytic cracking. The catalyst powder flows to the vertical reactor where cracking takes place. The cracked vapours pass to a fractionating column while the catalyst flows to the regenerator*

The catalyst is a synthetic sodium aluminium silicate belonging to a class of compounds called zeolites. A zeolite has a three-dimensional structure in which the silicon and oxygen atoms form tunnels and cavities into which small molecules can fit. Cracking takes place on the surface of the catalyst which is polar and causes the bond to break to form ionic intermediates.

Synthetic zeolites make excellent catalysts because they can be developed with active sites to favour the required reactions by acting on molecules with particular shapes and sizes.

Catalytic cracking is a continuous process. The finely-powdered catalyst gradually gets coated with carbon so it circulates through a regenerator where the carbon burns away in a stream of air.

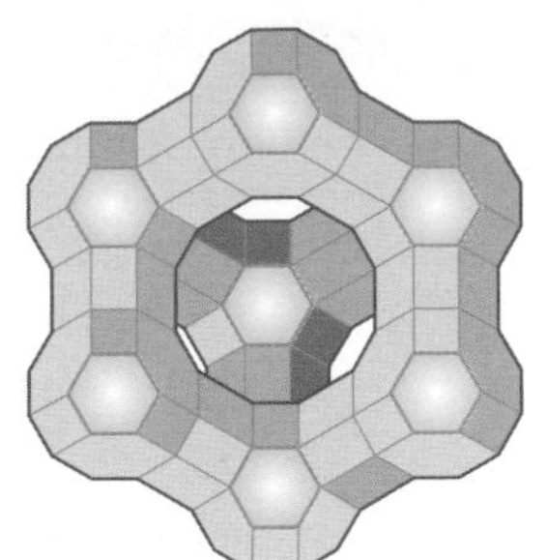

Figure 5.12.5 ▲
A model of a zeolite crystal structure

Isomerisation

Isomerisation converts straight-chain alkanes into branched-chain compounds. Typically compounds such as pentane are changed into branched isomers such as 2-methylbutane. The value of the process is that branched alkanes increase the octane number of petrol.

Isomerisation happens when the hot vapour of the hydrocarbon passes over a platinum catalyst. The isomerisation reactions do not go to completion but produce equilibrium mixtures containing the unchanged alkane and its isomers. The products flow from the catalyst to a bed of a special zeolite with pores just the right size to acts as a 'molecular sieve'. The branched compounds cannot enter the pores of the zeolite but the straight-chain compounds can do so. The branched isomers are collected while the straight-chain compounds are recycled over the catalyst.

Reforming

Reforming converts alkanes to arenes (aromatic hydrocarbons, see page 196) such as benzene and methylbenzene. It can also convert alkanes to cyclic alkanes. Hydrogen is a valuable by-product of the process. The gas can be used in processes elsewhere in the refinery.

The catalyst for this process is often one or more of the precious metals such as platinum and rhodium supported on an inert material such as aluminium oxide. The process operates at about 500 °C.

Test yourself

1 Write structures to show how:
   a) catalytic cracking converts dodecane, $C_{12}H_{16}$ to 2,5-dimethylhexane and but-1-ene
   b) isomerisation converts pentane to 2-methylbutane and 2,2-dimethylpropane
   c) reforming converts hexane to cyclohexane.

$$CH_3CH_2CH_2CH_2CH_2CH_2CH_3 \xrightarrow[\text{heat, pressure}]{\text{Pt catalyst}} C_6H_5CH_3 + 4H_2(g)$$

heptane — methylbenzene (CH_3 on a benzene ring)

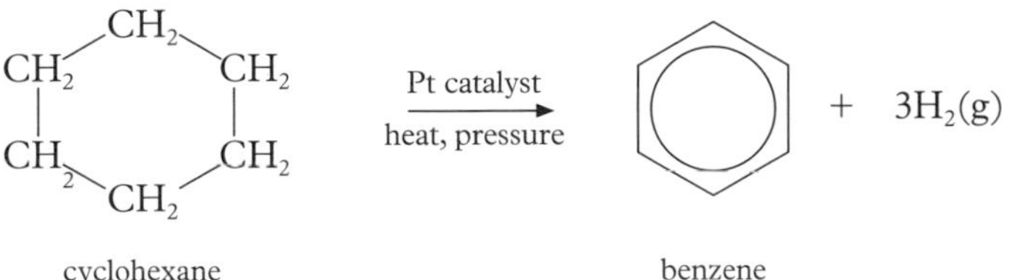

Figure 5.12.6 ◄
Examples of reforming. Note that the formation of hydrogen as a by-product means that these are not isomerisation reactions

Test yourself

2 Draw a diagram to show an ethane molecule splitting into two methyl radicals.

Chemicals from oil

Crude oil is not only a source of fuels. It is also a valuable source of chemicals, especially alkenes which make excellent building blocks for the synthesis of new chemicals.

The feedstocks for the petrochemical industry include ethane from natural gas and a so-called naphtha fraction from distilling oil which contains hydrocarbons with six to 10 carbon atoms.

Steam cracking

Steam cracking is the preferred method for converting hydrocarbons to alkenes. Steam cracking converts ethane to ethene which is a very important starting compound for chemical synthesis in industry.

A mixture of the hydrocarbon vapour and steam passes under pressure through tubes in a furnace where it is heated to about 1000 °C. This is thermal cracking as opposed to catalytic cracking. Under these conditions, cracking involves free-radical chain reactions (see page 194). The first step in the thermal cracking of ethane is for the molecules to split into methyl radicals.

Ethanol from ethene

The chemical industry makes ethanol by hydrating ethene in the presence of an acid catalyst (see page 204). An alternative source of the alcohol is fermentation (see page 230).

Ethanol is a useful solvent. Methylated spirit is ethanol with about 5% methanol to make it undrinkable. Industrial methylated spirit is used as a solvent commercially and in laboratories. Surgical spirit is industrial methylated spirit with other additives including castor oil. The methylated spirit ('meths') sold in hardware stores as a solvent and fuel is additionally coloured blue by a dye.

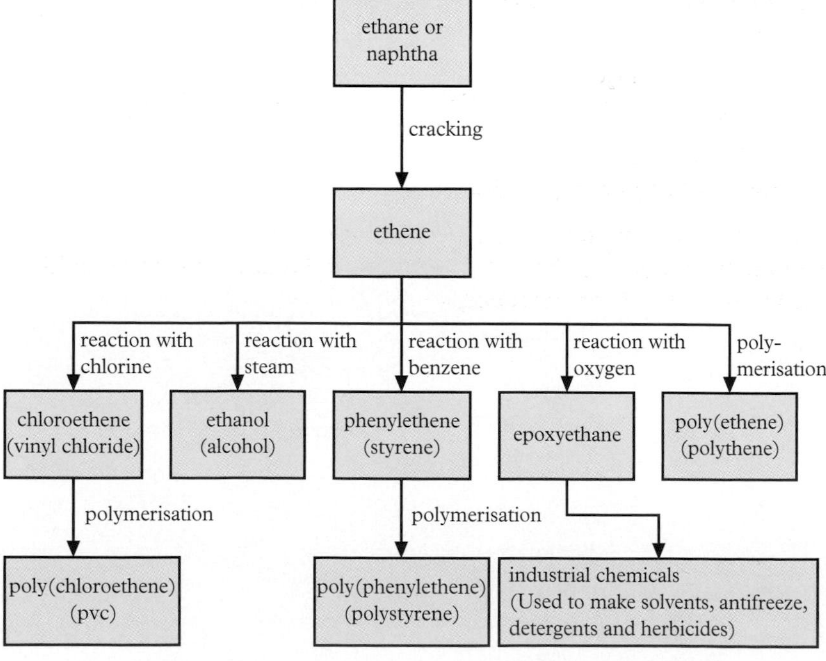

Figure 5.12.7 ► *Ethene is an important product of thermal cracking. Ethene can be converted into many other useful organic chemicals*

Epoxyethane from ethene

Epoxyethane is an important chemical intermediate used to manufacture surfactants, solvents and lubricants. Epoxyethane is made by mixing ethene with air or oxygen and passing the mixture at about 300 °C under pressure over a heterogeneous catalyst. The catalyst is finely-divided silver spread on the surface of an inert material such as alumina.

$$2CH_2{=}CH_2(g) + O_2(g) \rightleftharpoons 2\ \overset{\overset{O}{\diagup\ \diagdown}}{CH_2{-}CH_2}(g)$$

Epoxyethane is a reactive and hazardous chemical so it is generally made and stored where it will be used. It reacts vigorously with water producing a range of products depending on the conditions.

The main use of epoxyethane is to make ethane-1,2-diol, the antifreeze in motor vehicle engines. Antifreeze lowers the freezing point of water well below 0 °C. Ethane-1,2-diol is used as antifreeze because it mixes freely with water but has a high boiling point (198 °C) so that it does not evaporate from the coolant when the engine is hot. Hydrogen bonding between the two —OH groups in ethan-1,2-diol molecules and water molecules also helps to prevent the antifreeze evaporating. A further advantage of ethan-1,2-diol is that it limits corrosion of the metal parts in the cooling system.

Test yourself

3 Suggest reasons why the silver catalyst used to make epoxyethane is finely divided on the surface of aluminium oxide.

4 Draw the displayed formula of epoxyethane showing bond angles and bond polarities. Suggest reasons why it is a very reactive compound

5 Draw the structure of ethan-1,2-diol and show how the —OH groups can form hydrogen bonds with water.

$$n\ \underset{\diagdown O \diagup}{CH_2{-}CH_2} + H_2O \longrightarrow HO{+}CH_2CH_2O{+_n}H$$

Figure 5.12.8 ◀ *Reaction of epoxyethane with water. When n = 1, the product is ethane-1,2-diol which is the main ingredient of antifreeze. When n is greater than 4, the product may be useful as a solvent or as a non-ionic surfactant*

Epoxyethane is also the main source of non-ionic surfactants. Surfactants are surface-active agents. They are one of the ingredients of detergents because they help to separate greasy dirt from surfaces. They also keep dirt dispersed in water so that it rinses away. Non-ionic surfactants are used in many household cleaners because they allow smooth drainage and leave no deposit even when they are not fully rinsed away. Non-ionic surfactants also make less stable foams so they are included in washing powders for dishwashers and washing machines.

$$n\ \underset{\diagdown O \diagup}{CH_2{-}CH_2} + \underset{\text{alcohol}}{ROH} \longrightarrow HO{+}CH_2CH_2O{+_n}R$$

Figure 5.12.9 ◀ *Epoxyethane reacts with alcohols. With a primary alcohol, when n = 1 or n = 2 the product is a solvent*

Addition polymers from ethene

Ethene is the starting point for making most of the monomers used to manufacture addition polymers. Many molecules of the monomer add together to form a long chain polymer.

Monomer	Polymer	Notes
ethene $H_2C{=}CH_2$	poly(ethene) or polythene $\left[-\mathrm{CH_2}-\mathrm{CH_2}-\right]_n$	A high-pressure, high-temperature process in the presence of a peroxide initiator produces low-density poly(ethene) with branched chains. A low-pressure, low-temperature process with a Ziegler catalyst produces high-density poly(ethene); the polymer chains pack closer because they have no side branches.
propene $H_2C{=}CH(CH_3)$	poly(propene) or polypropylene $\left[-\mathrm{CH_2}-\mathrm{CH(CH_3)}-\right]_n$	A polymer with a higher softening temperature than poly(ethene). It is also tougher. It is used to make packaging materials and fibres for carpets and ropes
chloroethene $H_2C{=}CHCl$	poly(chloroethene) or polyvinylchloride (PVC) $\left[-\mathrm{CH_2}-\mathrm{CHCl}-\right]_n$	Unplasticised uPVC is a rigid polymer suitable for guttering and window frames. Plasticised PVC is flexible and used for packaging, flooring and cable insulation.
tetrafluoroethene $F_2C{=}CF_2$	poly(tetrafluoroethene) or PTFE $\left[-\mathrm{CF_2}-\mathrm{CF_2}-\right]_n$	The polymer used to coat non-stick pans. Engineers use ptfe to provide low-friction surfaces to allow bridges and other engineering structures to move slightly as the metals expand and contract with temperature change.

Figure 5.12.10 ▲ CD-ROM

Test yourself

6 Draw the structure of the polymer you would expect to form from

$H_2C{=}CH(C_6H_5)$

CD-ROM

Figure 5.12.11 ▲
This young boy is wearing a PVC jacket

5.13 Chemicals from plants

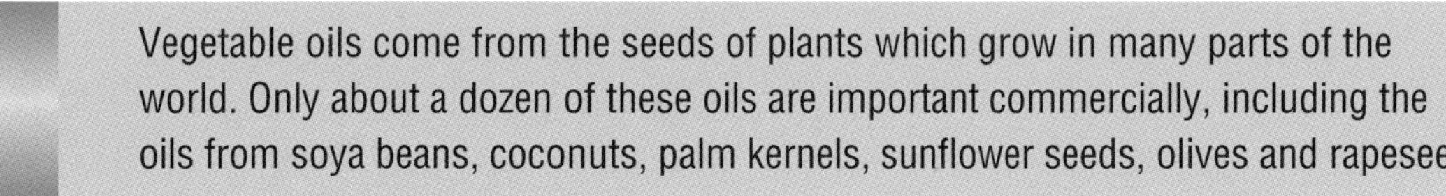

Vegetable oils come from the seeds of plants which grow in many parts of the world. Only about a dozen of these oils are important commercially, including the oils from soya beans, coconuts, palm kernels, sunflower seeds, olives and rapeseed.

Food from vegetable oils

Vegetable oils and animal fats are vital ingredients in a healthy diet. Chemically fats and oils are very similar. Some cultures favour liquid oils while others are more used to solid fats.

In many European countries people have traditionally eaten the animal fats in butter spread on bread. The search for an alternative to butter was prompted by Napoleon III in 1869. He suggested a competition to find a cheap fat which would keep well and be an alternative to butter for the working classes and in the Navy.

The presence or absence of double bonds is the main chemical difference which accounts for the different states of fats and oils at room temperature. The fatty acids in vegetable oils contain a higher proportion of unsaturated fatty acids than animal fats.

The molecules of unsaturated oils have a less regular structure than saturated fats. The molecules do not pack together so easily to make solids so they have lower melting points and have to be cooler before they solidify.

Hydrogenation is used industrially to add hydrogen to double bonds in oils. This produces saturated fats which are solid at room temperature. So the process is sometimes called 'hardening'. Manufacturers bubble hydrogen though the oil in the presence of finely-divided nickel which is the catalyst.

$$\text{---CH}{=}\text{CH---} \xrightarrow[\text{Ni catalyst at 140°C}]{H_2(g)} \text{---CH}_2{=}\text{CH}_2\text{---}$$

unsaturated fatty acid → saturated fatty acid

Figure 5.13.1 ◄ *Hydrogenation of a double bond in unsaturated fat with hydrogen in the presence of a nickel catalyst*

Margarine and non-dairy spreads are solid emulsions consisting of water finely dispersed in vegetable oils with a high enough proportion of hardened oils to make the product a spreadable solid.

Fuels from vegetable oils

Vegetable oils have always been used domestically for heating and lighting on a domestic scale though in industrialised countries they have been largely replaced by natural gas and other fossil fuels.

Today there is increasing commercial interest in biofuels. Countries such as Italy and Australia are making significant use of diesel fuel manufactured from the oils from crops such as rape.

Chemicals from vegetable oils

The most important use of vegetable oils as a chemical raw material is for the manufacture of soaps; but they are also used to make paints, varnishes, lubricants and plastics. Vegetable oils are esters and hydrolysis with solutions of potassium or sodium hydroxide converts them to soap.

5.14 Organic chemicals and the environment

People value crude oil and natural gas as sources of energy and chemicals, but extracting these raw materials from the Earth, transporting and burning fuels, using the chemicals and then throwing them away can damage the environment in a number of serious ways. In this section the main topics are the effect of fuels and chemicals on the atmosphere and the problems arising from the disposal of solid wastes.

Test yourself

1 Identify some of the ways that extracting and transporting crude oil can damage the environment.

Pollution of the atmosphere

Pollutant gases from motor vehicles, power stations and industrial processes affect the lower parts of the atmosphere (troposphere). The most serious consequence of organic halogen compounds, on the other hand, is to affect the ozone layer in the upper atmosphere (stratosphere).

Ozone often features in stories about environmental problems. The ozone layer in the upper atmosphere is vital for living things. The ozone high in the stratosphere protects living things by absorbing harmful UV radiation from the Sun. In the lower atmosphere, ozone is harmful because it damages living things and helps to cause photochemical smog.

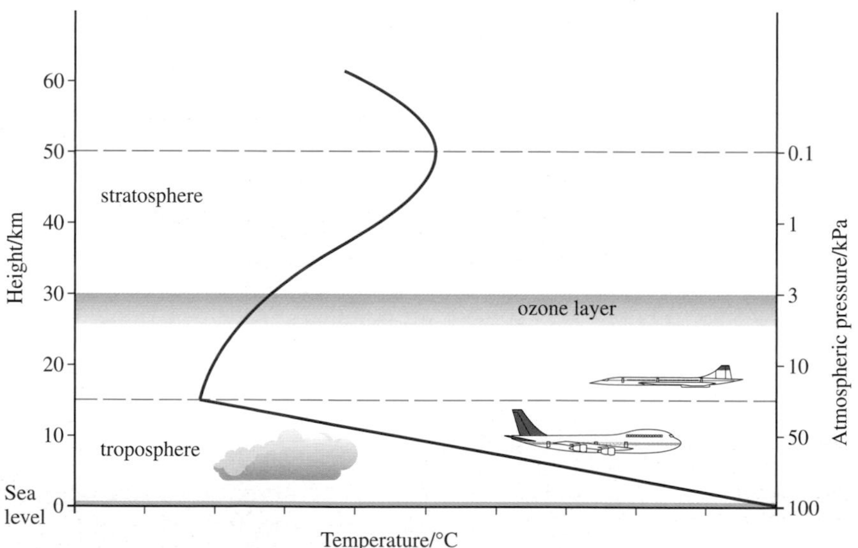

Figure 5.14.1 ► *The Earth's atmosphere*

Definition

Environmental chemists use the abbreviation $\mathbf{NO_x}$ to refer to a polluting mixture of nitrogen monoxide, NO, and nitrogen dioxide, NO_2.

Types of air pollution

Photochemical smog

Photochemical smog is produced by sunlight and the pollutants from the exhaust gases of motor vehicles. This type of smog forms on still sunny days when there is no wind to blow away the gases. It is severe in cities such as Los Angeles where weather conditions and the local geography tend to trap the pollutants.

The primary pollutants are nitrogen oxides and unburnt hydrocarbons. They are emitted in large amounts during the morning rush hour in cities. Bright sunlight during the middle of the day sets off photochemical reactions involving oxygen in the air. The products are the secondary pollutants which create the smog.

The level of ozone in the air rises and oxidising free radicals form. The unburnt hydrocarbons are then oxidised to aldehydes, ketones and other chemicals such as organic nitrates which irritate the eyes and lungs.

Acid rain

Acid rain is a type of pollution produced when burning fuels or industrial processes release acidic oxides into the air. Sulfur dioxide, SO_2, and nitrogen oxides, NO_x, form when fuels burn in engines, furnaces and power stations. These primary pollutants are converted to secondary pollutants by chemical reactions in the air. Among the secondary pollutants are sulfuric acid, nitric acid and ammonium sulfate. The pollutants cause acidification by being deposited in the environment as gases or particles (dry deposition) or in the form of rain or mist (wet deposition).

Fossil fuels contain varying amounts of sulfur compounds. The proportion of sulfur compounds in crude oil, for example, varies from less than 1% up to 7% depending on the origin of the crude. When crude oil is distilled, the sulfur compounds tend to be concentrated in the heavy fractions such as the fuel oils. Oil refining removes much of this sulfur and

Test yourself

2 Give examples to explain the difference between a primary pollutant and a secondary pollutant.

3 Why does air pollution by sulfur dioxide and oxides of nitrogen speed up both the corrosion of metals and the decay of building stones such as limestone?

Figure 5.14.2 ▼ *The effects of acid rain and acid gases*

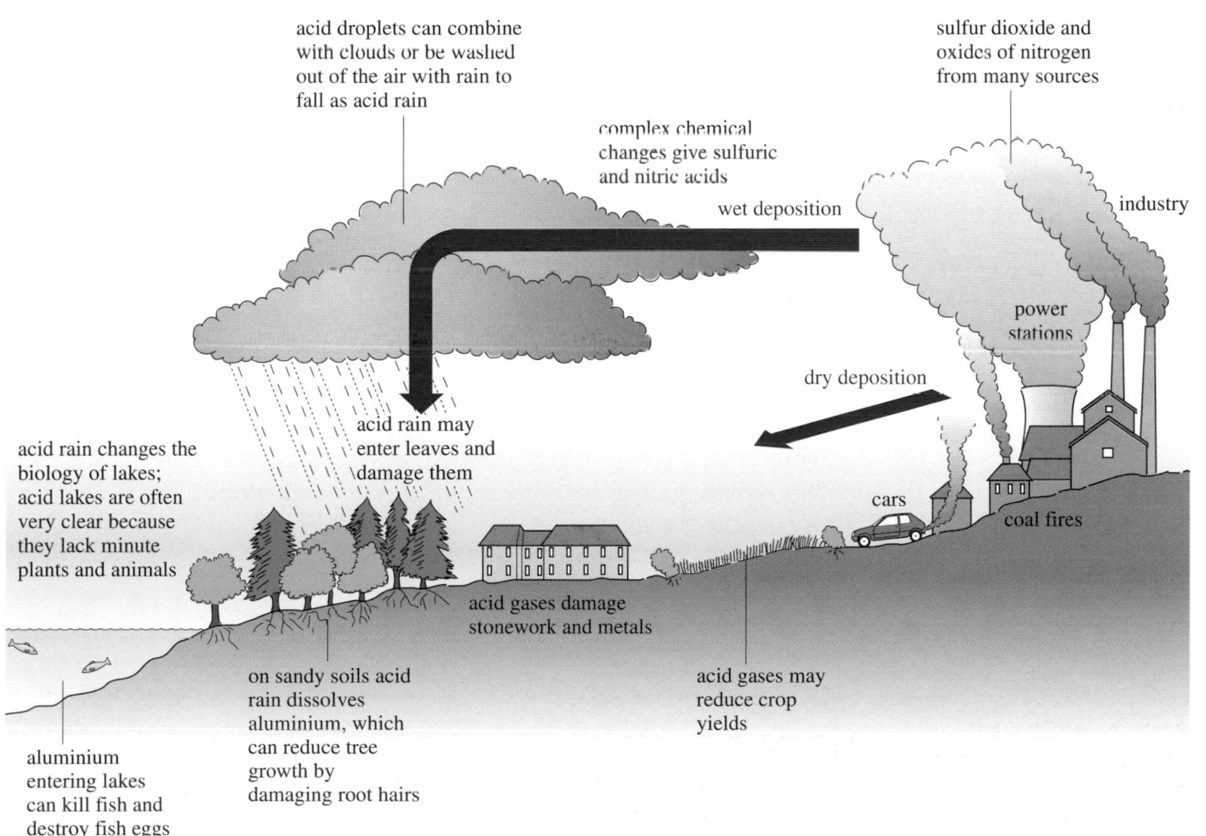

acid droplets can combine with clouds or be washed out of the air with rain to fall as acid rain
sulfur dioxide and oxides of nitrogen from many sources
complex chemical changes give sulfuric and nitric acids
wet deposition
industry
power stations
dry deposition
acid rain may enter leaves and damage them
acid rain changes the biology of lakes; acid lakes are often very clear because they lack minute plants and animals
cars
coal fires
acid gases damage stonework and metals
on sandy soils acid rain dissolves aluminium, which can reduce tree growth by damaging root hairs
acid gases may reduce crop yields
aluminium entering lakes can kill fish and destroy fish eggs

Test yourself

4 Estimate the volume of carbon dioxide produced at atmospheric pressure when a car travels about 100 miles using 12 dm^3 petrol. Assume that petrol consist of the alkane octane, C_8H_{18}, with a density of 0.8 g cm^{-3}.

5 Why is ozone in the stratosphere essential to living things while ozone in the troposphere is a threat to health?

6 What would you say to someone who believes that the hole in the ozone layer causes global warming by letting in more energy from the Sun. Is there any connection between global warming and the destruction of the ozone layer by CFCs?

there is very little in petrol. Sulfur compounds burn to produce sulfur dioxide.

Nitrogen oxides form by direct combination of nitrogen with oxygen in the air as fuels burn at high temperatures in furnaces or engine cylinders.

The greenhouse effect and global warming

The greenhouse effect keeps the surface of the Earth about 30 °C warmer that it would be if there were no atmosphere. Without the greenhouse effect there would be no life on Earth.

When radiation from the Sun reaches the Earth's atmosphere, about 30% is reflected into space, 20% is absorbed by gases in the air and about half reaches the surface of the Earth.

The warm surface radiates energy back into space but at longer, infra-red wavelengths. Some of the infra-red radiation is absorbed and warms up the atmosphere. This is the greenhouse effect.

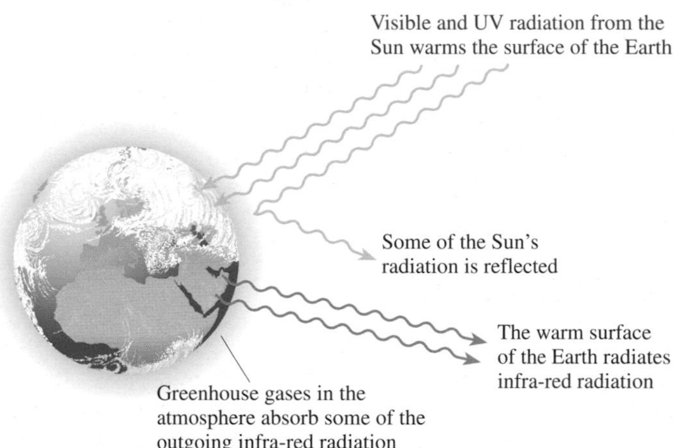
Visible and UV radiation from the Sun warms the surface of the Earth
Some of the Sun's radiation is reflected
The warm surface of the Earth radiates infra-red radiation
Greenhouse gases in the atmosphere absorb some of the outgoing infra-red radiation

Figure 5.14.3 ► The greenhouse effect

The gases in the air which absorb infra-red radiation are called greenhouse gases. Nitrogen and oxygen make up most of the air but they are not greenhouse gases. The main natural greenhouse gases are carbon dioxide, methane, dinitrogen oxide and water vapour.

The concentrations of greenhouse gases in the air are rising because of human activity such as burning fossil fuels and agriculture. This is enhancing the greenhouse effect. There is growing evidence that this is responsible for global warming and climate change.

The hole in the ozone layer

The ozone layer is a concentration of the gas in the stratosphere at about 10–50 km above sea level. At this altitude ultra-violet (UV) light from the Sun splits oxygen molecules into oxygen atoms.

$$O_2 + \text{UV light} \rightarrow O + O$$

These are free radicals which then convert oxygen molecules to ozone.

$$O + O_2 \rightarrow O_3$$

The ozone formed also absorbs UV radiation. This splits ozone molecules back into oxygen molecules and oxygen atoms destroying ozone.

$$O_3 + \text{UV light} \rightarrow O_2 + O$$

In the absence of pollutants there is a steady state with ozone being destroyed as quickly as it forms. Normally the steady-state concentration of ozone is sufficient to absorb most of the dangerous UV light from the Sun.

The problem with CFCs (chlorofluorocarbons) and some other halogen compounds is that they are chemically very unreactive. They escape into the atmosphere where they are so stable that they last for many years, long enough for them to diffuse up to the stratosphere. In the stratosphere, the intense ultra-violet light from the Sun splits CFCs into free radicals including chlorine atoms. Chlorine atoms react with ozone.

$$Cl\bullet + O_3 \rightarrow ClO\bullet + O_2$$

$$ClO\bullet + O\bullet \rightarrow Cl\bullet + O_2$$

The first reaction is much faster than other reactions in the stratosphere. The second reaction involves oxygen atoms which are common in the stratosphere and it recreates the chlorine atom. This means in effect that one chlorine atom can rapidly destroy many ozone molecules. This effect was noticed in the early 1980s when scientists in the Antarctic noticed that the ozone concentration in the stratosphere was much lower than expected. Since then the ozone layer has been monitored by satellites which have confirmed that there is a 'hole' in the ozone layer not only over the Antarctic but in other regions too.

Tackling pollution problems

One approach to tackling the problems of air pollution is to find ways of preventing the pollutants escaping into the air. Another approach is to develop alternative technologies which cause less pollution. A third approach is to change lifestyles and ways of working to cut the rate at which people use energy and consume valuable resources.

Cutting emissions from vehicles

Growing concern about acid rain and photochemical smog has led to many governments regulating new cars and insisting that they are fitted with catalytic converters. A catalytic converter is a device in the exhaust system of a car which contains a catalyst to covert pollutants in the exhaust gases to less harmful substances. Car exhausts pollute the air because the engine does not burn all the fuel and because the temperature and pressure in the cylinders are high enough for nitrogen from the air to react with oxygen.

Pollutant gas	Origin of the pollutant
carbon dioxide, CO_2	complete combustion of hydrocarbons in petrol
carbon monoxide, CO	incomplete combustion of fuel
hydrocarbons, C_xH_y	unburnt fuel
nitrogen oxides, NO_x	reaction of nitrogen and oxygen from the air in the hot engine
lead compounds	from anti-knock additives in leaded petrol

Figure 5.14.4 ◄ *Pollutants from petrol engines*

Unleaded petrol must be used in cars fitted with a catalytic converter because lead would poison the catalyst and stop it working.

The catalyst is a finely-divided alloy of platinum and rhodium supported on an inert ceramic pierced with many fine tubes to give a very large surface area. Once the catalyst is hot enough, it converts the pollutants to steam, carbon dioxide and nitrogen. So it helps to cut out the pollutants that cause acid rain and smog but it does nothing to limit the amount of carbon dioxide into the atmosphere. Catalytic converters can increase fuel consumption so they may mean that motor vehicles make an even bigger contribution to the level of greenhouse gases in the atmosphere.

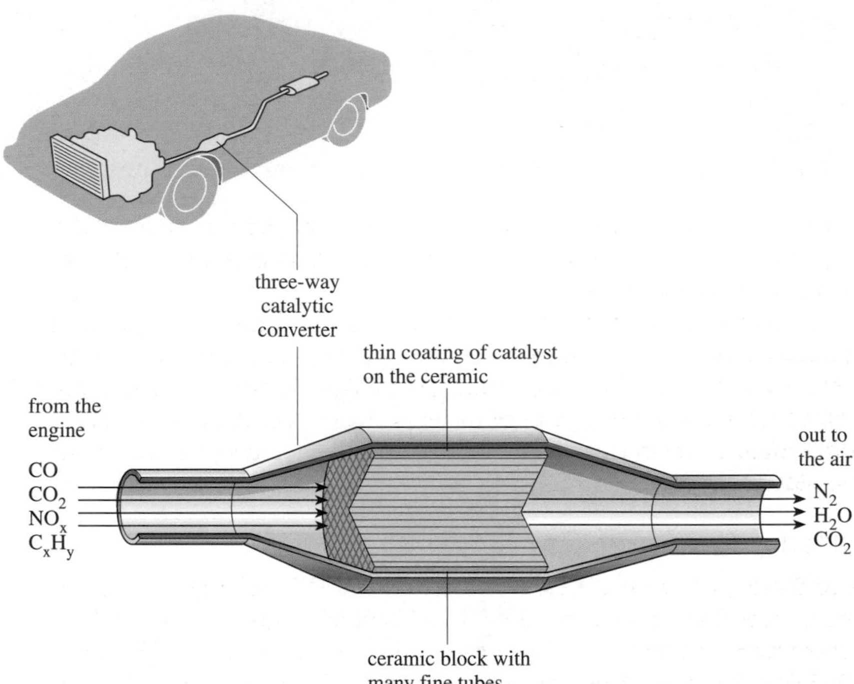

Figure 5.14.5 ► *Diagram of a catalytic converter. Catalytic converters remove 80–90% of polluting emissions*

Alternative fuels

Concern about the enhanced greenhouse effect is encouraging people to seek out alternative fuels such as biofuels. Crops take in carbon dioxide from the air through photosynthesis as they grow to make sugars or oil. The carbon dioxide returns to the air as the biofuels burn. It appears that the use of biofuels would have no overall effect on the level of carbon dioxide in the atmosphere. This, however, ignores the energy from burning fossil fuels during the planting, harvesting and processing of the crop. Fertilisers may also be needed for growth and these need fossil fuels for their manufacture (see pages 162–166).

Brazil is the leading country using ethanol as a biofuel. The ethanol is manufactured by fermenting sugars from sugar cane. Fermentation converts sugars to alcohol (ethanol) and carbon dioxide. Fermentation is an example of anaerobic respiration. This reaction is catalysed by an enzyme from yeast.

$$\underset{\text{glucose sugar}}{C_6H_{12}O_6(aq)} \rightarrow \underset{\text{carbon dioxide}}{2CO_2(g)} + \underset{\text{ethanol}}{2C_2H_5OH(aq)}$$

Figure 5.14.6 ◀
A petrol station in Sao Paulo, Brazil, with ethanol and petrol pumps

In Brazil some cars have specially adapted engines and fuel tanks so that they can run on pure ethanol. Other vehicles run on gasohol which is a blend of petrol and ethanol.

Ethanol is in many ways a good fuel. It has good anti-knock properties and the emissions of nitrogen oxides and carbon monoxide are lower than petrol when it burns. The disadvantages are that the fuel contains about 5% water so it is corrosive to parts of the engine; also, combustion produces some ethanal, which is an irritant. The long-term effects of emissions of this aldehyde are not yet known.

Limiting the damage to the ozone layer

Worldwide production of CFCs has fallen sharply since their impact on the ozone layer was discovered. At a United Nations conference in 1987, governments agreed on a 50% reduction in the production of CFCs by 1999 (the Montreal Protocol). The European Union has agreed to a total ban by the year 2000.

The problem with implementing a ban is that CFCs have proved to

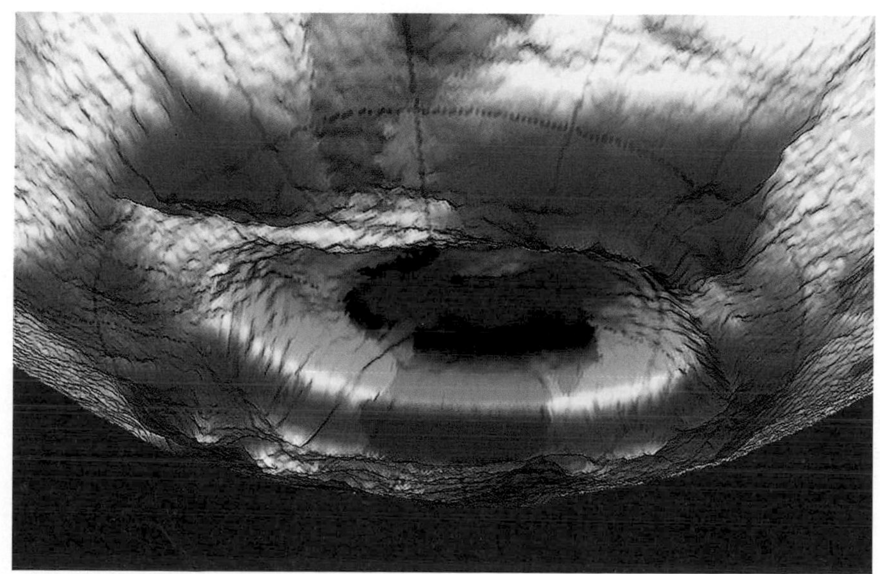

Figure 5.14.7 ◀
A satellite image showing ozone depletion over Antartica

have very desirable properties. CFCs are compounds containing carbon, chlorine and fluorine such as CCl_3F, CCl_2F_2 and CCl_2FCClF_2. They have advantages: they are unreactive, do not burn and are not toxic. Also it is possible to make CFCs with different boiling points to suit different applications. These properties make CFCs ideal as the working fluid in refrigerators and air conditioning units. They also act as blowing agents to make the bubbles in expanded plastics and insulating foams. CFCs make good solvents for dry cleaning and for removing grease from electronic equipment.

The hunt is on for alternative compounds with the desirable properties of CFCs but with fewer environmental problems. One possibility is to use hydrofluorocarbons (HFCs) instead. These compounds have no chlorine atoms and are less persistent in the environment because of the hydrogen atoms.

Managing and disposing of organic wastes

Waste causes pollution problems if it is not carefully managed. More waste is buried in landfill around the world than is managed by any other method such as recycling, composting or energy recovery. Much domestic waste is organic, including plastics.

Recycling

Plastics are much more difficult to recycle than iron or steel (see pages 173–174). One problem arises from the low density of plastic waste. It can take as many as 20 000 bottles to make up a tonne of waste. Even so, the typical European throws away about 36 kg of plastic waste each year.

There are many types of plastic so they have to be sorted which is difficult. The industry has introduced codes for labelling plastic products to help consumers separate them for recycling.

Recycling rates are increasing as automatic machines for sorting plastic bottles have been developed. Separation methods include the use of infra-red detectors, chemical scanning for chlorine in pvc, flotation to separate flaked plastics according to their densities and electrostatic separation which can classify and sort polyester, PET and pvc.

Plastics vary greatly in their value. The material used to make sparkling drink bottles, PET, is generally the most valuable. PET can be re-melted and spun into fibres for carpets, bedding and clothing. Another plastic worth recycling is pvc which can be recycled to make sewage pipes, flooring and

Test yourself

7 Why are recycling rates much higher for metals, such as aluminium and iron, than they are for plastics?

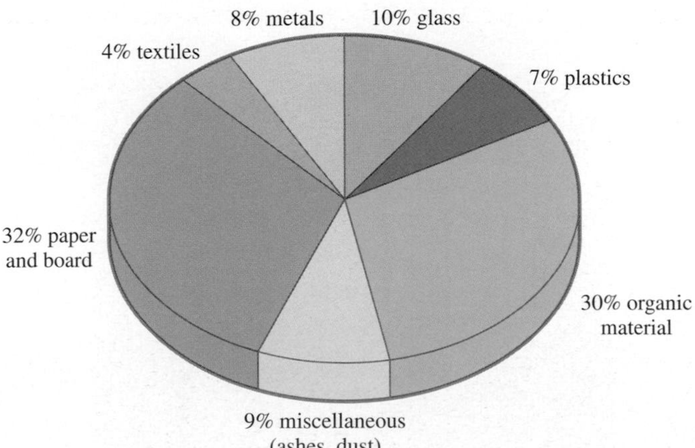

Figure 5.14.8 ▶
Typical composition of household waste in Europe

even the soles of shoes. An alternative approach to recycling plastics is to use cracking to break the polymer molecules into small molecules which can be added to the feedstock for oil refineries and chemical plants.

Recovery of plastic waste becomes easier when designers plan plastic products with recycling in mind. Engineers in the motor industry, for example, are putting much effort into designing components so that they can be recycled at the end of the life of a car.

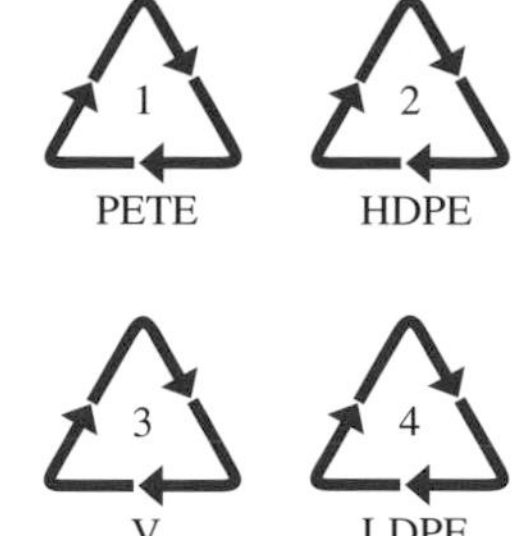

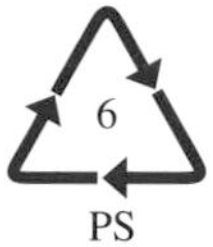

Figure 5.14.9 ▲
The above numbered codes are proposed by the American plastics manufacturers in a move to make different plastics readily identifiable by the general public and therefore more easily separated for recycling

Fuel from waste

An option worth considering is to burn plastic waste or domestic refuse and to use the energy to generate electricity. Just dumping rubbish in landfill is a waste of resources. Burning all the domestic waste from homes in Europe could generate 5% of European energy needs.

Material	Energy value/MJ kg^{-1}
oil	40
mixed plastic waste	25–40
domestic refuse	10
fuel made from refuse	16

Figure 5.14.10 ◀

There has been strong opposition to the burning of waste. Old incinerators were inefficient, operated at too low a temperature and often were not designed to generate electricity. Incomplete combustion produces carbon monoxide which is toxic.

Another worry is that if waste contains chlorine compounds such as pvc, an incinerator will give off corrosive or toxic chemicals such as hydrogen chloride or dioxins. High temperature combustion at 800–1250 °C, however, helps to achieve complete combustion and limit the emission of toxic chemicals. Modern plants have elaborate gas cleaning systems to control the emission of harmful gases and dust particles.

There are benefits too. Incineration greatly reduces the volume of waste to be dumped in landfill sites. It also replaces fossil fuels thus cutting the overall emission of greenhouse gases.

Definition

Dioxins are a family of stable compounds which persist in the environment. Some dioxins are toxic and can cause skin disease, cancer and birth defects.

Landfill

Most waste ends up in landfill. A modern landfill site is engineered to contain waste and the products of decomposition. Some landfills are for general domestic waste but others are planned to contain more hazardous wastes.

Figure 5.14.11 ◀
A waste site designed to collect landfill gas

Definitions

An **aerobic** change happens in the presence of oxygen from the air.

An **anaerobic** process happens in the absence of oxygen.

Water trickling through a landfill site leaches organic acids and metal ions from the waste. The **leachate** is the toxic solution flowing from the bottom of the decomposing waste.

Heavy vehicles compact the waste and the layers of waste are covered with soil. Micro-organisms break down the organic materials which make up more than half of household waste. At first there is oxygen present and micro-organisms convert complex carbohydrates to simple sugars. Fermentation begins as the conditions become anaerobic, turning carbohydrates, proteins and fats into organic acids, carbon dioxide and hydrogen. In time the micro-organisms start to produce methane and carbon dioxide. This landfill gas can be a hazard because methane is highly flammable, but the gas can also be collected from pipes buried in the waste and burnt to generate electricity.

Liquids flow from the bottom of the landfill. These leachates can be very toxic because they contain organic chemicals and heavy metals. The organic chemicals come partly from the breakdown of the waste and partly from chemicals discarded in the waste.

Acids in the liquids dissolve metals from the solid waste so that the leachate contains toxic compounds of metals such as mercury, chromium, lead and cadmium. Landfill sites have to be monitored to ensure that, so far as is possible, the toxic leachate does not contaminate underground water supplies.

Figure 5.14.12 ► *Monitoring effluent at a landfill site*

Test yourself

8 Consider the question: 'Chlorine – friend or foe?'. Some environmentalists would like to see a ban on the manufacture and use of many organic chlorine compounds. CFCs, the insecticide DDT and the polymer pvc are among the products that have given chlorine a bad name. The chemical industry argues that people need chlorine chemistry and that the environmental problems can be solved. Chlorine compounds purify drinking water, help to cure disease and protect crops from attack by pests. Draw up a table to compare the risks and benefits of chlorine and its compounds (see pages 154, 159, 207 and 229). What is your answer to the question – friend or foe?

Anaerobic digestion

Letting biodegradable wastes rot and ferment in landfill can be hazardous and throws away much of a possible fuel even if some of the landfill gas is collected. An alternative is to separate the organic materials and digest them under anaerobic conditions in heated containers. This speeds up breakdown by micro-organisms and makes it possible to collect all the biogas. Waste paper is an ideal material for anaerobic digestion along with vegetable and kitchen wastes. About 30% of the energy from the biogas is used to the heat the digesters. The rest is available to generate electricity.

Review

This guidance will help you to organise your notes and revision. Check the terms and topics against the specification you are studying. You will find that some topics are not required for your course.

Key terms

Show that you know the meaning of these terms by giving examples. Consider writing the key term on one side of an index card and the meaning of the term with an example on the other side. Then you can easily test yourself when revising. Alternatively use a computer database with fields for the key term, the definition and the example. Test yourself with the help of reports which just show one field at a time.

- Homologous series
- Isomers
- Functional group
- Saturated compounds
- Unsaturated compounds
- Double bond
- Pi(π)-bond
- Geometrical isomerism
- Addition reaction
- Substitution reaction
- Elimination reaction
- Dehydration
- Hydrolysis
- Oxidation
- Reduction
- Hydrogenation
- Polymerisation
- Addition polymerisation
- Fermentation
- Homolytic bond breaking
- Free radicals
- Heterolytic bond breaking
- Nucleophiles
- Leaving groups
- Electrophiles
- Carbocations

Symbols and conventions

Make sure that you understand the symbols and conventions which chemists use when writing equations. Illustrate your notes with examples.

- Names of carbon compounds: alkanes, alkenes, halogenoalkanes, alcohols, aldehydes, ketones, carboxylic acids, esters, nitriles.
- Types of formula: empirical, molecular, structural and displayed (or graphical).
- The use of the terms primary, secondary and tertiary as applied to alcohols and halogenoalkanes.
- Use of curly arrows to show how bonds break and form when describing reaction mechanisms.

Facts, patterns and principles

Use charts, annotated equations and spider diagrams to summarise key facts and ideas. Brighten your notes with colour to make them memorable. Make sure that you show clearly the reagents and conditions needed to bring about each reaction. Try writing the equation for a reaction on one side of an index card and the reagents and conditions on the other side. You can carry around a set of cards like this to revise from and test yourself with during any spare moments.

- The link between the physical properties of a series of organic compounds and the types of intermolecular forces between the molecules
- Reactions of alkanes
- Reactions of alkenes
- Reactions of halogenoalkanes
- Reactions of alcohols including the formation of aldehydes, ketones, carboxylic acids and esters
- Types of reaction mechanism: free radical chain reactions, nucleophilic substitution reactions, electrophilic addition reactions (and the explanation for Markovnikov's rule)

Laboratory techniques

Draw flow diagrams to show these key steps in practical procedures.

- Test-tube tests to identify functional groups.
- Simple distillation and fractional distillation.
- Recrystallisation to purify a solid product.
- Measuring boiling and melting points.

Calculations

Give your own worked examples, with the help of the Test Yourself questions to show that you can carry out calculations to work out the following from given data.

- Empirical formulae from percentage combustion and the results of combustion analysis.
- Molecular formula from the empirical formula and the molar mass.
- Theoretical and percentage yields for organic reactions.

Chemical applications

Use charts, tables and diagrams to summarise the principles and processes mentioned in the specification for the course you are studying.

- Refining and processing crude oil by fractional distillation, catalytic cracking, reforming, isomerisation.
- Formulation and combustion of fuels including petrol, biofuels (biogas, ethanol and biodiesel), methanol.
- Types of air pollution caused by burning fuels including the formation of greenhouse gases.
- Catalytic converters and how they help to cut pollution by acid rain and low-level ozone.
- Production of petrochemicals by thermal (steam) cracking and the conversion of alkenes to other useful materials for organic synthesis such as epoxyethane.
- Formation of addition polymers from unsaturated compounds such as ethene and related compounds.
- Valuable and harmful properties of halogenocompounds including CFCs, fluorocarbons and dioxins.
- The management of organic wastes including waste polymers by methods which include recycling, incineration (for energy recovery) and landfill.

Key skills

Communication

Reading, writing and talking about the applications of organic chemistry allows you to show that you can select and read material that contains information you need, pick out the main points, identify lines of reasoning and then synthesise the information in a form relevant to your purpose using specialist vocabulary when appropriate.

Information technology

When studying organic chemistry you can use chemical modelling software from a CD-ROM or the Internet to study the shapes of the molecules. In this way you can explore bond angles, molecular shapes and the effects of isomerism. CD-ROM

When summarising your knowledge and understanding you could record key facts and definitions in a database with fields such as organic reactant, reagents, conditions, products, type of reaction, mechanism. You could then test yourself during revision by displaying reports which show part of the information and prompt you to recall the rest.

Improving own learning and performance

You will have to learn how to make sense of a great deal of information about organic compounds, their reactions and their uses. Seek advice from your chemistry teacher about effective techniques for learning descriptive organic chemistry. Use the review section above to set realistic targets. Consider working with other students to test each other so that you can assess the quality of your learning.

Section six

Reference

Contents

Periodic table

Properties of elements and compounds

Solubility

Bond lengths and bond energies

Qualitative analysis tests

Periodic table

Key: Atomic number / Symbol / Name / Relative atomic mass

Period	1	2					Transition elements						3	4	5	6	7	8
1	1 H Hydrogen 1																	2 He Helium 4
2	3 Li Lithium 7	4 Be Beryllium 9											5 B Boron 11	6 C Carbon 12	7 N Nitrogen 14	8 O Oxygen 16	9 F Fluorine 19	10 Ne Neon 20
3	11 Na Sodium 23	12 Mg Magnesium 24											13 Al Aluminium 27	14 Si Silicon 28	15 P Phosphorus 31	16 S Sulphur 32	17 Cl Chlorine 35.5	18 Ar Argon 40
4	19 K Potassium 39	20 Ca Calcium 40	21 Sc Scandium 45	22 Ti Titanium 48	23 V Vanadium 51	24 Cr Chromium 52	25 Mn Magnesium 55	26 Fe Iron 56	27 Co Cobalt 59	28 Ni Nickel 59	29 Cu Copper 63.5	30 Zn Zinc 65.4	31 Ga Gallium 70	32 Ge Germanium 73	33 As Arsenic 75	34 Se Selenium 79	35 Br Bromine 80	36 Kr Krypton 84
5	37 Rb Rubidium 85	38 Sr Strontium 88	39 Y Yttrium 89	40 Zr Zirconium 91	41 Nb Niobium 93	42 Mo Molybdenum 96	43 Tc Technetium	44 Ru Ruthenium 101	45 Rh Rhodium 103	46 Pd Palladium 106	47 Ag Silver 108	48 Cd Cadmium 112	49 In Indium 115	50 Sn Tin 119	51 Sb Antimony 122	52 Te Tellurium 128	53 I Iodine 127	54 Xe Xenon 131
6	55 Cs Caesium 133	56 Ba Barium 137	57 ► La Lanthanum 139	72 Hf Hafnium 178	73 Ta Tantalum 181	74 W Tungsten 184	75 Re Rhenium 186	76 Os Osmium 190	77 Ir Iridium 192	78 Pt Platinum 195	79 Au Gold 197	80 Hg Mercury 201	81 Tl Thallium 204	82 Pb Lead 207	83 Bi Bismuth	84 Po Polonium	85 At Astatine	86 Rn Radon
7	87 Fr Francium	88 Ra Radium 226	89 ►► Ac Actinium 227	104 Rf Rutherfor dium	105 Db Dubnium	106 Sg Seaborgium	107 Bh Bohrium	108 Hs Hassium	109 Mt Meitnerium	110 Uun Ununnilium	111 Uuu Unununium	112 Uub Ununbium						

► Lanthanoid elements	58 Ce Cerium 140	59 Pr Praseo dymium 141	60 Nd Neo-dymium 144	61 Pm Promethium	62 Sm Samarium 150	63 Eu Europium 152	64 Gd Gadolinium 157	65 Tb Terbium 159	66 Dy Dysprosium 163	67 Ho Holmium 165	68 Er Erbium 167	69 Tm Thulium 169	70 Yb Ytterbium 173	71 Lu Lutetium 175
►► Actinoid elements	90 Th Thorium 232	91 PA Protactinium 231	92 U Uranium 238	93 Np Neptunium 237	94 Pu Plutonium	95 Am Americium	96 Cm Curium	97 Bk Berkelium	98 Cf Californium	99 Es Einstein -ium	100 Fm Fermium	101 Md Mendel -evium	102 No Nobelium	103 Lr Lawrencium

Note: Relative atomic masses are shown only for elements which have stable isotopes or isotopes with a very long half-life.

Properties of elements and compounds

The densities of gases are at 25 °C. Atomic and ionic radii are in picometres (1 pm = 10^{-12} m). The atomic radii are metallic radii for metals and covalent radii for non-metals (except for the noble gases for which the radii are van der Waals radii). Ionic radii are quoted for the common simple ions.

Properties of selected elements

Compound	Density/ g cm^{-3}	Molar mass/ g mol^{-1}	Melting point/°C	Boiling point/°C	Ionisation energies/kJ mol^{-1}			Radii	
					1st	2nd	3rd	atomic/ pm	ionic/ pm
Group 1									
Lithium, Li	0.53	7	180	1327	520	7298	11815	157	74
Sodium, Na	0.97	23	98	900	496	4563	6913	191	102
Potassium, K	0.86	39	63	777	419	3051	4412	235	138
Rubidium, Rb	1.53	85	39	705	403	2632	3900	250	149
Caesium, Cs	1.88	133	29	669	376	2420	3300	272	170
Group 2									
Beryllium, Be	1.85	9	1285	2470	900	1757	14890	112	27
Magnesium, Mg	1.74	24	650	1100	738	1451	7733	160	72
Calcium, Ca	1.53	40	840	1490	590	1145	4912	197	100
Strontium, Sr	2.60	88	769	1384	550	1064	4210	215	113
Barium, Ba	3.59	137	710	1640	503	965		224	136
Group 3									
Boron, B	2.47	11	2030	3700	801	2427	3660	98	12
Aluminium, Al	2.70	27	660	2350	578	1817	2745	143	53
Gallium, Ga	5.91	70	30	2070	579	1979	2963	153	62
Group 4									
Carbon, C (diamond)	3.53	12	3550	4827	1086	2353	4621	77	–
Silicon, Si	2.33	28	1410	2620	789	1577	3232	118	40
Germanium, Ge	5.32	73	959	2850	762	1537	3302	139	54
Group 5									
Nitrogen, N	0.00117	14	–210	–196	1402	2856	4578	75	171
Phosphorus, P (white)	1.82	31	44	280	1012	1903	2912	110	190
Group 6									
Oxygen, O	0.00133	16	–219	–183	1314	3388	5301	73	140
Sulfur, S	2.07	32	113	445	1000	2251	3361	102	185
Group 7									
Fluorine, F	0.00158	19	–220	–188	1681	3374	6051	71	133
Chlorine, Cl	0.00299	35.5	–101	–34	1251	2297	3822	99	180
Bromine, Br	3.12	80	–7	59	1140	2100	3500	114	195
Iodine, I	4.95	127	114	184	1008	1846	3200	133	215
Group 8									
Helium, He	0.00017	4	–270	–269	2372	5251	–	180	–
Neon, Ne	0.00084	20	–249	–246	2081	3952	6122	160	–
Argon, Ar	0.00166	40	–189	–186	1521	2666	3931	190	–
Krypton, Kr	0.00346	84	–157	–153	1351	2368	3565	200	–
Xenon, Xe	0.0055	131	–112	–108	1170	2047	3100	220	–
d-block elements									
Chromium, Cr	7.19	52	1860	2600	653	1592	2987	129	62
Manganese, Mn	7.47	55	1250	2120	717	1509	3249	137	67
Iron, Fe	7.87	56	1540	2760	759	1561	2958	126	55
Copper, Cu	8.93	64	1084	2580	746	1958	3554	128	73
Zinc, Zn	7.14	65	420	913	906	1733	3833	137	75

Properties of selected inorganic compounds

Compound	Melting point/°C	Boiling point/°C	$\Delta H_f^{\ominus}$/ kJ mol^{-1}	Solubility mol/100 g water
Aluminium chloride, $AlCl_3$	sublimes	–	–704	0.52
Aluminium oxide, Al_2O_3	2015	2980	–1676	insoluble
Ammonia, NH_3	–78	–34	–46.1	3.11
Barium chloride, $BaCl_2$	963	1560	–859	0.15
Barium oxide, BaO	1917	2000	–554	0.023
Barium hydroxide, $Ba(OH)_2$	408	decomposes	–945	0.015
Barium sulfate, $BaSO_4$	1580	decomposes	–1473	insoluble
Caesium fluoride, CaF	682	1250	–554	3.84
Caesium chloride, CsCl	645	1300	–443	1.13
Calcium chloride, $CaCl_2$	782	2000	–443	0.54
Calcium oxide, CaO	2600	3000	–635	reacts
Calcium hydroxide, $Ca(OH)_2$	decomposes	–	–986	0.0015
Calcium sulfate, $CaSO_4$	decomposes	–	–1434	0.0045
Carbon monoxide, CO	–250	–191	–110	insoluble
Carbon dioxide, CO_2	sublimes	–	–393	0.0033
Copper(II) oxide, CuO	1326	–	–157	insoluble
Hydrazine, N_2H_4	2	114	+50.6	very soluble
Hydrogen fluoride, HF	–83	20	–271	0.043
Hydrogen chloride, HCl	–114	–85	–92.3	5.97
Hydrogen bromide, HBr	–87	–67	–36.4	2.39
Hydrogen iodide, HI	–51	–35	+26.5	0.056
Lithium fluoride, LiF	845	1676	–616	0.005
Lithium chloride, LiCl	614	1382	–409	2.00
Lithium iodide, LiI	449	1171	–270	1.21
Magnesium chloride, $MgCl_2$	714	1418	–641	0.56
Magnesium oxide, MgO	2800	3600	–602	reacts
Magnesium hydroxide, $Mg(OH)_2$	dehydrates	–	–924	0.00002
Magnesium sulfate, $MgSO_4$	decomposes	–	–1285	0.18
Manganese(IV) oxide, MnO_2	decomposes	–	–520	insoluble
Nickel(II) chloride, $NiCl_2$	1001	sublimes	–305	0.51
Phosphorus(III) chloride, PCl_3	–112	76	–320	reacts
Phosphorus(V) chloride, PCl_5	sublimes	–	–444	reacts
Potassium chloride, KCl	776	1500	–437	0.48
Potassium iodide, KI	686	1330	–328	0.89
Silicon(IV) chloride, $SiCl_4$ (silicon tetrachloride)	–70	58	–687	reacts
Silicon dioxide, SiO_2	1610	2230	–911	insoluble
Sodium fluoride, NaF	993	1695	–574	0.098
Sodium chloride, NaCl	808	1465	–411	0.62
Sodium hydroxide, NaOH	318	1390	–426	1.05
Sodium oxide, Na_2O	sublimes	–	–414	reacts
Strontium chloride, $SrCl_2$	911	1250	–829	0.01
Strontium oxide, SrO	2430	3000	–592	0.008
Strontium hydroxide, $Sr(OH)_2$	375	decomposes	–959	0.003
Strontium sulfate, $SrSO_4$	1605	–	–1453	insoluble
Sulfur dioxide, SO_2	–75	–10	–297	0.17
Sulfur trioxide, SO_3	17	43	–441	very soluble
Zinc chloride, $ZnCl_2$	283	732	–415	3.0
Water, $H_2O(l)$	273	373	–286	–
Water, $H_2O(g)$	273	373	–242	–

Properties of selected organic compounds

Compound	Formula	Melting point/°C	Boiling point /°C	$\Delta H_c^\ominus$ / kJ mol⁻¹	$\Delta H_f^\ominus$ / kJ mol⁻¹
Alkanes					
Methane	CH_4	−182	−161	−890	−75
Ethane	C_2H_6	−183	−88	−1560	−85
Propane	C_3H_8	−188	−42	−2219	−104
Butane	C_4H_{10}	−138	−0.5	−2876	−126
Pentane	C_5H_{12}	−130	36	−3509	−173
Hexane	C_6H_{14}	−95	69	−4163	−199
Decane	$C_{10}H_{22}$	−30	174	−6778	−301
Eicosane	$C_{20}H_{42}$	−37	344		
2-methylpropane	C_4H_{10}	−159	−12	−2868	−134
2-methylbutane	C_5H_{12}	−160	28	−3503	−179
2-methylpentane	C_6H_{14}	−154	60	−4157	−204
2,2-dimethylpropane	C_5H_{12}	−16	10	−3492	−190
Alkenes					
Ethene	C_2H_4	−169	−104	−1411	+52
Propene	C_3H_6	−185	−48	−2058	+20
Arenes					
Benzene	C_6H_6	6	80	−3267	+49
Halogenoalkanes					
Chloromethane	CH_3Cl	−98	−24	−764	−82
Tetrachloromethane	CCl_4	−23	77	−360	−130
Bromomethane	CH_3Br	−178	4	−770	−37
1-chlorobutane	C_4H_9Cl	−123	79	−2704	−188
1-bromobutane	C_4H_9Br	−112	102	−2716	−144
2-bromobutane	C_4H_9Br	−112	91	−2705	−155
1-iodobutane	C_4H_9I	−103	131		
Alcohols					
Methanol	CH_3OH	−98	65	−726	−239
Ethanol	C_2H_5OH	−114	78	−1367	−277
Propan-1-ol	C_3H_7OH	−126	97	−2021	−303
Propan-2-ol	C_3H_7OH	−89	83	−2006	−318
Butan-1-ol	C_4H_9OH	−89	118	−2676	−327
Aldehydes					
Methanal	$HCHO$	−92	−21	−571	−109
Ethanal	CH_3CHO	−121	20	−1167	−191
Propanal	C_2H_5CHO	−81	49	−1821	−217
Ketones					
Propanone	CH_3COCH_3	−95	56	−1816	−248
Butanone	$C_2H_5COCH_3$	−86	80	−2441	−276
Carboxylic acids					
Methanoic acid	HCO_2H	9	101	−254	−425
Ethanoic acid	CH_3CO_2H	17	118	−874	−484
Propanoic acid	$C_2H_5CO_2H$	−21	141	−1527	−511

Solubility

	Soluble	Insoluble
Acids	All common *acids* are soluble	
Bases	The *alkalis*: the hydroxides of sodium and potassium, calcium hydroxide which is slightly soluble, ammonia, plus the carbonates of sodium and potassium	All other metal oxides, hydroxides and carbonates
Salts	All nitrates	
	All chlorides	*except* silver and lead chlorides
	All sulfates	*except* barium sulfate, lead sulfate, and calcium sulfate which is slightly soluble
	All sodium and potassium salts	All other carbonates, chromates, sulfides and phosphates

Bond lengths and bond energies

1 pm = 10^{-3} nm = 10^{-12} m

Bond	Bond length/pm	Average bond enthalpy/kJ mol^{-1}
H–H	74	435
Cl–Cl	199	243
Br–Br	228	193
I–I	267	151
O–H	96	464
C–H	109	435
C–C	154	347
C=C	134	612
C=C	120	838
C–Cl	177	346
C–Br	194	290
C–I	214	228
N=N	110	945
N–N	145	158
N–H	101	391

Qualitative analysis tests

Tests for cations

Positive ion in solution	Observations on adding sodium hydroxide solution drop by drop and then in excess	Observations on adding ammonia solution drop by drop and then in excess
Calcium, Ca^{2+}	White precipitate but only if the calcium ion concentration is high	No precipitate
Magnesium, Mg^{2+}	White precipitate insoluble in excess reagent	White precipitate insoluble in excess reagent
Barium, Ba^{2+}	No precipitate	No precipitate
Aluminium, Al^{3+}	White precipitate which dissolves in excess reagent	White precipitate insoluble in excess reagent
Chromium(III), Cr^{3+}	Green precipitate which dissolves in excess to form a dark green solution	Green precipitate insoluble in excess reagent
Manganese(II), Mn^{2+}	Off-white precipitate insoluble in excess reagent	Off-white precipitate insoluble in excess reagent
Iron(II), Fe^{2+}	Green precipitate insoluble in excess reagent	Green precipitate insoluble in excess reagent
Iron(III), Fe^{3+}	Browny-red precipitate insoluble in excess reagent	Browny-red precipitate insoluble in excess reagent
Copper(II), Cu^{2+}	Pale blue precipitate insoluble in excess	Pale blue precipitate dissolving in excess to form a dark blue solution
Zinc, Zn^{2+}	White precipitate which dissolves in excess reagent	White precipitate which dissolves in excess reagent
Lead, Pb^{2+}	White precipitate which dissolves in excess reagent	White precipitate insoluble in excess reagent
Ammonium, NH_4^+	Alkaline gas (ammonia) given off on heating	–

Tests for anions

Test	Observations	Inference
Test for carbonate, sulfite and nitrite: Add dilute hydrochloric acid to the solid salt. Warm gently if there is no reaction at first.	Gas which turns limewater milky white Gas which is acidic, has a pungent smell and turns acid-dichromate paper from orange to green Colourless gas given off which turns brown where it meets the air	Carbon dioxide from a carbonate Sulfur dioxide from a sulfite Nitrogen oxide (NO) from a nitrite turning to nitrogen dioxide (NO_2)

Test for halide ions: Make a solution of the salt. Acidify with nitric acid, then add silver nitrate solution. Test the solubility of the precipitate in ammonia solution	White precipitate soluble in dilute ammonia solution Cream precipitate soluble in concentrated ammonia solution Yellow precipitate insoluble in excess ammnia	Chloride Bromide Iodide
Test for sulfate and sulfite ions: Make a solution of the salt. Add a solution of barium nitrate or chloride. If a precipitate forms add dilute nitric acid.	White precipitate which redissolves on adding acid White precipitate which does not redissolve in acid	Sulfite Sulfate
Test for nitrates: Make a solution of the salt, add sodium hydroxide solution and then a piece of aluminium foil or a little Devarda alloy. Heat.	Alkaline gas evolved which turns red litmus blue and has a pungent smell	Ammonia from a nitrate (or nitrite)
Test for chromate(VI) ions: Make a solution of the salt. Divide into three: ■ add dilute acid ■ add a solution of barium nitrate. ■ add a solution of lead nitrate.	Yellow solution turns orange Yellow precipitates form with barium ions and lead ions	Yellow chromate ions turning to orange dichromate Precipitates of insoluble barium and lead chromates Chromate

Tests for gases

Gas	Test	Observations
Hydrogen	Burning splint	Burns with a 'pop'
Oxygen	Glowing splint	Splint bursts into flame (relights)
Carbon dioxide	Limewater (aqueous calcium hydroxide)	Turns milky white
Hydrogen chloride (hydrogen bromide and iodide react in the same way)	Smell Blue litmus Ammonia vapour (from a drop of concentrated ammonia solution on a glass rod)	Pungent Turns red Thick white smoke
Chlorine	Colour Smell Effect on moist blue litmus paper Moist starch–iodide paper	Greenish–yellow Pungent – bleach-like Turns the paper red and then bleaches it Turns blue–black

Sulfur dioxide	Smell Blue litmus Acid dichromate(II) paper	Pungent Turns red Turns green
Hydrogen sulfide	Smell Burning splint Lead(II) ethanoate paper	'Bad eggs' Gas burns – yellow deposit of sulfur Turns brown–black
Ammonia	Smell Red litmus	Pungent Turns blue
Nitrogen dioxide	Colour Blue litmus	Orange–brown Turns red
Water vapour	Appearance Anhydrous cobalt(II) chloride paper	'Steams' in the air Turns from blue to pink

Tests for organic functional groups

Functional group	Test	Observations
>C=C< in an alkene	Shake with a dilute solution of bromine Shake with a very dilute, acidic solution of potassium manganate(VII)	Orange colour of the solution decolourised Purple colour fades and the solution turns colourless
—C—X where X = Cl, Br or I in a halogenoalkane	Warm with a solution of sodium hydroxide. Cool. Acidify with nitric acid. Then add silver nitrate solution	Precipitate is white from a chloro compound, creamy yellow from a bromo compound and yellow from an iodo compound. (Hydrolysis produces ions from the covalent molecules)
H—C(H)—OH (—CH_2—OH) in a primary alcohol	Add a solution of sodium carbonate Add PCl_5 to the **anhydrous** compound Warm with an acidic solution of potassium dichromate(VI)	No reaction, unlike acids, alcohols do not react with carbonates Colourless, fuming gas forms (hydrogen chloride) Orange solution turns green and gives a fruity smelling vapour. Alcohols are oxidised by dichromate(VI) but not by Fehling's solution
>C=O in aldehydes and ketones	Warm with fresh Fehling's solution Warm with fresh Tollen's reagent	Solution turns greenish and then an orange–red precipitate forms with aldehydes, but not ketones Silver mirror forms with aldehydes but not with ketones
—CO_2H in carboxylic acids	Measure the pH of a solution Add a solution of sodium carbonate Add a neutral solution of iron(VII) chloride	Acidic solution with pH about 3–4 Mixture fizzes giving off carbon dioxide Methanoic and ethanoic acids give a red colouration

Answers to 'Test yourself' questions

In a few instances it is not possible to give a sensible short answer – so there is no answer included. There are also no answers to questions asking for large diagrams, charts or graphs.

Section one Studying chemistry

1.2 Why study chemistry?

1 All nitrates are soluble in water.
2 A more reactive metal displaces a less reactive metal from a solution of one of its salts. Zinc goes into solution forming zinc sulfate. Copper metal appears.
3 The metal in the compound forms at the cathode.
4 The carbonate fizzes giving off carbon dioxide gas.
5 They are all colourless, crystalline solids.

1.3 Laboratory investigations

1 a) Adding hydrochloric acid to a reactive metal such as magnesium.
b) Warming ammonium chloride with alkali giving a pungent smell of ammonia.
c) Adding acid to blue litmus which then turns red.
d) Adding silver nitrate solution to a solution of sodium chloride.
e) Methane, in natural gas, burning in air.
2 a) Useful in diagnosis to see if a patient has diabetes.
b) Useful to the police to see if they should prosecute for drunk driving.
c) Useful to a mining company to assess whether an ore is worth mining.
d) Useful to those monitoring air pollution in cities to see if there is a risk to the health of the population.

1.5 Units and measurement

1 a) 1, b) 3, c) 3, d) 1

1.6 Key skills

- See page 27
- See page 88
- See page 137
- See pages 158–159
- See pages 64–65
- See pages 229–231

Section two Foundations of chemistry

2.2 States of matter

1 decane(l), eicosane(s), krypton(g), gallium(s), 1-bromobutane(l), methanal(g), methanoic acid(l), silicon tetrachloride(l), hydrogen fluoride(g)

2.3 Matter and chemical change

1 Oxides of nitrogen split into oxygen and nitrogen on heating. Any ionic salt, such as sodium chloride or aluminium oxide, can be split into elements by electrolysis when molten.
3 Metals: usually solids with high melting points, shiny when polished, malleable and ductile, good conductors of electricity, form positive ions, form basic oxides, form solid (ionic) chlorides.
Non-metals: mostly gases at room temperature, dull not shiny if solid, brittle and usually do not conduct electricity, form negative ions, form acidic oxides, form covalent molecular chlorides which are usually liquids or gases.
4 Electrons in main shells: Be 2.3, F 2.7, Na 2.8.1
5 HCl, H_2S, CS_2, CCl_4, NH_3
6 All these insoluble: SiO_2, CH_4, CCl_4, NO
All these react with water becoming soluble: HCl, NH_3, SO_2, NO_2
7 KI, $CaCO_3$, Na_2SO_4, $Ca(OH)_2$, $AlCl_3$
8 Any ionic salt including: $NaCl$, $PbBr_2$, $MgCl_2$, $CuSO_4$, $ZnSO_4$
9 Molecules: H_2SO_4, PCl_3
Ions: CuO, $MgCl_2$, LiF

2.4 Solutions

1 Gas exchange in the lungs and transport in blood to and from the air and living cells requires the gases involved in respiration to be soluble.
2 $NaOH$ – soluble, CuO – insoluble, KI – very soluble, $NaCl$ – soluble, MnO_2 – insoluble, $ZnSO_4$ – very soluble, $NiCl_2$ – soluble.

2.5 Chemical equations

2 $2Mg(s) + O_2(g) \rightarrow 2MgO(s)$
$2Na(s) + 2H_2O(l) \rightarrow 2NaOH(aq) + H_2(g)$
$Ca(OH)_2(s) + 2HCl(aq) \rightarrow CaCl_2(s) + 2H_2O(aq)$

2.6 Types of chemical change

1 *oxidised* / *reduced*
a) magnesium — water (steam)
b) hydrogen — copper(II) oxide
c) aluminium — iron(III) oxide
d) carbon — carbon dioxide
2 a) $2Na \rightarrow 2Na^+ + 2e^-$
$Cl_2 + 2e^- \rightarrow 2Cl^-$
b) $2Zn \rightarrow 2Zn^{2+} + 4e^-$
$O_2 + 4e^- \rightarrow 2O^{2-}$
c) $Ca \rightarrow Ca^{2+} + 2e^-$
$Br_2 + 2e^- \rightarrow 2Br^-$
3 a) $CaO(s) + 2HNO_3(aq) \rightarrow Ca(NO_3)_2(aq) + H_2O(l)$
b) $Zn(s) + H_2SO_4(aq) \rightarrow ZnSO_4(aq) + H_2(g)$
c) $Na_2CO_3(s) + 2HCl(aq) \rightarrow 2NaCl(aq) + H_2O(l) + CO_2(g)$
4 a) hydrogen, H^+; nitrate, NO_3^-
b) hydrogen, H^+; sulfate, SO_4^{2-}
5 Bases: MgO, CuO, ZnO, $Cu(OH)_2$,
Alkalis: $NaOH$, KOH, $Ca(OH)_2$, NH_3
6 a) Yes: barium sulfate, $BaSO_4$
b) No
c) Yes: calcium carbonate, $CaCO_3$
d) Yes: lead(II) chloride, $PbCl_2$
e) Yes: copper(II) hydroxide, $Cu(OH)_2$.
7 a) Redox, b) Acid–base,
c) Redox, d) Acid–base,
e) Precipitation, f) Hydrolysis.

2.7 Chemical quantities

1 a) × 2, b) × 2, c) × 8, d) × 4
2 a) 124, b) 170, c) 98

3 a) 74.5, b) 267, c) 400
4 a) 0.5 mol, b) 0.05 mol, c) 1 mol, d) 0.1 mol, e) 0.25 mol
5 a) 12.7 g, b) 17.75 g, c) 36 g, d) 0.535 g, e) 8.0 g
6 a) 2 mol, b) 1 mol, c) 4 mol
7 a) 3.01×10^{23} b) 24.1×10^{23} c) 54.2×10^{23}

2.8 Finding formulae

1 a) Mg_3N_2, b) CH_4, c) Na_2SO_4
2 a) H_2SO_4, b) C_2H_6O
3 a) 88.9%, b) 57.7%, c) 63.5%

2.9 Calculations from equations

1 14 g
2 8.4 tonnes, 11.2 tonnes
3 2.0 g
4 a) 10 mol, b) 0.002 mol, c) 0.125 mol
5 a) 48 000 cm^3 b) 4.8 cm^3 c) 3000 cm^3
6 a) 1 dm^3 b) 150 dm^3, 100 cm^3
7 a) 600 cm^3 b) 240 cm^3 c) 90 cm^3
8 a) 0.2 mol dm^{-3}, b) 0.5 mol dm^{-3}
9 a) 9.8 g, b) 0.158 g
10 a) 0.29 g, b) 2.0 g
11 a) 10 cm^3, b) 1200 cm^3

2.10 Titrations

1 0.108 mol dm^{-3}
2 0.22 mol dm^{-3}
3 $H_3PO_4 + 2NaOH \rightarrow Na_2HPO_4 + 2H_2O$

Section three Physical chemistry

3.2 Atomic structure

1

	protons	neutrons	electrons
a)	4	5	4
b)	19	20	19
c)	92	143	92
d)	53	74	54
e)	20	20	18

2 a) $^{16}_{8}O$, b) $^{40}_{18}Ar$, c) $^{23}_{11}Na^+$, d) $^{32}_{16}S^{2-}$
3 79% magnesium-24, 10% magnesium-25 and 11 % magnesium-26.

Average relative atomic mass = $\dfrac{(79 \times 24) + (10 \times 25) + (11 \times 26)}{100} = 24.3$

4 a) & b) $^{28}_{14}Si$ – 14 protons, 14 neutrons
$^{29}_{14}Si$ – 14 protons, 15 neutrons
$^{30}_{14}Si$ – 14 protons, 16 neutrons
c) Average relative atomic mass = $\dfrac{(93 \times 28) + (5 \times 29) + (2 \times 30)}{100} = 28.1$

5 Relative molecular mass = 58
6 2, group 2
8 a) $\nu = \dfrac{\Delta E}{h} = \dfrac{2.18 \times 10^{-18}\ J}{6.6 \times 10^{-34}\ J\ Hz^{-1}} = 3.30 \times 10^{15}\ Hz$

b) Ionisation energy = $2.18 \times 10^{-18}\ J \times 6.02 \times 10^{23}\ mol^{-1}$
$= 13.1 \times 10^5\ J\ mol^{-1}$
$= 1310\ kJ\ mol^{-1}$

9 a) Be: $1s^22s^2$,
b) O: $1s^22s^22p^4$,
c) Si: $1s^22s^22p^63s^23p^2$,
d) P: $1s^22s^22p^63s^23p^3$
10 a) Be, b) Al, c) Cl

3.3 Kinetic theory and gases

1 a) Nitrogen: $T_b = 77$ K,
b) Butane: $T_b = 272.5$ K,
c) Sucrose: $T_m = 459$ K,
d) Iron: $T_m = 1813$ K
2 a) n, P and R constant so V = constant $\times$ T, so $V \propto T$.
b) P, T and R constant so V = constant $\times$ n, so $V \propto n$.
3 Molar mass = 74 g mol^{-1}
4 a) The average energy of the molecules is unchanged but twice as many molecules now collide with a given area of the wall of the container per second.
b) The kinetic energy of the molecules is proportional to the temperature. So the higher the temperature, the more energetic the collisions between the molecules and the walls of the container.

3.4 Structure and bonding

1 Be – giant, B – giant, F – molecular, Si – giant, P_{white} – molecular, S – molecular, Ca – giant, Co – giant, I – molecular.

3.5 Structure and bonding in metals

1 a) shiny – mercury and silver
b) conducts electricity – all metals especially copper and aluminium
c) bend and stretch – most metals including steel and copper
d) high tensile strength – steel
2 a) d-block metals generally have high densities and high melting points.
b) s-block metals have relatively low densities and low melting points.
3 Close packed: group 2 metals except barium, aluminium and many d-block metals.
Body centered cubic: examples are the group 1 metals plus barium, chromium and vanadium.

3.6 Ionic giant structures

2 1+, 2+, 3+, 2–, 1–
3 Across a period, such as Li to Ne, or Na to Ar, the nuclear charge progressively increases as electrons go into the same shell
metals: outer electrons held relatively weakly by low effective nuclear charge
non-metals: outer electrons held relatively strongly by high effective nuclear charge
4 a) Both have the same electron configuration but an Na nucleus has one more proton than an Ne nucleus.
b) Both have the same electron configuration but a Cl nucleus has one less proton than an Ar nucleus.
5 The charges on the ions in the oxides are twice as big as the charges on the ions of the chlorides. Also oxide ions are much smaller than chloride ions so the charges in MgO and CaO are closer together.

6 a) potassium metal and bromine gas
b) magnesium metal and chlorine gas

3.7 Covalent molecules and giant structures

1 Boron and germanium have giant structures. The others are molecular.

2 Carbon dioxide is molecular with weak forces between the molecules. Silicon dioxide has a giant structure so there is a continuous network of strong bonding throughout the crystals.

4 a) H:Br

b) H:S:H

c) H H / H:C:C:H / H H

d) :O::S::O:

e) :S::C::S:

5 NH_3 – one lone pair on the N atom
H_2O – two lone pairs on the O atom
HF – three lone pairs on the F atom

6 a) $[H_3O]^+$ (H:O:H with H)

b) BF_4 (F:B:F with F above and below)

7 a) BCl_3; PH_3

b) O—C—O; SO_2

c) NH_3; NH_3

d) SO_3^{2-}; CO_3^{2-}

3.8 Intermediate types of bonding

1 δ+ δ− δ+ δ− δ+ δ− δ+ δ−
H–S N–O C–Cl I–Cl

2 a) NaF > NaCl > NaBr > NaI
b) $Na_2O > MgO > Al_2O_3 > SiO_2$
c) CsI > KI > NaI > LiI

3.9 Intermolecular forces

1 polar: HBr, $CHCl_3$, SO_2 – the rest non-polar

2 The C=O bond in propanone is a permanent dipole. There are no permanent dipoles in butane.

3 HBr – attractions between permanent dipoles
ethane – attractions between transient dipoles
methanol – hydrogen bonding

4 a)

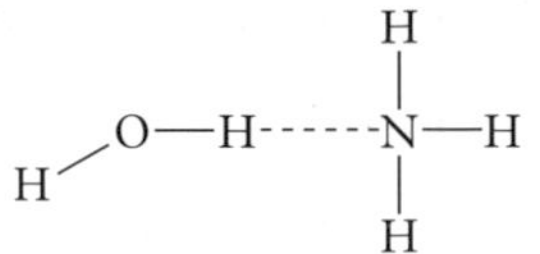

b)

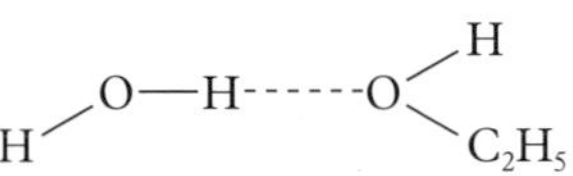

5 Down the group from HCl to HI the number of electrons increases so the molecules become more polarisable. Hydrogen bonding in HF.

3.10 Structure and bonding in forms of carbon

1 a) sodium chloride – conducts electricity when molten
b) methane gas – easily compressed
c) graphite – conducts electricity as a solid
d) ice – lower density than water at 0 °C
e) any metal, such as magnesium – conducts electricity when solid
f) ethanol – relatively low boiling point

3.13 Enthalpy changes

1 a) A→B Molecules lose energy and so move more slowly
b) B→C Molecules assemble into a crystal lattice and during this stage the energy given out is the energy released as more hydrogen bonds form between the molecules.
c) & d) After C: all the water is solid ice which cools until it is at the same temperature as the freezing mixture, about −10 °C.

2 $\Delta H_{vap} = 39.2\ kJ\ mol^{-1}$

3 a) molecular: HBr, Cl_2, Br_2, CCl_4, H_2O
b) metal giant structures: Zn, Na, K, Mg, Ca, Pb
c) ionic giant structures: LiCl, NaCl, AgCl

4 $\Delta H_{combustion} = -1680\ kJ\ mol^{-1}$

5 $\Delta H_{reaction} = -185\ kJ\ mol^{-1}$

6 $\Delta H_{neutralisation} = -55\ kJ\ mol^{-1}$

7 $\Delta H_{solution} = +25\ kJ\ mol^{-1}$

8 $\Delta H_{reaction} = -200\ kJ\ mol^{-1}$

9 $C(s) + O_2(g) \rightarrow CO_2(g)$
$\Delta H^{\ominus}_{reaction} = \Delta H^{\ominus}_c[C(s)] = \Delta H^{\ominus}_f[CO_2(g)]$

10 $\Delta H^{\ominus}_f\ [CH_3OH(l)] = -239\ kJ\ mol^{-1}$

11 $C(s) + 2H_2(g) + \frac{1}{2}O_2(g) \rightarrow CH_3OH(l)$

12 There is an enthalpy change for $H_2O(l) \rightarrow H_2O(g)$ (the enthalpy change of vaporisation of water).

13 $\Delta H^{\ominus}_{reaction} = -534\ kJ\ mol^{-1}$

14 Average bond enthalpies = + 463 kJ mol^{-1}

15 Bond lengths: C–C > C=C > C≡C
Bond strength: E(C≡C) > E(C=C) > E(C–C)
But the bond enthalpy for a double bond is not twice the bond enthalpy of a single bond.

16 $\Delta H_{reaction} = -125\ kJ\ mol^{-1}$

17 Enthalpies of formation give the more accurate value because they are specific for the substances in the reaction. Bond energies are mean values averaged over a range of elements and compounds.

3.14 Reversible reactions

1 a) Raising the temperature will make the ice melt again.
b) Adding alkali will turn the litmus blue again.
c) Allowing the white solid to cool and then adding water.
2 $I_2(s) \rightleftharpoons I_2(g)$

3.15 Chemical equilibrium

1 a) At 0 °C
b) At 100 °C
c) When the solution is saturated
3 $Ag^+(aq) + Fe^{2+}(aq) \rightleftharpoons Ag(s) + Fe^{3+}(aq)$

Disturbance	How does the equilibrium mixture respond	The result
Increasing $Fe^{2+}(aq)$	It moves to the right	More Ag(s) precipitates
Lowering the concentration of $Ag^+(aq)$	It moves to the left	Some Ag(s) reacts

4 $2CrO_4^{2-}(aq) + 2H^+(aq) \rightleftharpoons Cr_2O_7^{2-}(aq) + H_2O(l)$

a) Adding acid: increases $[H^+(aq)]$, shifts equilibrium to the right, turns from yellow to orange
b) Adding alkali: lowers $[H^+(aq)]$, shifts equilibrium to the left, turns from orange to yellow.

5 $CaCO_3(s) \rightleftharpoons CaO(s) + CO_2(g)$

Allowing the carbon dioxide to escape stops the back reaction so the system cannot reach equilibrium. The forward reaction continues until all the limestone decomposes.

3.17 Rates of chemical change

1 a) Keep the moisture level as low as possible.
b) Turn off the toaster to let the bread cool.
c) Keep the milk cool in a refrigerator.
2 a) Warm the dough
b) Blow air onto the fuel to keep up the oxygen concentration
c) Warm the object being repaired
d) Use a catalyst in a catalytic converter
3 a) Rate = 4.8 $cm^3\ s^{-1}$
b) Rate = 0.0002 mol s^{-1} (given that the volume of 1 mol gas is 24 000 cm^3)
c) $Mg(s) + 2HCl(aq) \rightarrow MgCl_2(aq) + H_2(g)$
Rate of removal of Mg(s) = rate of formation of $MgCl_2(aq)$ = 0.0002 mol s^{-1}
Rate of removal of HCl(aq) = 0.0004 mol s^{-1}
5 The rate of reaction is directly proportional to the concentration of thiosulfate ions.
Rate $\propto [S_2O_3^{2-}(aq)]$, or
Rate = constant $\times [S_2O_3^{2-}(aq)]$
6 b) The reaction with the small marble chips.
c) Small chips – 480 s; larger chips – 600 s. The marble was in excess, so the reactions stopped when all the acid was used up.
d) The smaller the pieces, the larger the surface area for a given mass of marble. The larger the surface area the faster the reaction where the acid and marble are in contact.
7 a) $Zn(s) + H_2SO_4(aq) \rightarrow ZnSO_4(aq) + H_2(g)$
b) See Figure 3.17.1 for a suitable apparatus.
c) i) B (or possibly C), ii) C, iii) D, iv) A

3.18 Collision theory

1 a) Speeds up – the powder has a larger surface area where hydrogen ions from the acid can collide with zinc atoms.
b) Slows down – sodium carbonate neutralises some of the acid reducing the concentration of the acid.
c) Slows down – the ice both cools the mixture and dilutes the acid as it melts.
d) Speeds up – a little zinc displaces copper from the copper(II) sulfate and the copper acts as a catalyst speeding up the formation of hydrogen.
2 This reaction tends to go but it has a high activation energy. At room temperature the molecules do not collide with enough energy to break the bonds in the reactants and start the reaction. A flame or spark heats the gas mixture and increases the energy of the colliding methane and oxygen molecules. The reaction is exothermic so once it starts, it keeps the gas mixture hot enough to react.

3.19 Stability

1 Thermodynamic stability: water not decomposing on boiling
Kinetic stability: methane not burning at room temperature, diamond not changing into graphite.

Section four Inorganic chemistry

4.2 The Periodic table

1 See Figure 4.2.1.
3 Metals form positive ions
Group 1: 1+; group 2: 2+; group 3: 3+
d-block elements can often form more than one positive ion.
Non-metals form negative ions.
Group 7: 1–; group 6: 2–, group 5: 3–.
4 K > Na > Al > B >
5 a) $Cl^- > Cl$, b) Al > N
6 N: $1s^22s^22p^3$
O: $1s^22s^22p^4$
In a nitrogen atom the three p electrons are unpaired – one in each p orbital (so that the p sub-shell is half full). In an oxygen atom the fourth p electron is paired up. The charge on the oxygen nucleus is one greater than the charge on a nitrogen nucleus. A possible explanation for the lower first ionisation energy of oxygen is that the electrostatic repulsion between the two electrons in the same p orbital makes it easier to remove the fourth electron in the 2p sub-shell. Another explanation used by chemists is to suggest that the half-filled sub-shell has greater stability so that it takes more energy to remove one electron from a nitrogen atom than from an oxygen atom.
7 Density is related to the size of atoms, the mass of atoms and the way that they are packed together. Across the second and third periods, density rises from group 1 to group 4 (in the second period) or group 3

(in the third period) as the atoms get both heavier and smaller. Then there is a drop in density with the switch from giant to molecular structures. Even in the liquid or solid states the molecular non-metals have lower densities because the bonding between molecules is relatively weak.

8 As a general rule: metals on the left of each period conduct electricity; non-metals on the right are non-conductors.

4.3 Oxidation numbers

1 a) +3, b) −3, c) +2, d) −3, e) +5

2 a) oxidised, b) reduced c) oxidised, d) reduced, e) oxidised

3 a) SnO, b) SnO_2, c) $NaClO_3$, d) $Fe(NO_3)_3$, e) K_2CrO_4.

4 a) $2Fe(s) + 3Br_2(l) \rightarrow 2FeBr_3(s)$

iron oxidised, bromine reduced

b) $2F_2(g) + 2H_2O(l) \rightarrow 4HF(g) + O_2(g)$

fluorine reduced, oxygen oxidised

c) $IO_3^-(aq) + 6H^+(aq) + 5I^-(aq) \rightarrow 3I_2(s) + 3H_2O(l)$

iodine in iodate ions reduced, iodine in iodide ions oxidised

d) $2S_2O_3^{2-}(aq) + I_2(aq) \rightarrow 2I^-(aq) + S_4O_6^{2-}(aq)$

iodine reduced, sulfur oxidised

e) $Cl_2(aq) + 2OH^-(aq) \rightarrow Cl^-(aq) + ClO^-(aq) + H_2O(l)$

chlorine oxidised and reduced (disproportionation)

4.4 Group 1

1 Li: $1s^22s^1$
Na: $1s^22s^22p^63s^1$
K: $1s^22s^22p^63s^23p^64s^1$

2 Na^+: $1s^22s^22p^6$

3 Same: soft metals, shiny when freshly cut, good conductors of electricity, form 1+ ions, form ionic chlorides (M^+Cl^-), form basic oxides, react with water giving hydrogen and forming an alkaline solution, MOH(aq).
Trends down the group: become softer, atomic radii increase, first ionisation energies decrease, react more violently with water.

4 a) $2Na(s) + 2H_2O(l) \rightarrow 2NaOH(aq) + H_2(g)$
b) $2K(s) + Cl_2(g) \rightarrow 2KCl(s)$
c) $Li_2O(s) + H_2O(l) \rightarrow 2LiOH(aq)$
d) $Na_2O(s) + 2HCl(aq) \rightarrow 2NaCl(aq) + H_2O(l)$
e) $2KOH(aq) + H_2SO_4(aq) \rightarrow K_2SO_4(aq) + 2H_2O(l)$

4.5 Group 2

1 Mg: $1s^22s^22p^63s^2$
Ca: $1s^22s^22p^63s^23p^64s^2$
Sr: $1s^22s^22p^63s^23p^63d^{10}4s^24p^65s^2$

2 a) $Ba(s) + O_2(g) \rightarrow BaO_2(s)$
b) $Ca(s) + 2HCl(aq) \rightarrow CaCl_2(aq) + H_2(g)$
c) $Sr(s) + Cl_2(g) \rightarrow SrCl_2(s)$
d) $Ba(s) + 2H_2O(g) \rightarrow Ba(OH)_2(s) + H_2(g)$

3 a) $MgO(s) + 2HCl(aq) \rightarrow MgCl_2(aq) + H_2O(g)$
b) $MgO(s) + H_2O(g) \rightarrow Mg(OH)_2(s)$
c) $Ca(OH)_2(aq) + CO_2(g) \rightarrow CaCO_3(s) + H_2O(g)$

4 The solubility increases from 2.0 × 10^{-5} mol/100g water for $Mg(OH)_2$ to 1500 × 10^{-5} mol/100g water for $Ba(OH)_2$.

6 a) $MgCO_3(s) \rightarrow MgO(s) + CO_2(g)$
b) $MgCO_3(s) + 2HCl(aq) \rightarrow MgCl_2(aq) + CO_2(g) + H_2O(g)$
c) $2Ca(NO_3)_2(s) \rightarrow 2CaO(s) + 4NO_2(g) + O_2(g)$
d) $Ba^{2+}(aq) + SO_4^{2-}(aq) \rightarrow BaSO_4(s)$

7 The solubility decreases from 1830 × 10^{-4} mol/100g water for $MgSO_4$ to 0.009 × 10^{-4} mol/100g water for $BaSO_4$.

4.6 Group 7

1 a) Cl: $1s^22s^22p^63s^23p^5$
b) Cl^-: $1s^22s^22p^63s^23p^6$
c) Br: $1s^22s^22p^63s^23p^63d^{10}4s^24p^5$
d) Br^-: $1s^22s^22p^63s^23p^63d^{10}4s^24p^6$

2 a) $Mg(s) + Br_2(l) \rightarrow MgBr_2(s)$
b) $2Fe(s) + 3Cl_2(g) \rightarrow 2FeCl_3(s)$
c) $Fe(s) + I_2(s) \rightarrow FeI_2(s)$

3 $SiCl_4$ is molecular. The Si–Cl bonds are strong but the forces between the molecules are weak, too weak to hold the molecules in the solid state at room temperature but strong enough to prevent them evaporating.

4 a) $2P(s) + 3Cl_2(g) \rightarrow 2PCl_3(s)$
P in oxidation state zero oxidised to oxidation state +3.
b) $Cl_2(g) + 2I^-(aq) \rightarrow 2Cl^-(aq) + I_2(s)$
Cl in oxidation state zero reduced to oxidation state −1.

5 a) The colourless solution turns to the orange-brown colour of bromine.
b) The colourless solution turns to yellow-brown and then grey specks of insoluble iodine appear.

7 $\overset{-1}{2NaBr(s)} + \underset{+6}{H_2SO_4(l)} \rightarrow \overset{0}{Br_2(l)} + \underset{+4}{SO_2(g)} + 2H_2O(g)$

8 a) $Ag^+(aq) + I^-(aq) \rightarrow AgI(s)$
b) $Ag^+(aq) + Br^-(aq) \rightarrow AgBr(s)$

4.7 Water treatment

1 $Cl_2(g) + 2NaOH(aq) \rightarrow NaOCl(aq) + NaCl(aq) + H_2O(l)$

2 Adding acid to bleach solution reverses the reaction of chlorine with alkali and releases toxic chlorine gas.

3 Adding acid to lower the pH raises the concentration of hydrogen ions. The equilibrium for this reaction shifts to the right increasing the concentration of HOCl(aq).

$$OCl^-(aq) + H^+(aq) \rightleftharpoons HOCl(aq)$$

Adding alkali to raise the pH has the reverse effect. Alkali neutralises hydrogen ions removing them from the equilibrium which therefore shifts to the left lowering the concentration of HOCl(aq).

4 Concentration of chlorine in the bleach = 0.103 mol dm^{-3}

4.8 Inorganic chemicals in industry

2 Chlorine is reduced as it oxidises bromide ions (stages 1 and 4)
Bromine is reduced as it oxidises sulfur dioxide (stage 3)

3 $Cl_2(aq) + H_2O(l) \rightleftharpoons HOCl(aq) + H^+(aq) + Cl^-(aq)$
Adding acid increases the concentration of hydrogen ions on the right-hand side of the equation.

Le Chatelier's principle predicts that this will cause the equilibrium to shift to the left. This reduces the tendency for chlorine to react with water.
4 Many new products have been developed which include chlorine such as the plastic PVC and a variety of chlorinated solvents.
5 Chlorine and sodium hydroxide are formed in fixed proportions by electrolysis of sodium chloride solution. For every tonne of chlorine, the process makes 2.25 tonnes of 50% sodium hydroxide.
6 a) Increasing yield at equilibrium: cooling the gas mixture between each catalyst bed, also adding more oxygen from the air and removing SO_3 from the gas mixture before it passes through the final bed of catalyst.
b) Increasing the rate: using a catalyst and keeping the temperature high enough for the catalyst to be active.
7 $CO_3^{2-}(aq) + CO_2(g) + H_2O(l) \rightarrow 2HCO_3^-(aq)$
8 a) Increasing yield at equilibrium: raising the pressure and keeping the temperature as low as is consistent with a reasonable rate of reaction.
b) Increasing the rate: using a catalyst and keeping the temperature high enough for the reaction to proceed fast enough.
9 Ammonia is the only gas with hydrogen bonding between the molecules in the liquid state. Hydrogen bonding is much stronger than other intermolecular forces.
10 According to the equation, 9 moles of gas react to give 10 moles and the reaction is exothermic. So in theory a lower pressure and a lower temperature will increase the yield. In practice raising the pressure is an advantage because it speeds up the reaction and forces more of the gas mixture through the catalyst in a given time. The catalyst glows red hot and under these conditions the product is formed fast enough.
11 According to the equation, 2 moles of gas react to give 1 mole and the reaction is exothermic. Le Chatelier's principle predicts a higher yield of N_2O_4 at equilibrium if the pressure is higher and the temperature is lower.
12 Nitrogen starts in the +4 state. In the products it is in the +5 and +2 states so it is both oxidised and reduced.
13 $NH_3(aq) + HNO_3(aq) \rightarrow NH_4NO_3(aq)$
14 28 g N in 80 g NH_4NO_3, 35%
28 g N in 60 g NH_2CONH_2, 46.7%
28 g N in 132 g $(NH_4)_2SO_4$, 21.2%
15
$N_2 \rightarrow NH_3 \rightarrow NO \rightarrow NO_2 \rightarrow N_2O_4 \rightarrow HNO_3 \rightarrow NH_4NO_3$
0 −3 +2 +4 +4 +5 −3 +5

4.9 Extraction of metals

1 Al(III) reduced to Al(0). The Al^{3+} ions gain electrons.
2 The process uses large currents and is generally only economic with low-cost electricity.
3 It has a high melting point, conducts electricity and does not dissolve in the hot electrolyte.
4 Grinding powders the solids and mixes them together well. This increases the areas of surfaces in contact and so speeds up the process.
5 $2Fe_2O_3(s) + 3C(s) \rightarrow 4Fe(s) + 3CO_2(g)$
6 The furnace operates at a high temperature and contains reactive chemicals. Refractories have high melting points and are relatively inert.
7 $CaCO_3(s) \rightarrow CaO(s) + CO_2(g)$
8 Exothermic: reactions of coke with oxygen and iron(III) oxide with carbon monoxide.
Endothermic: reaction of coke with carbon dioxide.
9 Using the hot gases leaving the top of the furnace to preheat the incoming blast of air.
10 Cold scrap steel takes in energy as it melts and so prevents the iron becoming too hot as the oxygen blast oxidises impurities.
11 $S(s) + Mg(s) \rightarrow MgS(s)$
$Si(s) + O_2(g) \rightarrow SiO_2(g)$
$CaO(s) + SiO_2(s) \rightarrow CaSiO_3(s)$
12 Aluminium combines very rapidly and strongly with oxygen.
13

Continuous method	Batch method
Raw materials fed in continuously	Raw materials mixed in batches and placed in reactor
Products tapped off at regular intervals day and night	Products removed from the reactor in batches
Furnace stays hot	Furnace regularly heated and then cooled
Process runs day and night	Process in episodes

14 $2Na^+ + 2e^- \rightarrow 2Na$
$2Cl^- \rightarrow Cl_2 + 2e^-$
15 Sodium chloride is cheap but electricity is expensive. Energy is also needed to melt the solid sodium chloride and keep it hot. Also, special equipment and procedures are needed when handling a very reactive metal such as sodium.
16 Hot sodium reacts very violently with oxygen in the air. Argon is inert and does not react with sodium. The argon atmosphere keeps out the air.
17 The reaction is highly exothermic.

4.10 Environmental issues

1 $CaO(s) + SO_2(g) \rightarrow CaSiO_3(s)$
2 Calcium sulfate can be used to make plaster and plasterboard.
3 Recycling cuts the use of raw materials, saves energy and reduces the quantity of waste dumped in landfill.

Section five Organic chemistry

5.2 Organic molecules

1 a) CH, b) $C_5H_{10}O$
2 a) Empirical: CH_3O, molecular: $C_2H_6O_2$
b) Empirical: C_2H_5, molecular: C_4H_{10}
4 Alkanes: C_3H_8, C_7H_{16}.
5 C_nH_{2n}

8

```
    H   H
    |   |
H — C — C — OH
    |   |
    H   H
```

alcohol

```
    H       H
    |       |
H — C — O — C — H
    |       |
    H       H
```

ether

9

```
    H   H   H   H               H   H   OH  H
    |   |   |   |               |   |   |   |
H — C — C — C — C — OH      H — C — C — C — C — H
    |   |   |   |               |   |   |   |
    H   H   H   H               H   H   H   H
```

```
          H                           H
          |                           |
    H H — C — H H               H H — C — H H
    |     |     |               |     |     |
H — C ——— C ——— C — OH      H — C ——— C ——— C — H
    |     |     |               |     |     |
    H     H     H               H     OH    H
```

10

```
Br        H
  \      /
   C = C
  /      \
Br        H
```

```
Br        Br
  \      /
   C = C
  /      \
H         H
```

cis

```
Br        H
  \      /
   C = C
  /      \
H         Br
```

trans

11 Non-polar: C_5H_{12} and CCl_4. The other two formulae give polar molecules.

12 The two polar molecules with oxygen atoms that can form hydrogen bonds with water molecules.

5.3 Names of carbon compounds

1 2-methylbutane
1-bromopropane
methylpropene
pentan-1-ol
propanoic acid

5.4 Types of organic reaction

1 a) $2CH_3CO_2H(aq) + Na_2CO_3(aq) \rightarrow 2CH_3CO_2Na(aq) + CO_2(g) + H_2O(l)$

b) $2CH_3CO_2H(aq) + Mg(aq) \rightarrow Mg(CH_3CO_2)_2(aq) + H_2(g)$

2 a) oxidising agent
b) oxidising agent
c) base
d) reducing agent

3 a) elimination
b) substitution
c) addition
d) substitution

5.5 Mechanisms of organic reactions

1 a) nucleophile
b) electrophile
c) free radical
d) nucleophile

5.7 Alkanes

1 The more branching the lower the boiling point.

3 Branched molecules are more compact and so the area over which intermolecular forces can operate is smaller. So the attraction between the molecules is less, making it easier to separate the molecules. So the branched compounds boil at a lower temperature.

4 $C_4H_{10}(g) + 7^1/_2O_2(g) \rightarrow 4CO_2(g) + 5H_2O(l)$

5 Incomplete combustion produces the toxic gas carbon monoxide and very fine particles of soot (carbon) which can be damaging to the lungs.

6 $C_2H_6(g) \rightarrow C_2H_4(g) + H_2(g)$

7 Cracking turns a liquid to a gas. Hydrocarbon gas molecules are generally smaller than the molecules of a liquid hydrocarbon.
The product burns more readily and decolourises bromine solution.

9 The light provides the energy to split bromine molecules to produce the free radicals that initiate the reaction. One of the products is hydrogen bromide, a colourless gas which fumes in moist air.

5.8 Alkenes

1 Buta-1,3-diene and pent-2-ene have geometric isomers.

2

```
C3H7        CH3          H           CH3
    \      /              \         /
     C = C                 C  =  C
    /      \              /         \
H           H         C3H7           H
```

cis *trans*

3

```
H   120°    H
  \        /
   C  =  C    120°
  /        \
H           H
```

4 Alkenes form by cracking alkanes, dehydrating alcohols and by eliminating hydrogen halides from halogenoalkanes.

5 a) Propane
b) 1, 2-dichloropropane
c) Propan-1, 2-diol

6

```
C2H5                      CH3
    \                    /
     C = O          O = C
    /                    \
H                         H
```

These aldehydes are then oxidised to carboxylic acids.

5.9 Halogenoalkanes

1

a)
```
       H   H   H
       |   |   |
   H—C—C—C—I
       |   |   |
       H   H   H
```

b)
```
       H  CH3 H   H
       |   |   |   |
   H—C—C—C—C—H
       |   |   |   |
       H   Cl  H   H
```

c)
```
       H   H   Br  H   H
       |   |   |   |   |
   H—C—C—C—C—C—H
       |   |   |   |   |
       H   H   H   H   H
```

2 Ethene plus hydrogen bromide at room temperature to form bromoethane.
Butan-1-ol with phosphorus pentachloride at room temperature to form 1-chlorobutane.
Ethane with bromine in sunlight to form bromoethane.

3 Only CCl_4 is non-polar.

4 The boiling point rises as the halogen atom gets larger and has more electrons. In this series the molecules become more polarisable and so intermolecular forces increase.

5 Branched molecules are more compact and so the area over which intermolecular forces can operate is smaller. Therefore there is less attraction between the molecules making it easier to separate the molecules. Hence the tertiary compound boils at the lowest temperature.

6 Hydrolysis means splitting apart with water. Hydrolysis of a halogenoalkane produces an alcohol. The reaction goes faster with an alkali so that the nucleophile is the hydroxide ion rather than a water molecule.

7 a) Silver ions react with halide ions not with covalently bonded halogen atoms.
b) The alkali used to hydrolyse the halogenoalkane must be neutralised, otherwise a precipitate of silver oxide forms on adding silver nitrate. Nitric acid is suitable because nitrate ions do not interfere with the test.
c)
$CH_3CH_2CH_2CH_2Br + OH^-$
$\rightarrow CH_3CH_2CH_2CH_2OH + Br^-$
$OH^-(aq) + H^+(aq) \rightarrow H_2O(l)$
$Ag^+(aq) + Br^-(aq) \rightarrow AgBr(s)$

8 $CH_3CH_2CH_2CH_2Br + CN^-$
$\rightarrow CH_3CH_2CH_2CH_2CN + Br^-$

9
```
         ••
   C4H9—N
         |
        C4H9
```

This has a lone pair and can react as a nucleophile.

```
         ••
   C4H9—N—C4H9
         |
        C4H9
```

11 Using aqueous alkali favours the formation of an alcohol. Using alkali in a non-aqueous solvent (such as ethanol) favours formation of an alkene.

12 a) See Figure 5.9.11
b) See Figure 5.9.9

13 a)
```
           CH3
           |
   CH3—C=CH2
```

b)
```
       H   H   H
       |   |   |
   H—C—C—C—OH
       |   |   |
       H   H   H
```

5.10 Alcohols

1 Hydrogen bonding between the –OH groups in alcohols means that the intermolecular forces are stronger than in alkanes. So alcohols have higher boiling points than comparable alkanes.

2 H–O–H + Na $\rightarrow$ Na^+OH^- + $^1/_2H_2$

with propan-1-ol one of the hydrogen atoms in water is replaced with C_3H_7–.

3 The gas is acidic, it fumes in moist air and gives a white smoke with ammonia gas.

4 $3C_3H_7–OH + PI_3 \rightarrow 3C_3H_7–I + H_3PO_3$

5 The gas decolourises an orange solution of bromine and also decolourises a dilute, acidified solution of potassium manganate(VII).

6 Adding more concentrated sulfuric acid than is needed to catalyse the reaction removes water as it forms because it is a dehydrating agent. This limits the back reaction.

7 a) O–H, b) C–O, c) C–O

8 $CH_3CH_2CHO + 2[H] \rightarrow CH_3CH_2CH_2OH$

9 Ethanol and propan-2-ol only.

5.11 Theoretical and percentage yields

1 Figure 5.11a

2 59%

5.12 Fuels and chemicals from crude oil

2
```
       H   H                      H          H
       |   |                      |          |
   H—C—C—H    ——→    H—C•      •C—H
       |   |                      |          |
       H   H                      H          H
```

3 Silver is expensive. Using small particles of silver increases the surface area for reaction. Supporting the silver grains on inert aluminium oxide helps to keep the pieces of silver apart while holding them in place. This all helps to make a little silver go a long way.

5
```
       H                  H
       |                  |
   H—C—O—H------O—H
       |
   H—C—O—H------O—H
       |                  |
       H                  H
```

5.14 Organic chemicals and the environment

1. Fires and blow outs at oil wells can pollute soil and air around oil fields. Pipelines can leak. Oil tankers release some oil into the sea and a wreck of a tanker can cause widespread pollution of the coastline.
2. The gases leaving the exhaust pipe of a car are primary pollutants. They turn to secondary pollutants when they react with air, water and each other in the atmosphere.
3. Sulfur dioxide and some nitrogen oxides are acidic oxides. They dissolve in water to make acids. Acids corrode metals and attack calcium carbonate in limestone.
4. About 16 000 dm^3 at room temperature.
5. Ozone in the stratosphere absorbs UV light from the Sun which can damage living organisms. Ozone in the troposphere is harmful because it is a toxic gas which can affect plant growth and the health of animals which breathe the polluted air.
6. CFCs both help to destroy ozone in the ozone layer and act as greenhouse gases, so they are damaging in two ways but these are separate effects. The destruction of ozone in the stratosphere cools that part of the atmosphere because less radiation is absorbed in those regions when there is less ozone. The greenhouse effect is a separate effect caused by any gas which absorbs infra-red radiation from the surface of the Earth.
7. Recycling of metals is well established. Metals can generally be recycled without loss of quality. Aluminium and steel are distinct and identifiable materials which can be separated relatively easily from other wastes.
   There are many types of plastic, each needing different treatment for processing. Separation of plastic from other wastes and from each other is not easy to automate. The arguments for recycling are less clear cut for plastics because incineration to generate electricity may be a better option.

Index

Page numbers in bold refer to illustrations.

accuracy and errors 11
acid-base equilibria 111
  Brønsted-Lowry theory 111
  acid-base reactions 30–3
  acids 30
  bases and alkalis 31
  neutralisation reactions 31
  organic 191
  proton transfer 111
  salts 32
  strong and weak acids 31
  what makes an acid an acid? 30–1
acid rain 227–8
  effect of **227**
acids 30
  proton donors 111
  reactions of 30–1
  strong and weak 31, 111
activation energy 118–9
addition 192
  polymers from ethene 223–4
  polymerisation 204
    of ethene **204**
air pollution, types of 226–9
  acid rain 227–8
  greenhouse effect and global warming 228
  hole in ozone layer 228–9
  photochemical smog 226–7
alcohol(s) 212–16
  chemical properties 212
    combustion 212
    elimination of water (dehydration) 213–15
    ester formation 215
    oxidation 215–16
    reaction with sodium **212,** 212–13
    substitution of halogen atom for an –OH group 213
    tri-iodomethane reaction 216
  physical properties 212
  reduction of ketone to **192**
  structure and names 212, **212**
aldehyde, oxidation of primary alcohol to **216**
alicyclic hydrocarbons 196, **196**
aliphatic hydrocarbons 196
alkali(s)
  and bases 31
  reactions of group seven elements with 153
alkanes 197–200
  chemical properties 197–200
    burning 198
    cracking 198–9
    reactions with chlorine and bromine 199–200
  physical properties 197
  structures and names 197
alkenes 201–6
  addition to unsymmetrical alkenes 206
  chemical reactions 202–5
    addition of bromine or chlorine 203
    addition of hydrogen 202–3
    addition of hydrogen halides 204
    addition of water 204
    addition polymerisation 204
    oxidation 205
  double bond in 201–2
  electrophilic addition 205
  physical properties 201
  structures and names 201, **201**
alkyl groups 188–90
  structures of **189**
allotropes 87
alternative fuels 230–1
aluminium **167**
  electrolysis cell for extracting **168**
  extraction 167–8
  recycling 174
amine, two-step synthesis of with one more carbon atom than starting material **209**
ammonia
  equilibrium yield varying with pressure and temperature **163**
  flow diagram for synthesis of **162**
  and hydrogen chloride, combining **105**
  manufacture 162–4
  oxidation of 164, **164**
  uses of **163**
ammonium chloride decomposing into two gases on heating **106**
ammonium nitrate **165**
amounts in moles 36–7
anaerobic digestion 234
analysis 4, 179
anions, tests for 244–5
answers to 'Test yourself' questions 247–56
application of number 12–13, 18, 123
arenes 196
atomic absorption spectrometer **4**
atomic masses, relative 35
atomic number 52, 127
atomic orbitals, shapes of s and p in **59**
atomic properties, periodicity 128–9
atomic radii **129**, 129–30, **130**
  group one 138
  group two 141
atomic spectra 56
atomic structure 52–60
  atomic spectra 56
    quantum theory 56–7
  diagram **21**
  electron configurations 59–60, **138**
  electrons in energy levels 57–9
    quantum numbers 59
  ionisation energies 55–6
  isotopes 52
  mass spectrometry 53–5
    average relative atomic masses 54–5
    relative molecular masses **55**
atoms 15
  chlorine **55**
  of elements 21–2
  hydrogen **52**
  into ions 72–3
  for metals and non-metals **126**
  structures of **22**
average bond enthalpies 104
average relative atomic masses 54–5
Avogadro constant 37
Avogadro's law 41

balanced symbol equations 27, 136
Balmer series in emission spectrum **58**
barium **144**
bases 31
  as proton acceptors 111
benzene molecule **196**
beryllium 142, **144**
beryllium chloride, structure of **144**
big molecules 179
biofuels 230–1
blast furnace
  for extracting iron **169**
  in steel foundry **171**
bleach, analysing 155–6
body-centered cubic structure 71, **71**
boiling 19
  point of elements, periodicity of **132**
bomb calorimeter **97**
bond angles 79–80
bond breaking and bond forming 97–8
bond enthalpies 97–98, 103–104, 243
bonding
  in carbon dioxide **22**
  enthalpy changes and 103–4
  multiple 77–8
  three types of strong 68–9
    covalent 68
    ionic 69
    metallic 68
bond lengths and bond energies 243
bonds
  between molecules, weak 186–7
  within molecules, strong 182–3
bromine 148–9, **150, 159**
  addition to alkenes 203
  addition to ethene **192**
  addition to propene **203**
  covalent and van der Waals radii for **85**
  electrophilic addition to ethene **205**
  effect of light on solution of bromine in hexane **200**
  manufacture 158–9
  oxidation numbers for **134**
  reactions with alkanes 199
  structure 68
Brønsted-Lowry theory 111
building-up principle **60**
burning, alkanes and 198, **199**

C_{60} buckyball **5**, 88
caesium chloride, structure of 73
calcium **141**
calcium fluoride
  fluorite or fluorspar **33, 126**
  in Derbyshire Blue John **143**
calculations 123
  from equations 40–4
  calculating masses of reactants and products 40
  gas volume calculations 41–3
  gas volumes 40–1
    Avogadro's law 41
    Gay-Lussac's law of combining volumes 41
  molar volume of a gas 41
  solution calculations 43–4
  yields 217
carbocations 206
carbon
  atoms, electron configuration of **142**
  compounds, series of 178
      names of 188–90
      alkyl groups 188–90
      and IUPAC 188
      systematic names 188
      ways of representing bonding and shape of **182**
  as special element 178
  structure and bonding in forms of 87–8
    diamond 87
    fullerenes 88
    graphite 87–8
carbonates 140, 144–5
  thermal stability of 146–7
carbon dioxide, bonding in **22**
catalysts 116, 119
catalytic converter **230**
catalytic cracking **220**, 220–1
cations, tests for 34, 244
caustic soda (sodium hydroxide) **31**, 140
change
  direction of, and enthalpy changes 103
changes of state 18–19
  boiling 19
    energy changes accompanying **19**
  and enthalpy changes 94–5
  evaporation 18
  melting and freezing 18
  subliming 19
  vapours 18
chemical amounts 35
CFCs 207, 229
chemical elements, organising 127
chemical equations 26–7
  balancing redox equations 136
  balanced symbol equations 27
  ionic 34
  molecular models 26–7
  word equations 26
chemical equilibrium 107–10
  dynamic equilibrium 108–9
  factors affecting equilibrium 109
    changing concentration 109–10
    changing pressure and temperature 110
    predicting direction of change 109
  reaching an equilibrium state 107–8
chemical industry (inorganic) 157–66
  ammonia manufacture 162–4
  bromine manufacture 158–9
  chlor-alkali industry 159–60
  fertilisers 165–6
  major sectors of **158**
  nitric acid manufacture 164–5
  sulfuric acid manufacture 160–2
chemical plants **157**
chemical quantities 35–7
  amount in moles 36
  Avogadro constant 37
  chemical amounts 35
  molar mass 36
  relative atomic masses 35
  relative formula mass 36
  relative molecular mass 36
chemical reactions
  acid–base reactions 30–3
    acids 30
    bases and alkalis 31
    neutralisation reactions 31
    salts 32
    strong and weak acids 31
    what makes an acid an acid? 30–1
  hydrolysis 34
  ionic precipitation 33–4
    spectator ions 34
    tests for anions 33
    tests for cations 34
  oxidation and reduction 28–30
    electron transfer 28–9
    half-equations 30
    oxidising and reducing agents 29
  thermal decomposition 28
  see also matter and chemical change
chemical reactions and enthalpy changes 96–7
chemicals
  and origins 15–16
    purity 15–16
    raw materials 15
chlor-alkali industry 159–60
chlorine 148–9
  addition to alkenes 203
  and aqueous potassium iodide, solution of **151**
  atoms **55**
  gas **149**
  reactions with
    alkanes 199
    alkenes 203
    group one elements 139
    group two elements 143

chlorine *continued*
  metals 150
  non-metals 150-1
  substitution reactions with methane **199**
  uses of **159**
close-packed structures 70–1
cobalt(II) chloride paper to test for water **105**
collision model 117–19
  catalysts 119
    concentration, pressure and surface area 117
    heterogeneous reaction of solid with liquid or gas **117**
  temperature 118
combustion
  analysis **180**
  enthalpies of 96
  of alcohols 212
  of hydrogen, energy level diagram for **98**
communication 12, 176, 237
composition
  percentage 38–9
compounds 22–4
  names of and oxidation states 135–6
  of non-metals with non-metals 22–3
  properties of 241-2
concentration 43-4
  and rates of chemical change 113–14, 117
Contact process 160–1
control of substances hazardous to health 8
copper(II) sulfate crystals **17**
corrosive substances **9**
COSHH regulations 8
covalent bonding 68, 76–7
  dative covalent bonds 78–9
  in hydrogen chloride **151**
  multiple bonding 77–8
covalent giant structures 76
cracking 198-9
  catalytic 220–1
  steam 222
crude oil *see* fuels and chemicals from crude oil
  crystals 17
  metal 70
cyanide ions, substitution by 208–9
cyclohexanol, converting to cyclohexene **214**

dative covalent bonds 78–9, **79**
dehydration 213–15
  small-scale, of ethanol to ethene **214**
delocalised electrons
  in graphite 88
  in metals 68, 71
diamond(s) 67, 76, 87
dioxins 233
dipoles
  interactions between 83-5
direction of change, enthalpy changes and 103
disinfection 154
displacement reactions of halogens 151, 158–9
displayed formula 182
disproportionation reaction 153
distillation **213, 214, 216**
distribution of molecular energies 65, 118-9
dot-and-cross diagrams 72
double bond in alkenes 201–2
dynamic equilibrium 108–9

Earth
  atmosphere **226**
  crust, proportion of elements in **20**
electrolysis of molten sodium chloride **23**
electron(s)
  configurations 59–60, 126, **138**
    of magnesium and calcium atoms **142**
  in energy levels 57–9
  lone pairs of 78
  in shells for sodium atom **56**
  transfer 28–9
electron affinity 74
electronegativity 81
  periodicity of 130–1
electrophiles 195
electrophilic addition reactions in alkenes 205
elements 20
  atoms of 21–2
  group one 137
  group seven 148–9
  group two 141
  organising 127
  properties of 240
  from the stars 52–3
elimination 193, 210
  of water (dehydration) from alcohols 213–15, **215**
emissions from vehicles, cutting 229–30
empirical formulae 180–1
endothermic and exothermic changes 93
energy
  change 50–1
  changes accompanying changes of state **19**
energy levels in atoms 56–60
energy level diagrams 93
enthalpy changes 92–104
  and bonding 103–4
  and changes of states 94–5
  and chemical reactions 96–7
    bond breaking and bond forming 97–8
    enthalpy changes in solution 98–9
    enthalpies of combustion 96
  and direction of change 103
environment, organic chemicals and 226–34
  managing and disposing of organic wastes 231–4
  pollution of atmosphere 226
  tackling pollution problems 229–31
  types of air pollution 226–9
environmental issues 172–4
  recycling metals 173–4
  waste from mining 172–3
epoxyethane 223
equations
  balancing redox 136
  balanced symbol 27
  chemical 26–7
  half- 30
  ionic 34
  molecular models 26–7
  word 26
equilibrium *see* chemical equilibrium
errors and accuracy 11
ester formation 215
  hydrolysis 192
ethanol
  formula of **182**
  from ethene 204, 222
  from fermentation 230
  infra-red spectra **91**
  reaction with sodium **212**
  structure of **178, 183**
ethene
  addition of hydrogen bromide **204**
  addition polymerisation of **204**, 223-4
  electrophilic addition **205**
  epoxyethane from 223
  ethanol from 204
  pi (π)-bond in **202**
  and thermal cracking **222**
ether MTBE, structure of **220**
evaporation 18
exothermic and endothermic changes 93
  diagrams of **93**
experimental formulae 38
explosive substances **9**
extraction of metals 167–71
    aluminium 167–8
    iron making 168–9
    steel making 169–71
    titanium 171
    waste from 172–3

Fehling's solution 216
fermentation 230
fertilisers 165–6
flame colours
  group one 138
  group two 142
flame tests **138**
flammable substances **9**
fluorine 148-9
  bonding in **76**
fluorite crystals **33**
fluorspar, crystalline specimen of **126**
food from vegetable oils 225
formulae
  empirical 180–1
finding 38–9
  molecular 181
  skeletal 188
  structural 181–2
formula mass, relative 36
forward reaction **106**
foundations of chemistry 14–47
fractional distillation **218,** 218–19
free radicals 194, 199
freezing and melting 18
fuel(s)
  alternative fuels 230–1
  from waste 233
fuels and chemicals from crude oil 218–24
  chemicals from oil 222–4
    addition polymers from ethene 223–4
    epoxyethane from ethene 223
    ethanol from ethene 222
    steam cracking 222
  fractional distillation **218,** 218-19
  fuels from oil 219–21
    catalytic cracking **220**, 220–1
    isomerisation 221
    reforming 221
  refining crude oil 218
fuels and chemicals from vegetable oils 225
fullerenes (C_{60}) 88
functional groups 183
  test for 246
fusion of helium nuclei **53**

gas(es) 17
  effect of the amount of gas on its volume 62
  effect of pressure on gas volumes 62
  effect of temperature on gas volumes 61–2
  tests for 245
  volume calculations 41–3
Gay-Lussac's law of combining volumes 41
geometric isomers 185–6, **186**
giant structures **67**
  ionic 72–4
  of diamond 67, 76
  of graphite **87**
  of metals 70-1
glass, covalent and ionic bonding in **69**
global warming and greenhouse effect 228
granite, polished **15**
graphite 87–8
  giant structure of **87**
greenhouse effect **228**
group one 137–40
  atomic and ionic radii 138
  elements 137
    lithium 137, **137**
    potassium 137, **137**
    sodium 137, **137**
  flame colours 138, **142**
  ionisation energies 138
  oxidation states 139
  properties of compounds 139–40
    carbonates 140
    hydroxides 140
    nitrates 140
    oxides 139–40
  reactions of elements 139
    with chlorine 139
    with oxygen 139
    with water 139
group seven 148–53
  elements 148–9
  oxidation state –1 151–3
    displacement reactions 151
    hydrogen halides 152–3
    reactions of halides with concentrated sulfuric acid 151–2
  oxidation states +1, +3 and +5 153
    reactions with alkali 153
    reactions with water 153
  reactions of elements 149–51
    with metal elements 150
    with non-metal elements 150–1
group two 141–7
  atomic and ionic radii 141
  elements 141
    barium 141
    beryllium 141
    calcium **141**
    magnesium **141**
  flame colours 142
  ionisation energies 141–2
  oxidation states 142
  properties of compounds 143–6
    carbonates 144–5
    chlorides 144
    hydroxides 144
    nitrates 145
    oxides 143
    sulfates 145–6
  reaction of elements 142–3
    with chlorine 143
    with oxygen 143
    with water and dilute acids 143
  thermal stability of carbonates and nitrates 146–7
gypsum, crystals of **146**

Haber process 162–4
half-equations 30
halides
  displacement reactions 151
  with concentrated sulfuric acid 151–2
  identification 153, 245
halogenoalkanes 207–11
  chemical reaetions **207**, 208
    elimination reaction 210
    nucleophilic substitution 209–10
    substitution by hydroxide ions 208
    substitution by cyanide ions 208–9
    substitution reaction with
  ammonia 209
  hydrolysis of **208**
  physical properties 208
  structures and names 207, **207**
  uses **207**
halogens 148–51
harmful substances **9**
hazards
  and risks 8
  warning symbols **9**
health, control of substances
  hazardous to 8

Hess's law 100–3
  enthalpies of formation from enthalpies of combustion 101–2
  standard enthalpies of reaction from standard enthalpies of formation 102–3
heterogeneous
  catalyst 116
  equilibrium 110
heterolytic bond breaking 194–5, **195**
hexagonal close packing of metal atoms **71**
Hodgkin, Dorothy **66**
hole in ozone layer 228–9
homogeneous
  catalyst 116
  equilibrium 110
homologous series 184
homolytic bond breaking 194, **194**
household waste, typical composition of **232**
hydration of ions 89
hydrocarbon(s) 196
  mass spectrum of **55**
  types of 196
    alicyclic 196, **196**
    aliphatic 196
    arenes 196
    saturated compounds 196
    unsaturated compounds 196, **196**
hydrogen
  addition to alkenes 202–3
  addition to propene **202**
  atoms **52**
  burning to produce water **26**
  line spectrum for **57**
  uses **160**
hydrogenation
  of an alkene in presence of catalyst **203**
  of double bond in unsaturated fat **225**
hydrogen bonding 85–6
  in hydrogen fluoride **86**
  in water **85**
hydrogen bromide
  addition to ethene **204**
  electrophilic addition to ethene **205**
  elimination of from 2 bromopropane **210**
hydrogen chloride
  covalent bonding in **151**
  reaction with water **152**
hydrogen halides
  addition to alkenes 204
  boiling points **86**
  properties 152–3
hydrolysis 34, 192
  of ester **192**
  of halogenoalkanes **208**
hydroxide ions
  acting as a base 111, 210
  in alkalis 31
  substitution by 208
hydroxides 140, 144

ice, structure 86
ideal gas equation 63
incineration of waste 233
industry *see* chemical industry
information technology 13, 48, 123, 176, 237
infra-red spectroscopy 90–1, **91**
inorganic chemistry 124–76
  definition of 125–6
inorganic compounds, properties of 241
intermediates in reactions 206
intermediate types of bonding 81–2
  polar covalent bonds 81–2
  polarisation of ions 82
intermolecular forces 83–6, **186**
  attractions between temporary dipoles 84–5
    van der Waals radius 85
  dipole–dipole interactions 84
  hydrogen bonding 85–6
  polar molecules 83
internal combustion engine **219**
International Union of Pure and Applied Chemistry (IUPAC) 188
iodine **148**
  structure of 75
  subliming **19**
iodine-thiosulfate titration 156
ionic bonding 69, **69**, 73–4, **82**
  caesium chloride structure 73
  energy changes 73–4
  ionic radii 74
  sodium chloride structure 73
ionic equation 34
ionic giant structures 72–4
  atoms into ions 72–3
  properties of ionic compounds 74
  see also ionic bonding
ionic precipitation 33–4
ionic radii
  group one 138
  group two 141
  for metal ions in the third period **82**
  trend in group two metals in comparison with group one metals **141**
ionic salts in water, solutions of 89
ionisation energies 55–6, 130, **130**
  group one 138
  group two 141–2
ions 23–4
  and oxidation numbers 136
iron
  blast furnace for extracting **169**
  cycle of extraction and corrosion for **28**
  making 168–9
irritants 9
isomerisation 221
isomerism 185
  geometric isomers 185–6
  structural isomers 185, **185**
isotopes 52
IUPAC (International Union of Pure and Applied Chemistry) 188

Kelvin temperature scale 62
ketone to alcohol, reduction of **192**
key skills 12–13, 237
  application of number 12–13, 123
    standard form 13
  communication 12, 176, 237
  information technology 13, 123, 176, 237
  learning, problem solving and collaboration 14, 123
kinetic theory of gases 64–6
  Maxwell–Boltzmann distribution 65, 118-9
  real and ideal gases 63–4
    ideal gas equation 63
kinetic stability 120
knocking 219
Kroto, Professor Harry **5**

laboratory investigations 6–7
lactic acid, structure of **183**
landfill 233–4
Le Châtelier principle 109, **109**
limestone, products from **145**
line spectrum for hydrogen **57**
liquid crystal(s) 19
liquids 17
lithium 137, **137**
lone pairs of electrons 78
Lyman series **58**

Magnesium **141**
Markovnikov's rule 206, **206**
mass number 52
mass spectrometry 53–5
Maxwell–Boltzmann distribution 65
  of molecular kinetic energies in a gas at two temperatures **65, 118**
mechanisms of reactions
  electrophilic addition 205
  elimination 210
  free radical substitution 199
  heterolytic bond breaking 194–5, 195
  homolytic bond breaking 194, 194
  nucleophilic substitution 209–210
melting
  and freezing 18
  point of elements, periodicity of **132**
membrane cell **160**
metallic bonding 68, **68**
metals
  extraction of 167–71
    aluminium 167–8
    iron making 168–9
    steel making 169–71
    titanium 171
    waste from 172–3
  group one **138**
  group two 141
  reactions of group seven elements with 150
  samples **20**
  structure and bonding 70–1
    metal crystals 70
    metal properties 71
    metal structures 70–1
      body-centered cubic structure 71
      close-packed structures 70–1
methane
  covalent bonding in **77**
  molecule of **22**
  substitution reactions with chlorine **199**
minerals, major chemicals from 25
models, molecular 26–7
molar mass(es) 36
measuring 63
  of volatile liquids, syringe method for determining **63**
molecular formulae 181
molecular models 26–7
molecule(s) 15
  big 179
  with dipoles, attractions between 84, **84**
  with double bonds **78**
  of methane **22**
  octahedral 80
  organic *see* organic molecules
    and oxidation numbers 134–5
  polar 83, **83**
  shapes of 79–80
    with multiple bonds **80**
  single covalent bonding in **77**
  with triple bonds **77**
  of water **22**
molecular mass, M_r, relative 36
molecular structures 75–6
moles 36
  of atoms of elements **36**
  of ionic compounds **37**
  of molecules of compounds **37**
molar volume of a gas 41
multiple bonding 77–8

names
  of carbon compounds 188–190
  of inorganic compounds 135–6
negative ions 24
neutralisation reactions 31
nitrates 140, 145
  thermal stability of 146–7
nitric acid manufacture 164–5
nitrile formation **209**
noble gases, trend in boiling point of **85**
non-metals, samples **21**
nucleophiles 195, **195**
nucleophilic substitution 209–10

octahedral molecule 80
octane number 220
octet rule and its limitations 79
oil *see* fuels and chemicals from crude oil
ores, waste from processing 172
organic acids, two-step synthesis of with one more carbon atom than starting material **209**
organic chemicals and environment 226–34
organic chemistry 177–237
  big molecules 179
  carbon as special element 178
  definition of 178–9
  types of organic reaction 191-3
organic compounds, properties of 242
organic functional groups, tests for 246
organic molecules 180–7
  empirical formulae 180–1
  functional groups 183
  homologous series 184
  isomerism 185
    geometric isomers 185–6
    structural isomers 185
  molecular formulae 181
  skeletal formulae 188
  strong bonds 182–3
  structural formulae 181–2
  weak bonds between molecules 186–7
organic reaction(s)
  mechanisms of 194–5
    heterolytic bond breaking 194–5, **195**
    homolytic bond breaking 194, **194**
  types of 191–3
    acid–base 191
    addition 192
    elimination 193
    hydrolysis 192
    redox 191–2
    substitution 193
organic wastes, managing and disposing of 231–4
origins of chemicals 15–16
oxidation
  of alcohols 215–16
  of alkenes 205
  numbers 133–6
    of atoms and ions **133**
    balancing redox equations 136
    and ions 133–4
    and molecules 134–5
    oxidation states and names of compounds 135–6
    periodicity of oxidation states 135
    rules 134, **134**
of primary alcohol
of propan-1-ol **215**
of propan-2-ol to propanone **215**
  and reduction 28–30
  electron transfer 28–9
  half-equations 30
  oxidising and reducing agents 29
  states
  group one 139
  group seven 151–3
  group two 142
oxides 139–40, 143
oxidising substances **9**
oxygen
  converter for making steel **170**
  reactions of group one elements with 139
  reactions of group two elements with 143
ozone
  depletion over Antarctica **232**
  layer
    hole in 228–9
  limiting damage to 231

particles
  in a gas, arrangement of **17**

in a pure solid as it melts **18**
percentage composition 38–9
percentage yield 217
periodic table 127–32, **128**, 239
periodicity 128
of atomic properties 128–9
atomic radii 129–30
electronegativity 130–1
ionisation energies 130
of physical properties 131–2
permanent dipoles, attractions between molecules with 84, **84**
petrochemical industry 218–221
petrol
engines, pollutants from **229**
knocking 219
octane number 220
phosphorus pentachloride vapour, molecule of 80
photochemical smog 226–7
pH scale **30**
physical chemistry 49–123
what is physical chemistry? 50–1
physical properties 131
pi (π)-bond in ethene **202**
plants, chemicals from 225
chemicals from vegetable oils 225
food from vegetable oils 225
fuels from vegetable oils 225
plastic product codes to identify polymer material **233**
polar covalent bonds 81–2
polarisation of ions 82
polar molecules 83
pollutants from petrol engines **229**
pollution
of atmosphere 226
tackling problems of 229–31
alternative fuels 230–1
cutting emissions from vehicles 229–30
limiting damage to ozone layer 231
polymerisation, addition 204, 223-4
polymers 179, 204, 223-4
polythene molecule **179**
positive ions 24
potassium 137, **137**
pressure, effect of on gas volumes 62
propene
addition of bromine to **203**
addition of hydrogen to **202**
properties of elements and compounds 240
purity 15
PVC **224**

qualitative analysis 6
tests 244
quantitative analysis 6
quantum
numbers 59
theory 56–7

radiation and matter 50
radii
of atoms and ions, comparison of **74**
ionic 74
for metal ions in the third period **82**
rates of reaction 51, 112–16
catalysts 116
concentration 113–14
measuring reaction rates 112–13
surface area of solids 114–15
temperature 115–16
raw materials 15
reactants and products, calculating masses of 40
reaction profiles
for decomposition of dinitrogen oxide (kinetic stability) **120**
showing activation energy for reaction **118**
showing effect of catalyst on activation energy of reaction **119**
reaction types 28–34
real and ideal gases 63–4
recycling 232–3
metals 173–4
aluminium 174
steel 173–4
plastics 232–3
redox
equations, balancing 136
reactions 191–2
reduction 28–30, 133
reduction of ketone to alcohol **192**
refining crude oil 218
reflux condenser **209**
reforming, examples of **221**
relative atomic masses 35
relative formula mass 36
relative molecular masses 55
reversible reactions 105–6
risk assessment 8–9
rotation about a single bond **185**

safety 8–9
control of substances hazardous to health 8
hazards and risks 8
hazard warning symbols **9**
of homes and swimming pools 154–5
risk assessment 8–9
salts 32
saturated compounds 196
saturated solutions 25
shapes of molecules 79–80
shielding 129, **129**
sigma bonds in molecules **202**
significant figures 11
silicon dioxide, structure of **76**
silver halides 153
SI units 10
skeletal formula 188
smog, photochemical 226–7
sodium 137, **137**
reaction with alcohols 212–13
reaction with ethanol **212**
representations of electron configuration of **60**
sodium atom, electrons in shells for **56**
sodium chloride,
crystal structure **24**
electrolysis of molten **23**
electrolysis of solution 159–60
sodium hydroxide (caustic soda) **31**, **140**, **160**
solids 17
solubility 243
patterns of 25, 89
solution(s) 25, 89
calculations 43–4
enthalpy changes in 98–9, **98**
patterns of solubility 25, 89
saturated solutions 25
of ionic salts in water 89
spectator ions 34
spectrometer, atomic absorption **4**
spectrometry, mass 53–5
stability 120
standard form 13
standard solution 43
states of matter 17–19
changes of state 18–19
steam cracking 222
steel
making 169–71
recycling 173–4
strong and weak acids 31
structural formulae 181–2
structural isomers 185
structure and bonding 66–9, 131–2
in forms of carbon 87–8
diamond 87
fullerenes 88
graphite 87–8
investigation of 66–7
in metals 70–1
metal crystals 70
metal properties 71
metal structures 70–1
body-centered cubic structure 71
close-packed structures 70–1
three types of strong bonding 68–9
covalent bonding 68
ionic bonding 69
metallic bonding 68
two types of structure 67–8
subliming 19
substitution reaction 193, 199-200, 208–10, 213
sulfates 145–6
sulfur 161
sulfur dioxide 161
sulfuric acid **162**
manufacture 160–2
oxidation states of elements in **135**
sulfur trioxide 161
surface area of solids, and rates of chemical change 114–15
swimming pools, sterilisation of 155
symbols and conventions 47, 121, 175, 235
codes on plastic products to identify polymer material **233**
warning symbols **9**
synthesis 4, 7, 179
syringe method for determining molar masses of volatile liquids **63**
system and its surroundings **92**
systematic names 135, 188

temporary dipoles, attractions between 84–5
temperature
and collision model 118
and rates of chemical change 115–16
at which group two carbonates begin to decompose **146**
effect of on gas volumes 61–2
tests
for anions 33, 245
for cations 34, 244
for gases 245–6
for organic functional groups 246
theoretical yield 217
thermal cracking 222
thermal stability of carbonates and nitrates 146–7
thermochemistry 92
thermometer
liquid crystal **19**
scale **11**
titanium
extraction 171
titrations 45–6
calculation 45
analysing solutions 45–6
investigating reactions 46
procedure 45
see also iodine-thiosulfate titration
toxic substances **9**
tri-iodomethane reaction 216

units and measurements 10–11
errors and accuracy 11
significant figures 11
unsaturated compounds 196, **196**
unsymmetrical alkenes, addition of alkenes to 206

van der Waals radius 85
vaporisation
of liquid, apparatus for measuring enthalpy of **95**
plot of enthalpy of, against boiling point **95**
vapours 18
vegetable oils

wastes
fuel from 233
from mining 172–3
from processing ores 172
from metal extraction 172–3
managing and disposing of organic 231–4
water
addition to alkenes 204
addition to ethene to make ethanol **204**
absorption of nitric acid in 165
elimination (dehydration) by alcohols 213–15
molecule **22**
and oil **187**
reactions of group one elements with 139
reactions of group seven elements with 153
reactions of group two elements with 143
treatment 154–6
analysing bleach 155–6
disinfection 154
safe homes and swimming pools 154–5
weak bonds between molecules 186–7
word equations 26

X-ray diffraction **66**

yields, theoretical and percentage 217

zeolite crystal structure **221**